# Theory of Nonlinear Age-Dependent Population Dynamics

## MONOGRAPHS AND TEXTBOOKS IN PURE AND APPLIED MATHEMATICS

1. *K. Yano,* Integral Formulas in Riemannian Geometry (1970) *(out of print)*
2. *S. Kobayashi,* Hyperbolic Manifolds and Holomorphic Mappings (1970) *(out of print)*
3. *V. S. Vladimirov,* Equations of Mathematical Physics (A. Jeffrey, editor; A. Littlewood, translator) (1970) *(out of print)*
4. *B. N. Pshenichnyi,* Necessary Conditions for an Extremum (L. Neustadt, translation editor; K. Makowski, translator) (1971)
5. *L. Narici, E. Beckenstein, and G. Bachman,* Functional Analysis and Valuation Theory (1971)
6. *D. S. Passman,* Infinite Group Rings (1971)
7. *L. Dornhoff,* Group Representation Theory (in two parts). Part A: Ordinary Representation Theory. Part B: Modular Representation Theory (1971, 1972)
8. *W. Boothby and G. L. Weiss (eds.),* Symmetric Spaces: Short Courses Presented at Washington University (1972)
9. *Y. Matsushima,* Differentiable Manifolds (E. T. Kobayashi, translator) (1972)
10. *L. E. Ward, Jr.,* Topology: An Outline for a First Course (1972) *(out of print)*
11. *A. Babakhanian,* Cohomological Methods in Group Theory (1972)
12. *R. Gilmer,* Multiplicative Ideal Theory (1972)
13. *J. Yeh,* Stochastic Processes and the Wiener Integral (1973) *(out of print)*
14. *J. Barros-Neto,* Introduction to the Theory of Distributions (1973) *(out of print)*
15. *R. Larsen,* Functional Analysis: An Introduction (1973) *(out of print)*
16. *K. Yano and S. Ishihara,* Tangent and Cotangent Bundles: Differential Geometry (1973) *(out of print)*
17. *C. Procesi,* Rings with Polynomial Identities (1973)
18. *R. Hermann,* Geometry, Physics, and Systems (1973)
19. *N. R. Wallach,* Harmonic Analysis on Homogeneous Spaces (1973) *(out of print)*
20. *J. Dieudonné,* Introduction to the Theory of Formal Groups (1973)
21. *I. Vaisman,* Cohomology and Differential Forms (1973)
22. *B.-Y. Chen,* Geometry of Submanifolds (1973)
23. *M. Marcus,* Finite Dimensional Multilinear Algebra (in two parts) (1973, 1975)
24. *R. Larsen,* Banach Algebras: An Introduction (1973)
25. *R. O. Kujala and A. L. Vitter (eds.),* Value Distribution Theory: Part A; Part B: Deficit and Bezout Estimates by Wilhelm Stoll (1973)
26. *K. B. Stolarsky,* Algebraic Numbers and Diophantine Approximation (1974)
27. *A. R. Magid,* The Separable Galois Theory of Commutative Rings (1974)
28. *B. R. McDonald,* Finite Rings with Identity (1974)
29. *J. Satake,* Linear Algebra (S. Koh, T. A. Akiba, and S. Ihara, translators) (1975)

30. *J. S. Golan,* Localization of Noncommutative Rings (1975)
31. *G. Klambauer,* Mathematical Analysis (1975)
32. *M. K. Agoston,* Algebraic Topology: A First Course (1976)
33. *K. R. Goodearl,* Ring Theory: Nonsingular Rings and Modules (1976)
34. *L. E. Mansfield,* Linear Algebra with Geometric Applications: Selected Topics (1976)
35. *N. J. Pullman,* Matrix Theory and Its Applications (1976)
36. *B. R. McDonald,* Geometric Algebra Over Local Rings (1976)
37. *C. W. Groetsch,* Generalized Inverses of Linear Operators: Representation and Approximation (1977)
38. *J. E. Kuczkowski and J. L. Gersting,* Abstract Algebra: A First Look (1977)
39. *C. O. Christenson and W. L. Voxman,* Aspects of Topology (1977)
40. *M. Nagata,* Field Theory (1977)
41. *R. L. Long,* Algebraic Number Theory (1977)
42. *W. F. Pfeffer,* Integrals and Measures (1977)
43. *R. L. Wheeden and A. Zygmund,* Measure and Integral: An Introduction to Real Analysis (1977)
44. *J. H. Curtiss,* Introduction to Functions of a Complex Variable (1978)
45. *K. Hrbacek and T. Jech,* Introduction to Set Theory (1978)
46. *W. S. Massey,* Homology and Cohomology Theory (1978)
47. *M. Marcus,* Introduction to Modern Algebra (1978)
48. *E. C. Young,* Vector and Tensor Analysis (1978)
49. *S. B. Nadler, Jr.,* Hyperspaces of Sets (1978)
50. *S. K. Segal,* Topics in Group Rings (1978)
51. *A. C. M. van Rooij,* Non-Archimedean Functional Analysis (1978)
54. *L. Corwin and R. Szczarba,* Calculus in Vector Spaces (1979)
53. *C. Sadosky,* Interpolation of Operators and Singular Integrals: An Introduction to Harmonic Analysis (1979)
54. *J. Cronin,* Differential Equations: Introduction and Quantitative Theory (1980)
55. *C. W. Groetsch,* Elements of Applicable Functional Analysis (1980)
56. *I. Vaisman,* Foundations of Three-Dimensional Euclidean Geometry (1980)
57. *H. I. Freedman,* Deterministic Mathematical Models in Population Ecology (1980)
58. *S. B. Chae,* Lebesgue Integration (1980)
59. *C. S. Rees, S. M. Shah, and C. V. Stanojević,* Theory and Applications of Fourier Analysis (1981)
60. *L. Nachbin,* Introduction to Functional Analysis: Banach Spaces and Differential Calculus (R. M. Aron, translator) (1981)
61. *G. Orzech and M. Orzech,* Plane Algebraic Curves: An Introduction Via Valuations (1981)
62. *R. Johnsonbaugh and W. E. Pfaffenberger,* Foundations of Mathematical Analysis (1981)
63. *W. L. Voxman and R. H. Goetschel,* Advanced Calculus: An Introduction to Modern Analysis (1981)
64. *L. J. Corwin and R. H. Szcarba,* Multivariable Calculus (1982)
65. *V. I. Istrătescu,* Introduction to Linear Operator Theory (1981)
66. *R. D. Järvinen,* Finite and Infinite Dimensional Linear Spaces: A Comparative Study in Algebraic and Analytic Settings (1981)

67. *J. K. Beem and P. E. Ehrlich,* Global Lorentzian Geometry (1981)
68. *D. L. Armacost,* The Structure of Locally Compact Abelian Groups (1981)
69. *J. W. Brewer and M. K. Smith, eds.,* Emmy Noether: A Tribute to Her Life and Work (1981)
70. *K. H. Kim,* Boolean Matrix Theory and Applications (1982)
71. *T. W. Wieting,* The Mathematical Theory of Chromatic Plane Ornaments (1982)
72. *D. B. Gauld,* Differential Topology: An Introduction (1982)
73. *R. L. Faber,* Foundations of Euclidean and Non-Euclidean Geometry (1983)
74. *M. Carmeli,* Statistical Theory and Random Matrices (1983)
75. *J. H. Carruth, J. A. Hildebrant, and R. J. Koch,* The Theory of Topological Semigroups (1983)
76. *R. L. Faber,* Differential Geometry and Relativity Theory: An Introduction (1983)
77. *S. Barnett,* Polynomials and Linear Control Systems (1983)
78. *G. Karpilovsky,* Commutative Group Algebras (1983)
79. *F. Van Oystaeyen and A. Verschoren,* Relative Invariants of Rings: The Commutative Theory (1983)
80. *I. Vaisman,* A First Course in Differential Geometry (1984)
81. *G. W. Swan,* Applications of Optimal Control Theory in Biomedicine (1984)
82. *T. Petrie and J. D. Randall,* Transformation Groups on Manifolds (1984)
83. *K. Goebel and S. Reich,* Uniform Convexity, Hyperbolic Geometry, and Nonexpansive Mappings (1984)
84. *T. Albu and C. Năstăsescu,* Relative Finiteness in Module Theory (1984)
85. *K. Hrbacek and T. Jech,* Introduction to Set Theory, Second Edition, Revised and Expanded (1984)
86. *F. Van Oystaeyen and A. Verschoren,* Relative Invariants of Rings: The Noncommutative Theory (1984)
87. *B. R. McDonald,* Linear Algebra Over Commutative Rings (1984)
88. *M. Namba,* Geometry of Projective Algebraic Curves (1984)
89. *G. F. Webb,* Theory of Nonlinear Age-Dependent Population Dynamics (1985)
90. *M. R. Bremner, R. V. Moody, and J. Patera,* Tables of Dominant Weight Multiplicities for Representations of Simple Lie Algebras (1985)
91. *A. E. Fekete,* Real Linear Algebra (1985)

*Other Volumes in Preparation*

# Theory of Nonlinear Age-Dependent Population Dynamics

G. F. WEBB
Vanderbilt University
Nashville, Tennessee

MARCEL DEKKER, INC. New York and Basel

Library of Congress Cataloging in Publication Data

Webb, Glenn F., [date]
Theory of nonlinear age-dependent population dynamics.

(Monographs and textbooks in pure and applied mathematics ; 89)
Bibliography: p.
Includes indexes.
1. Population biology--Mathematical models.
2. Population--Mathematical models. I. Title.
II. Title: Age-dependent population dynamics.
III. Series.
QH352.W43 1985 574.5'248'0724 84-23140
ISBN 0-8247-7290-3





MARCEL DEKKER, INC.
270 Madison Avenue, New York, New York 10016

Current printing (last digit):
10 9 8 7 6 5 4 3 2 1

PRINTED IN THE UNITED STATES OF AMERICA

# Preface

This monograph is devoted to the mathematical theory of continuous, deterministic, nonlinear models of populations with age structure. There is a long history and extensive literature of linear models of age-structured populations, but the theory of nonlinear models of age-structured populations is much more recent. This nonlinear theory has arisen in research articles published in the past 10 years and has undergone a rapid development. The principal value of nonlinear models is that they allow consideration for the effects of crowding, resource limitation, and interaction. The inclusion of nonlinearities in the equations of age-dependent population models increases considerably their mathematical difficulty, but increases also their reliability for physical description and behavior prediction.

The purpose of this monograph is to provide a mathematically complete treatment for a general class of models of nonlinear age-dependent population dynamics. A general model allowing for nonlinear birth, mortality, and immigration processes is formulated, and its mathematical theory is developed. Emphasis is given to the basic theory of the solutions, such as the properties of existence, uniqueness, positivity, regularity, representation, and numerical approximation. Emphasis is also given to the analysis of time-independent equilibrium states and the ultimate behavior of the solutions as time evolves. Some specific examples of biological populations are presented to illustrate the theoretical results. The underlying theme that unites the mathematical methods in this monograph is that the solutions of the equations of nonlinear age-dependent population dynamics constitute the mathematical structure of a nonlinear semigroup of operators, or a dynamical system. It is this vantage point from which the problem is viewed and this approach that allows the application of the general methods and techniques from the modern theory of nonlinear analysis.

This monograph is written at an advanced mathematical level, and is intended for advanced students, specialists, and researchers in differential equations, dynamical systems, applied mathematics, mathematical biology, and mathematical demography. The level of exposition assumes a basic knowledge of integration theory, linear operator theory, functional analysis, and the theory of semigroups of linear operators in Banach spaces. An effort has been made throughout to provide mathematically rigorous proofs and complete mathematical details.

The author wishes to express his thanks to the many friends and colleagues who have provided encouragement, advice, and assistance in the writing of this monograph. In particular, the support of S. Busenberg, G. Da Prato, G. Di Blasio, M. Gurtin, M. Iannelli, F. Kappel, K. Kunisch, R. MacCamy, J. Prüss, W. Schappacher, E. Sinestrari, and R. Villella-Bressan is gratefully acknowledged. The author also wishes to express his thanks to Ruby J. Moore for her careful and diligent typing of the manuscript.

G. F. Webb

# Contents

# Theory of Nonlinear Age-Dependent Population Dynamics

# 1
# Statement of the Problem

## 1.1 HISTORY OF THE PROBLEM

In 1798 T. R. Malthus proposed a model of population dynamics in which the rate of population growth was proportional to the size of the population. In the model of Malthus the function $P(t)$, which represents the total population size at time $t$, satisfies the differential equation

$$(1.1) \qquad \frac{d}{dt} P(t) = \lambda_1 P(t) \qquad t \geq 0$$

where $\lambda_1$ is the *malthusian parameter* of the given population. The solution of (1.1) is the exponential function $P(t) = e^{\lambda_1 t}P(0)$, which makes apparent the famous exponential growth rate of a malthusian population.

A malthusian population makes no allowance for the effects of crowding or the limitations of resources. A more realistic model of population growth would allow the malthusian parameter to depend upon the size of the total population itself. Such a model was proposed by P. F. Verhulst in 1838. In the model of Verhulst it was assumed that the population function $P(t)$ satisfied the differential equation

$$(1.2) \qquad \frac{d}{dt} P(t) = \lambda_1 \left[1 - \frac{P(t)}{K}\right] P(t) \qquad t \geq 0$$

Equation (1.2) is known as the *logistic equation*. The constants $\lambda_1$ and $K$ are known as the *intrinsic growth constant* and the *environmental carrying capacity*, respectively. Equation (1.2) is easily solved by the method of separation of variables to yield the explicit formula

$$(1.3) \qquad P(t) = \frac{K}{1 + [K/P(0) - 1]e^{-\lambda_1 t}} \qquad t \geq 0$$

From the formula (1.3) it is evident that the solutions of equation (1.2) have the property that $\lim_{t\to\infty} P(t) = K$. Thus, in contrast to a malthusian population, such a population approaches a nontrivial equilibrium state as time becomes infinite.

The models of Malthus and Verhulst are examples of continuous or deterministic population models. The theory of continuous population dynamics has been extensively developed by mathematical demographers and population biologists. One of the most important theories in this development has been for models which allow for the effects of age structure. For many populations consideration of the age distribution within the population leads to a more realistic and useful mathematical model.

Among the first continuous models incorporating age effects were those of F. R. Sharpe and A. Lotka in 1911 and A. G. McKendrick in 1926. In the models of Sharpe-Lotka-McKendrick the birth and mortality processes were linear functions of the population densities, which meant that the equations of their models were linear. In this early period the theory of linear age-dependent population dynamics was extensively developed by many mathematicians. As with the malthusian models of age-independent population dynamics, the models of this theory were necessarily linear, and consequently, permitted no influence for crowding effects or environmental limitations.

In 1974 M. Gurtin and R. C. MacCamy and F. Hoppensteadt introduced the first models of nonlinear continuous age-dependent population dynamics. In the Gurtin-MacCamy study the effects of crowding were incorporated into the model by allowing the birth and mortality processes to be nonlinear functions of the population densities. Consequently, the equations of their models, as in the case of the verhulstian models, contained nonlinear terms involving the unknown solutions. Analogously with the verhulstian models of age-independent population dynamics, these nonlinearities provided a mechanism by which the population might stabilize to a nontrivial equilibrium state as time evolved.

In this chapter we will discuss in turn the classical linear model of Sharpe-Lotka-McKendrick and the nonlinear model of Gurtin-MacCamy. We will then present a general formulation of age-dependent population dynamics which allows general birth and mortality processes. In the chapters which follow it will be our objective to develop the mathematical theory of this formulation, and to then apply this theory to examples of biological populations.

## 1.2 THE CLASSICAL LINEAR MODEL

The classical model of linear age-dependent population dynamics of Sharpe-Lotka-McKendrick is formulated as follows: Let $\ell(a,t)$ be the density with respect to age a of a population at time t. The units of $\ell(a,t)$ are given in units of population divided by units of time. Accordingly, the total population at time t of members of the population between ages $a_1$ and $a_2$ is

$$\int_{a_1}^{a_2} \ell(a,t)\, da$$

and the total population at time t of all members of the population is

$$P(t) = \int_0^\infty \ell(a,t)\, da \tag{1.4}$$

This density function satisfies the so-called *balance law* (or aging process of the population)

$$D\ell(a,t) = -\mu(a)\ell(a,t) \tag{1.5}$$

where $\mu$ is a nonnegative function of age called the *age-specific mortality modulus* and the differentiation operation D is defined by

$$(1.6) \qquad D\ell(a,t) \overset{\text{def}}{=} \lim_{h\to 0^+} h^{-1}[\ell(a + h, t + h) - \ell(a,t)]$$

The birth process of the population (or the imput of population of age 0) satisfies the so-called *birth law*

$$(1.7) \qquad \ell(0,t) = \int_0^\infty \beta(a)\ell(a,t)\, da \qquad t > 0$$

where $\beta(a)$ is a nonnegative function of age known as the *age-specific fertility modulus*. The expression $\ell(0,t)$ may be interpreted as the birth rate at time t. Its units are given in units of population per units of time. Its value at a given time t depends on the age distribution of the population at that time, as determined by the integral of the density $\ell(a,t)$ weighted with the fertility modulus $\beta(a)$.

Last, the *initial age distribution* of the population is

$$(1.8) \qquad \ell(a,0) = \phi(a) \qquad a \geq 0$$

where $\phi$ is a known nonnegative function of age a. Notice that equation (1.7) is not required to hold at $t = 0$. If (1.7) does hold at $t = 0$, then (1.7) and (1.8) must be compatible in the sense that

$$(1.9) \qquad \int_0^\infty \beta(a)\phi(a)\, da = \phi(0)$$

Condition (1.9) is called the *compatibility condition*, and it is not, in general, required of the initial age distribution $\phi$.

The problem (1.5), (1.7), (1.8) constitutes the classical linear model of age-dependent population dynamics. The equation (1.5), which is known as the McKendrick or the Von Foerster equation, is implicit in the earlier models of Sharpe and Lotka. Since it is instructive for the nonlinear models which we will treat later, we will present here a brief discussion concerning this classical linear model. For a detailed treatment we refer the reader to W. Feller [104], N. Keyfitz [168], and M. Gurtin [127].

The problem (1.5), (1.7), (1.8) is a linear first order hyperbolic partial differential equation with initial and boundary conditions. It is easily solved by the method of characteristics. The main idea is to convert the problem to a Volterra integral equation involving the birth rate $\ell(0,t)$. The problem is thus transformed from a linear differential equation in the two independent variables a and t (with initial and boundary conditions) to a linear Volterra integral equation in the single independent variable t.

The method proceeds in the following way: Suppose that the solution $\ell(a,t)$ of (1.5), (1.7), (1.8) is known. The characteristic curves of the equation (1.5) are the lines $a - t = c$, where c is a constant. Let $c \in R$ and define

$$w_c(t) \stackrel{\text{def}}{=} \ell(t + c, t) \qquad t \geq t_c$$

where $t_c \stackrel{\text{def}}{=} \max\{0,-c\}$ (see Figure 1.1 below).

The function $w_c$ is called the *cohort function* corresponding to age c. It keeps track of those members of the population who are initially of age c as time evolves. From (1.5) we obtain

$$\begin{aligned} \frac{d}{dt} w_c(t) &= \lim_{h \to 0^+} \frac{\ell(t + h + c, t + h) - \ell(t + c, t)}{h} \\ &= D\ell(t + c, t) = -\mu(t + c)w_c(t) \qquad t \geq t_c \end{aligned}$$

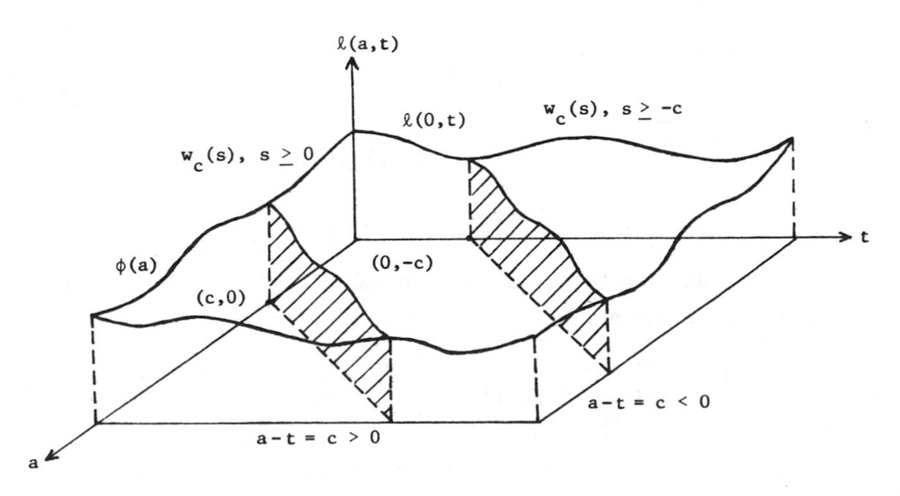


*Figure 1.1*

which implies that

$$w_c(t) = w_c(t_c) \exp\left[-\int_{t_c}^{t} \mu(s + c)\, ds\right] \qquad t \geq t_c$$

If we set $c = a - t$, where $a \geq t$, then

$$w_c(t) = w_c(0) \exp\left[-\int_{0}^{t} \mu(s + c)\, ds\right] \qquad t \geq 0$$

which yields

$$\text{(1.10)} \qquad \ell(a,t) = \ell(a - t, 0) \exp\left[-\int_{0}^{t} \mu(s + a - t)\, ds\right] \qquad a \geq t$$

If we set $c = a - t$, where $a < t$, then

$$w_c(t) = w_c(-c) \exp\left[-\int_{-c}^{t} \mu(s + c)\, ds\right] \qquad t \geq -c$$

which yields

$$\text{(1.11)} \qquad \ell(a,t) = \ell(0, t - a) \exp\left[-\int_{t-a}^{t} \mu(s + c)\, ds\right] \qquad a < t$$

Combine formulas (1.10), (1.11), and (1.8) to obtain

$$\text{(1.12)} \qquad \ell(a,t) = \begin{cases} \ell(0, t - a) \exp\left[-\int_{0}^{a} \mu(s)\, ds\right] & a < t \\ \phi(a - t) \exp\left[-\int_{a-t}^{a} \mu(s)\, ds\right] & a \geq t \end{cases}$$

The formulas in (1.12) allow a simple interpretation of the density function $\ell$ in terms of the mortality modulus $\mu$. Suppose that we define the function

$$\Pi(b,a) \overset{\text{def}}{=} \exp\left[-\int_{a}^{b} \mu(s)\, ds\right] \qquad 0 \leq a \leq b$$

as the probability that a member of the population of age $a$ survives to age $b$. Let the time $t$ be fixed. If $a < t$, then

$$\ell(a,t) = \ell(0, t - a)\Pi(a,0)$$

represents those members of the population born t - a units of time ago which survive to age a at time t. If $a \geq t$, then

$$\ell(a,t) = \phi(a - t)\Pi(a, a - t)$$

represents those members of the population which were of age a - t at the initial instant and which survived to age a at time t.

From (1.12) we see that if the birth rate $\ell(0,t)$ can be determined as a function of t, then the density function $\ell$ becomes known. Equations (1.7) and (1.12) can be used to arrive at a single equation for the birth rate $\ell(0,t)$. Set $B(t) \overset{\text{def}}{=} \ell(0,t)$ and substitute (1.12) into (1.7) to obtain

$$\text{(1.13)} \qquad \begin{aligned} B(t) &= \int_0^t \beta(a)\ell(a,t)\,da + \int_t^\infty \beta(a)\ell(a,t)\,da \\ &= \int_0^t \beta(a)B(t - a)\Pi(a,0)\,da \\ &\quad + \int_t^\infty \beta(a)\phi(a - t)\Pi(a, a - t)\,da \end{aligned}$$

The birth rate thus satisfies the integral equation

$$\text{(1.14)} \qquad B(t) = \int_0^t K(t - a)B(a)\,da + H(t) \qquad t \geq 0$$

where

$$K(b) \overset{\text{def}}{=} \beta(b)\Pi(b,0) \qquad b \geq 0$$

$$H(t) \overset{\text{def}}{=} \int_0^\infty \beta(a + t)\phi(a)\Pi(a + t, a)\,da \qquad t \geq 0$$

are known quantities.

Integral equations having the form (1.14) are known as renewal equations. The theory of renewal equations has been extensively developed by such authors as W. Feller [104], R. Bellman and K. Cooke [19], and R. K. Miller [220]. One aspect of this theory which has received considerable attention is the asymptotic behavior of the solutions of these equations. We will provide here a brief discussion of the theory of asymptotic behavior in classical linear age-dependent population dynamics.

We first define an equilibrium solution of the problem (1.5), (1.7), (1.8) as a solution $\ell(a,t) = \hat{\phi}(a)$ independent of time. An equilibrium solution $\hat{\phi}$ will satisfy the equations

$$(1.15) \qquad \frac{d}{da}\hat{\phi}(a) = -\mu(a)\hat{\phi}(a) \qquad a \geq 0$$

$$(1.16) \qquad \hat{\phi}(0) = \int_0^\infty \beta(a)\hat{\phi}(a)\, da$$

From (1.15) we see that $\hat{\phi}$ has the form

$$(1.17) \qquad \hat{\phi}(a) = \hat{\phi}(0) \exp\left[-\int_0^a \mu(\tau)\, d\tau\right] \qquad a \geq 0$$

and from (1.16) and (1.17) we see that if $\hat{\phi}(0) \neq 0$, then

$$(1.18) \qquad 1 = \int_0^\infty \beta(a) \exp\left[-\int_0^a \mu(s)\, ds\right] da$$

Thus, a nontrivial equilibrium solution of the problem (1.5), (1.7), (1.8) exists if and only if (1.18) holds, and if (1.18) holds, then a family of nontrivial equilibrium solutions is given by the formula (1.17) as $\hat{\phi}(0)$ ranges over all positive values. If (1.18) does not hold, then the zero solution is the only equilibrium solution of the problem.

Since it is a rare occurrence for equation (1.18) to hold, the investigators of the Sharpe-Lotka-McKendrick model turned their attention to another class of special solutions. These are the so-called *stable age distributions*, or *persistent solutions*. These solutions have the form

$$(1.19) \qquad \ell(a,t) = A(a)T(t) \qquad a \geq 0,\ t \geq 0$$

For a stable age distribution the proportion of the population within any age bracket $[a_1,a_2]$ remains constant for all time in the sense that

$$\frac{\int_{a_1}^{a_2} \ell(a,t)\, da}{\int_0^\infty \ell(a,t)\, da} = \frac{\int_{a_1}^{a_2} A(a)\, da}{\int_0^\infty A(a)\, da} \qquad t \geq 0$$

If we substitute (1.19) into (1.5), we obtain

$$\frac{\frac{d}{dt} T(t)}{T(t)} = - \frac{[\frac{d}{da} A(a) + \mu(a)A(a)]}{A(a)} = \lambda \tag{1.20}$$

where $\lambda$ is a constant. From (1.20) we see that a stable age distribution has the form

$$\ell(a,t) = A(0) \exp\left[\lambda(t - a) - \int_0^a \mu(b)\, db\right] \tag{1.21}$$

Since (1.7) must hold, the choice of $\lambda$ must satisfy

$$1 = \int_0^\infty \beta(a) \exp\left[-\lambda a - \int_0^a \mu(b)\, db\right] da \tag{1.22}$$

Equation (1.22), which is known as the *characteristic equation*, was discovered by A. Lotka in 1922. Under appropriate assumptions on the fertility modulus $\beta$ and the mortality modulus $\mu$, the right-hand side of (1.22) is a strictly decreasing function of $\lambda$ which assumes all values in $(0,\infty)$. In this case (1.22) has a unique solution $\lambda_1$, and for this choice of $\lambda_1$, formula (1.21) provides a stable age distribution for every positive value $A(0)$.

Stable age distributions for the Sharpe-Lotka-McKendrick model have attained considerable importance in mathematical demography. Under appropriate assumptions on the fertility modulus $\beta$, the mortality modulus $\mu$, and the initial age distribution $\phi$, it can be shown that every solution $\ell(a,t)$ of the problem (1.5), (1.7), (1.8) has the property that, uniformly in finite intervals of a,

$$\lim_{t\to\infty} \left| e^{-\lambda_1 t} \ell(a,t) - C(\phi) \exp\left[-\lambda_1 a - \int_0^a \mu(b)\, db\right]\right| = 0 \tag{1.23}$$

where $\lambda_1$ is the unique solution of the characteristic equation (1.22) and $C(\phi)$ is a value which depends on $\phi$. This remarkable result is originally due to A. Lotka, and was proved rigorously by W. Feller in 1941. We will give a proof of this result in Chapter 4 using operator semigroup methods.

Stable age distributions are of great interest in the classical linear theory of age-dependent population dynamics, but they are not actual equilibrium solutions unless the solution of the characteristic equation (1.22) is $\lambda_1 = 0$. It would be physically reasonable to expect that many biological populations have an equilibrium state to which the population stabilizes as time evolves. For the linear model of age-dependent population dynamics, as for the linear malthusian model of age-independent population dynamics, this expectation is not realized except in rare cases. In the verhulstian models of age-independent population dynamics a nontrivial equilibrium state does exist, and, in fact, attracts every solution as time becomes infinite. A verhulstian model possesses a mechanism to explain this phenomenon, namely, the effect of decreased population growth when the population becomes large. As with the nonlinear verhulstian models, we shall see that nonlinear age-dependent population models also allow the phenomena of stable time-independent states for age structured populations.

## 1.3 THE GURTIN-MACCAMY MODEL

In the Gurtin-MacCamy model of age-dependent population dynamics the fertility and mortality moduli are allowed to be density dependent. Let $\ell(a,t)$ and $P(t)$ be as before, so that

$$(1.24) \qquad P(t) = \int_0^\infty \ell(a,t)\, da \qquad t \geq 0$$

The *balance law* of the Gurtin-MacCamy model is

$$(1.25) \qquad D\ell(a,t) = -\mu(a,P(t))\ell(a,t)$$

where $D\ell(a,t)$ is as in (1.6) and the *mortality modulus* $\mu$ is a given nonnegative function of two independent variables. The *birth law* of the Gurtin-MacCamy model is given by

$$(1.26)\qquad \ell(0,t) = \int_0^\infty \beta(a,P(t))\ell(a,t)\,da \qquad t > 0$$

where the *fertility modulus* is also a given nonnegative function of two independent variables. The initial condition is again

$$(1.27)\qquad \ell(a,0) = \phi(a) \qquad a \geq 0$$

where $\phi$ is the known *initial age distribution*.

The model (1.25), (1.26), (1.27) was introduced by M. Gurtin and R. MacCamy in [130]. This model and similar nonlinear models have been investigated by many researchers, and we refer the reader to our references for a listing of relevant articles. The method employed by Gurtin and MacCamy was analogous to that of the classical linear case, and that was to apply the method of characteristics to convert the problem to a system of Volterra integral equations involving the birth rate. For the Gurtin-MacCamy model (1.25), (1.26), (1.27) this system consists of nonlinear Volterra integral equations.

The method proceeds as follows: As before we suppose that the solution $\ell(a,t)$ of (1.25), (1.26), (1.27) is known and we define the *cohort function*

$$w_c(t) \overset{\text{def}}{=} \ell(t + c,\ t) \qquad t \geq t_c \overset{\text{def}}{=} \max\{0,-c\}$$

for a fixed value $c \in R$. Since $\ell$ satisfies (1.25), we obtain

$$(1.28)\qquad \begin{aligned} \frac{d}{dt} w_c(t) &= D\ell(t + c,\ t) \\ &= -\mu(t + c,\ P(t))w_c(t) \qquad t \geq t_c \end{aligned}$$

Next, integrate (1.28) to obtain

$$(1.29)\qquad w_c(t) = \begin{cases} w_c(t - a)\ \exp\left[-\int_{t-a}^{t} \mu(s + c,\ P(s))\,ds\right] & a < t \\ w_c(0)\ \exp\left[-\int_0^t \mu(s + c,\ P(s))\,ds\right] & a \geq t \end{cases}$$

Now use (1.27) and substitute c = a - t to obtain

$$(1.30)\qquad \ell(a,t) = \begin{cases} \ell(0, t-a) \exp\left[-\int_{t-a}^{t} \mu(s+a-t, P(s))\, ds\right] & a < t \\ \phi(a-t) \exp\left[-\int_{0}^{t} \mu(s+a-t, P(s))\, ds\right] & a \ge t \end{cases}$$

Define $B(t) \overset{\text{def}}{=} \ell(0,t)$ and substitute the formula for $\ell(a,t)$ given by (1.30) into (1.24) and (1.26) to obtain

$$(1.31)\qquad P(t) = \int_0^t B(t-a) \exp\left[-\int_{t-a}^{t} \mu(s+a-t, P(s))\, ds\right] da + \int_t^\infty \phi(a-t) \exp\left[-\int_0^t \mu(s+a-t, P(s))\, ds\right] da$$

$$(1.32)\qquad B(t) = \int_0^t \beta(a,P(t))B(t-a) \exp\left[-\int_{t-a}^{t} \mu(s+a-t, P(s))\, ds\right] da + \int_t^\infty \beta(a,P(t))\phi(a-t) \exp\left[-\int_0^t \mu(s+a-t, P(s))\, ds\right] da$$

or equivalently,

$$(1.33)\qquad P(t) = \int_0^t B(a) \exp\left[-\int_a^t \mu(s-a, P(s))\, ds\right] da + \int_0^\infty \phi(a) \exp\left[-\int_0^t \mu(s+a, P(s))\, ds\right] da$$

$$(1.34)\qquad B(t) = \int_0^t \beta(t-a, P(t))B(a) \exp\left[-\int_a^t \mu(s-a, P(s))\, ds\right] da + \int_0^\infty \beta(a-t, P(t))\phi(a) \exp\left[-\int_0^t \mu(s+a, P(s))\, ds\right] da$$

The equations (1.33), (1.34) constitute a coupled pair of nonlinear Volterra integral equations in B(t), P(t). In fact, the problem (1.33), (1.34) is equivalent to the original problem (1.25), (1.26), (1.27), and if we can solve (1.33), (1.34) for B(t) and P(t), we can then obtain $\ell(a,t)$ from (1.30). Such a program was carried out by Gurtin and MacCamy in [130] under appropriate hypotheses on $\mu$, $\beta$, and $\phi$.

In [130] the existence and stability of equilibrium solutions for the model (1.25), (1.26), (1.27) were investigated. We will discuss briefly some of those results here. For this model an *equilibrium solution* $\hat{\phi}$ will satisfy

$$\frac{d}{da}\hat{\phi}(a) = -\mu(a,P\hat{\phi})\hat{\phi}(a) \qquad a \geq 0 \tag{1.35}$$

$$\hat{\phi}(0) = \int_0^\infty \beta(a,P\hat{\phi})\hat{\phi}(a)\, da \tag{1.36}$$

where

$$P\hat{\phi} \overset{\text{def}}{=} \int_0^\infty \hat{\phi}(a)\, da$$

The solution of (1.35) has the form

$$\hat{\phi}(a) = \hat{\phi}(0) \exp\left[-\int_0^a \mu(b,P\hat{\phi})\, db\right] \qquad a \geq 0 \tag{1.37}$$

Substitution of (1.37) into (1.36) yields

$$1 = \int_0^\infty \beta(a,P\hat{\phi}) \exp\left[-\int_0^a \mu(b,P\hat{\phi})\, db\right] da \tag{1.38}$$

if $\hat{\phi}(0) \neq 0$.

Thus, we see that the problem (1.25), (1.26), (1.27) has a nontrivial equilibrium solution if and only if the equation

$$1 = \int_0^\infty \beta(a,P) \exp\left[-\int_0^a \mu(b,P)\, db\right] da \tag{1.39}$$

has a nonzero solution $P = \hat{P}$. In this case a nontrivial equilibrium solution is given by (1.37), where

$$\hat{\phi}(0) = \frac{\hat{P}}{\int_0^\infty \exp\left[-\int_0^a \mu(b,\hat{P})\, db\right] da} \tag{1.40}$$

Moreover, the formula (1.37) with $\hat{\phi}(0)$ as in (1.40), gives the unique nontrivial equilibrium solution corresponding to this solution $\hat{P}$ of equation (1.39). As noted in [130], the equation (1.39) is much more likely to have a solution than the equation (1.18), and thus the nonlinear Gurtin-MacCamy model is much more likely to have a nontrivial equilibrium solution than the linear Sharpe-Lotka-McKendrick model. Analogously with the verhulstian age-independent model of population dynamics, the Gurtin-MacCamy age-dependent model of population dynamics provides a physically more realistic description of the behavior of biological populations.

## 1.4 FORMULATION OF A GENERAL MODEL

We have seen that it is advantageous to allow the birth and mortality processes of age-dependent population dynamics to be nonlinear functions of the population densities. It is our purpose now to describe a formulation of age-dependent population dynamics which allows such nonlinearities and which is applicable to many diverse population problems. Our formulation generalizes the formulation of Gurtin and MacCamy in that our birth and mortality processes are not required to have a special form. Our formulation will also include vector systems models, as well as scalar models, since many population problems involve interactions between population subclasses.

One of the most important considerations in developing a general formulation of age-dependent population dynamics is to find a mathematically tractable setting for the problem. In our context this setting will be a Banach space of functions having age a as their independent variable. We will choose as our setting the Banach space $L^1$. For many population problems $L^1$ is the natural choice for a mathematical setting in that the physical interpretation of the density function requires that it should be integrable, and the mathematical treatment of the problem requires that the density functions belong to a complete normed linear space. The $L^1$

norm of the density is a natural measure of the size of the population. In many cases the $L^1$ norm yields a priori estimates on the growth of the solutions, and these estimates are essential in order to analyze the behavior of the solutions for large time.

Our approach to age-dependent population dynamics has been influenced by our choice of the Banach space $L^1$ as a mathematically natural setting. It is also influenced by our intention to view age-dependent population dynamics from the vantage point of the theory of semigroups of operators in Banach spaces. The solution of an age-dependent population model evolves with time. The solution should have the property that if the initial age distribution $\phi$ belongs to $L^1$, then the density $\ell(\cdot,t)$ will also belong to $L^1$ at any later time t. Furthermore, the family of mappings $S(t)$, $t \geq 0$, in $L^1$, defined by the formula

$$S(t)\phi = \ell(\cdot,t) \qquad \phi \in L^1 \qquad t \geq 0$$

should form a strongly continuous semigroup of nonlinear continuous mappings (or a dynamical system) in the Banach space $L^1$. Investigation of this nonlinear semigroup and its infinitesimal generator will provide a convenient framework with which to view the structure and behavior of the solutions.

Still another consideration in formulating our problem consists of determining the proper balance law for the mortality process of the population and the proper birth law for the birth process. In what sense will the density function be uniquely determined as a solution of the mathematical equations of the model? This sense of solution will be influenced by our choice of $L^1$ as the mathematical setting for the problem. It should reduce to the sense of solution given in (1.5), (1.7), (1.8) for the classical linear formulation and in (1.25), (1.26), (1.27) for the Gurtin-MacCamy nonlinear formulation.

In order to formulate our approach to age-dependent population dynamics precisely, we now introduce some notation. Let $R^n$ denote n-dimensional vector-space with norm

$$|x| \overset{\text{def}}{=} \sum_{i=1}^{n} |x_i| \qquad x = (x_1, \ldots, x_n)^T \in R^n$$

(for population problems the $\ell^1$ norm of $R^n$ is more natural than the Euclidean norm). Let $B(R^n, R^n)$ denote the Banach algebra of bounded linear operators from $R^n$ to $R^n$ (or equivalently, the linear space of $n \times n$ matrices), with norm

$$|L| \overset{\text{def}}{=} \sup_{x \in R^n, |x|=1} |Lx| \qquad L \in B(R^n, R^n)$$

Let $L^1 \overset{\text{def}}{=} L^1(0,\infty;R^n)$ be the Banach space of equivalence classes of Lebesgue integrable functions from $[0,\infty)$ to $R^n$ which agree almost everywhere on $(0,\infty)$, with norm

$$\|\phi\|_{L^1} \overset{\text{def}}{=} \int_0^\infty |\phi(a)|\, da \qquad \phi \in L^1$$

(here $\phi$ is any representative of the equivalence class in $L^1$ containing $\phi$). If $\phi \in L^1$ and there is some representative of its equivalence class which is a continuous function on $[0,\infty)$, then we will identify $\phi$ as a continuous function defined on $[0,\infty)$.

Let $T > 0$ and let $L_T \overset{\text{def}}{=} C([0,T];L^1)$ be the Banach space of continuous $L^1$-valued functions on $[0,T]$ with the supremum norm

$$\|\ell\|_{L_T} \overset{\text{def}}{=} \sup_{0 \le t \le T} \|\ell(t)\|_{L^1} \qquad \ell \in L_T$$

In Lemma 2.1 in Chapter 2 we will show that each element of $L_T$ is identified in a natural way with an element of $L^1((0,\infty) \times (0,T);R^n)$, the Banach space of equivalence classes of Lebesgue integrable functions from $[0,\infty) \times [0,T]$ to $R^n$. We will use the symbol $\ell$ to denote both of these elements in that

$$\ell(t)(a) = \ell(a,t) \qquad 0 \le t \le T, \text{ a.e. } a > 0 \tag{1.41}$$

where $\ell \in L_T$ and $\ell \in L^1((0,\infty) \times (0,T);R^n)$. If $\ell \in L_T$ and $t \in [0,T]$, then we will also use the notation $\ell(\cdot,t)$ to denote the element of $L^1$ given by (1.41). In our context

$$\ell(\cdot,t) = (\ell_1(\cdot,t),\ldots,\ell_n(\cdot,t))^T$$

where $\ell_i(\cdot,t)$ denotes the density with respect to age at time t of the i-th subclass of a population divided into n subclasses.

Our formulation of age-dependent population dynamics is motivated in the following way: As in (1.24), let P(t) represent the total population at time t. The average rate of change in the total population size in the time interval (t, t + h) is

$$(1.42) \qquad h^{-1}(P(t+h) - P(t)) = h^{-1}\int_0^h \ell(a, t+h)\,da + \int_0^\infty h^{-1}[\ell(a+h, t+h) - \ell(a,t)]\,da$$

As $h \to 0$ in (1.42), the term on the left-hand side converges to the instantaneous rate of change of the total population size at time t, the first term on the right-hand side converges to the instantaneous birth rate at time t, and the second term on the right-hand side converges to the instantaneous rate of change of total population at time t due to causes other than births.

We are thus led to the following formulation of age-dependent population dynamics: Let $T > 0$, let $\ell \in L_{T_1}$, let F be a mapping from $L^1$ to $R^n$, let G be a mapping from $L^1$ into $L^1$, and let $\phi \in L^1$. The *balance law* of the population is given by

$$(1.43) \qquad \lim_{h\to 0^+} \int_0^\infty |h^{-1}[\ell(a+h, t+h) - \ell(a,t)] - G(\ell(\cdot,t))(a)|\,da = 0 \qquad 0 \le t \le T$$

The *birth law* of the population is given by

$$(1.44) \qquad \lim_{h\to 0^+} h^{-1}\int_0^h |\ell(a, t+h) - F(\ell(\cdot,t))|\,da = 0 \qquad 0 \le t \le T$$

The *initial age distribution* of the population is given by

$$(1.45) \qquad \ell(\cdot,0) = \phi$$

From (1.42), (1.43), (1.44) we see that the instantaneous rate of change of the total size of the population satisfies

$$\frac{d}{dt} P(t) = F(\ell(\cdot,t)) + \int_0^\infty G(\ell(\cdot,t))(a)\, da$$

where $F(\ell(\cdot,t))$ represents the birth rate at time t and

$$\int_0^\infty G(\ell(\cdot,t))(a)\, da$$

represents the rate of change of total population at time t due to causes other than births.

We will refer to F as the *birth function* and G as the *aging function*. The term aging function is used here rather than the term mortality function because we wish to allow the inclusion of migration processes as well as mortality processes in the balance law. We will say that the birth law (1.44) describes the *birth process* of the population and the balance law (1.43) describes the *aging process* of the population. The equations (1.43), (1.44), (1.45) constitute our formulation of age-dependent population dynamics in the Banach space $L^1$. The choice of $L^1$ as a setting and the generality of the birth and aging processes dictate the integral conditions in the formulation of the balance law and the birth law. We will refer to the equations (1.43), (1.44), and (1.45) as the *problem (ADP)*. We define

DEFINITION 1.1 Let $T > 0$ and let $\ell \in L_T$. We say that $\ell$ is a *solution of (ADP) on* $[0,T]$ provided that $\ell$ satisfies (1.43), (1.44), and (1.45).

Observe that the solution $\ell$ of (ADP) on $[0,T]$ has the properties that $\ell(\cdot,t) \in L^1$ for each $t \in [0,T]$, the mapping $t \to \ell(\cdot,t)$ is continuous from $[0,T]$ to $L^1$, and the total population between ages $a_1$ and $a_2$ at time t is given by

$$\int_{a_1}^{a_2} \ell(a,t)\, da$$

In Chapter 2 we will prove that the solution $\ell$ also has the property that for almost everywhere $c > -T$, the *cohort function*

$$w_c(t) \overset{\text{def}}{=} \ell(t + c, t) \qquad t_c \overset{\text{def}}{=} \max\{0,-c\} \le t \le T \tag{1.46}$$

is continuous from $[t_c,T]$ to $R^n$, differentiable almost everywhere on $(t_c,T)$, and satisfies

$$\frac{d}{dt} w_c(t) = G(\ell(\cdot,t))(t + c) \qquad \text{a.e. } t \in (t_c,T) \tag{1.47}$$

In the classical linear model (1.5), (1.7), (1.8) F and G have the forms

$$F(\phi) = \int_0^\infty \beta(a)\phi(a)\, da \qquad \phi \in L^1$$

$$G(\phi)(a) = -\mu(a)\phi(a) \qquad \phi \in L^1,\ a \ge 0$$

respectively. In the Gurtin-MacCamy model F and G have the forms

$$F(\phi) = \int_0^\infty \beta(a,P\phi)\phi(a)\, da \qquad \phi \in L^1$$

$$G(\phi)(a) = -\mu(a,P\phi)\phi(a) \qquad \phi \in L^1,\ a \ge 0$$

respectively, where

$$P\phi = \int_0^\infty \phi(a)\, da \qquad \phi \in L^1$$

In the problem (ADP) we allow much greater generality than these examples. As a special case, however, we will restrict our attention to aging functions G having the form

$$G(\phi)(a) = -\mu(a,\phi)\phi(a) \qquad \phi \in L^1,\ \text{a.e. } a > 0 \tag{1.48}$$

where $\mu$ is a function from $[0,\infty) \times L^1$ into $B(R^n,R^n)$. In Chapter 2 we will prove that if G has this form, then for every $\phi \in L^1$ the solution $\ell$ of (ADP) on $[0,T]$ is continuous on the triangle $0 \le a \le t \le T$, satisfies the balance law

$$D\ell(a,t) = -\mu(a,\ell(\cdot,t))\ell(a,t)$$

on this triangle (where D is defined as in (1.6)), and satisfies the birth law

$$\ell(0,t) = F(\ell(\cdot,t))$$

for $t \in [0,T]$.

An aging function G of the form (1.48) can be interpreted as a pure mortality process in the following sense: Suppose that $\Pi(a,b;\phi)$ is the probability that a member of the population of age a will survive to age b when the population density is $\phi$. Suppose that $\Pi(a,b;\phi)$ has the form

$$\Pi(a,b;\phi) = \exp\left[-\int_a^b \mu(\tau,\phi)\, d\tau\right] \qquad 0 \le a \le b$$

where $\mu$ is the age and density dependent mortality modulus. The change in the total population due to mortality in the time interval $[t, t + h]$ is

$$\int_0^\infty [\Pi(a, a + h;\ell(\cdot,t)) - 1]\ell(a,t)\, da$$

If we divide by h and take the limit as h goes to 0, then the instantaneous rate of change of the total population due to mortality (that is, the mortality rate of the total population) is

$$-\int_0^\infty \mu(a,\ell(\cdot,t))\ell(a,t)\, da$$

If we assume that the only reason for total population change other than births is mortality, then it is natural to take an aging function having the form (1.48).

In our development we will not assume that the aging function G always has the form in (1.48). In fact, our basic theory requires no assumptions on the form of the birth function F and the aging function G. For example, we may allow the aging function G to include a migration term, as well as a mortality term, as given by the form

$$G(\phi)(a) = -\mu(a,\phi)\phi(a) + \int_0^\infty \gamma(a,b,\phi(b))\, db$$

The function $\gamma$: $[0,\infty) \times [0,\infty) \times R^n \to R^n$ is associated with a density-dependent migration process into or out of the population, which in this example is nonlocal with respect to the population density.

The method we use to solve the general problem (ADP) is to convert the problem to an equivalent integral equation in which the unknown density function $\ell(a,t)$ is regarded as a function of two independent variables, namely, age a and time t. This approach to age-dependent population dynamics and to other types of functional differential equations has been studied by M. Marcus and V. Mizel in [214], [215], and [216]. This equivalent integral equation is

$$(1.49) \qquad \ell(a,t) = \begin{cases} F(\ell(\cdot,t-a)) + \int_{t-a}^{t} G(\ell(\cdot,s))(s+a-t)\, ds & \text{a.e. } a \in (0,t) \\ \phi(a-t) + \int_0^t G(\ell(\cdot,s))(s+a-t)\, ds & \text{a.e. } a \in (t,\infty) \end{cases}$$

In Lemma 2.2 in Chapter 2 we will show that integrals in (1.49) exist under appropriate assumptions on the mapping F from $L^1$ into $R^n$ and the mapping G from $L^1$ to $L^1$. We define

DEFINITION 1.2 Let $T > 0$ and let $\ell \in L_T$. We say that $\ell$ is a *solution of* (1.49) *on* $[0,T]$ provided that $\ell(\cdot,t)$ satisfies (1.49) for $t \in [0,T]$.

What is the motivation for considering the integral equation (1.49)? Suppose that we have a solution $\ell$ of (ADP) on $[0,T]$ which is sufficiently regular so that it satisfies

$$(1.50) \qquad D\ell(a,t) = G(\ell(\cdot,t))(a) \qquad t \in [0,T],\ a \geq 0$$

$$(1.51) \qquad \ell(0,t) = F(\ell(\cdot,t)) \qquad t \in (0,T]$$

We again consider the solution $\ell(a,t)$ along the characteristic curves $a - t$ = constant. As before, fix $c > -T$ and define the cohort function as in (1.46). From (1.50) we see that the cohort function $w_c$ satisfies (1.47).

We cannot now solve (1.47) explicitly for $w_c$ as we did in (1.11) and (1.29), because we have made no assumptions on the form of G. We can, however, integrate (1.47) to obtain

$$(1.52) \qquad w_c(t) = \begin{cases} w_c(t - a) + \int_{t-a}^{t} G(\ell(\cdot,s))(s + c)\, ds & a < t \\ w_c(0) + \int_{0}^{t} G(\ell(\cdot,s))(s + c)\, ds & a \geq t \end{cases}$$

Now substitute a - t for c and use (1.51) to show that $\ell$ satisfies (1.49) on [0,T]. We see, therefore, that a sufficiently regular solution of (ADP) on [0,T] provides a solution of the integral equation (1.49) on [0,T]. We will see in Chapter 2 that the two problems are, in fact, equivalent.

The nonlinear integral equation (1.49) will be the starting point for our development. In Chapter 2 we will prove the existence of unique local solutions to (1.49) under local Lipschitz continuity conditions on the birth function F and the aging function G. We will then establish the equivalence of the problem (ADP) and the integral equation (1.49). We will next consider the continuability of local solutions to global solutions, demonstrate the continuous dependence of solutions upon the initial age distribution $\phi$, the birth function F, and the aging function G, provide criteria for the positivity of solutions, and study the regularity properties of these solutions. In Chapter 3 we will show that the solutions of (ADP) form a strongly continuous semigroup of nonlinear operators in the Banach space $L^1$. We will analyze the infinitesimal generator of this nonlinear semigroup and give a representation of the nonlinear semigroup by means of an exponential formula involving the infinitesimal generator. Under appropriate assumptions we will prove that the trajectories of this nonlinear semigroup are compact in the Banach space $L^1$. We will then discuss several numerical schemes which may be used to approximate the trajectories. In

Chapter 4 we will consider the question of the existence and stability of equilibrium solutions of (ADP). Under appropriate hypotheses we will prove the existence of nontrivial equilibrium solutions by means of a fixed point argument. We will analyze the asymptotic behavior of the solutions of (ADP) by means of Lyapunov functionals and the Invariance Principle. We will also analyze the stability of stable age distributions for linear equations. We will then analyze the local stability of equilibria of nonlinear equations by the method of linearization. Last, in Chapter 5 we will apply the theory we have developed to various examples of biological population models.

## 1.5 NOTES

An interesting historical discussion of the early development of classical continuous linear age-dependent population dynamics is given by P. Samuelson in [253]. The classical theory of continuous linear age-dependent population dynamics has a discretized analog originating with the work of H. Bernardelli [20], P. H. Leslie [193], [194], and E. G. Lewis [196]. Detailed treatments of discrete linear age-dependent population dynamics, as well as stochastic versions, are given in the books of N. Keyfitz [168] and J. H. Pollard [235].

The theory of nonlinear continuous age-dependent population dynamics has undergone extensive development since the appearance of the first work of F. Hoppensteadt in [152], [153] and M. Gurtin and R. C. MacCamy in [130]. In our references we have assembled many of the relevant articles. We mention some of the directions of this development here.

(1) *Nonautonomous models and control models:* Nonlinear models with nonautonomous time dependence or models with a coupled control variable have been investigated by M. Chipot [37], C. V. Coffman and B. D. Coleman [44], J. Cushing [68], [69], J. Cushing and M. Saleem [70], R. H. Elderkin [100], M. Gyllenberg [137],

[138], A. Haimovici [140], K. Swick [267], and E. Sinestrari [261].

(2) *Models with delay:* Models of nonlinear age-dependent population dynamics with history dependence have been studied by J. Cushing [68], G. Di Blasio [84], K. Gopalsamy [118], P. Marcati [210], and K. Swick [268].

(3) *Diffusion models:* Nonlinear age-dependent population dynamics with spatial diffusion has been studied by S. Busenberg and M. Iannelli [33], [34], [35], G. Di Blasio [83], [85], G. Di Blasio, M. Iannelli, and E. Sinestrari [86], G. Di Blasio and L. Lamberti [88], M. G. Garroni and M. Langlais [115], M. Gurtin [126], M. Gurtin and R. C. MacCamy [131], [134], K. Kunisch, W. Schappacher, and G. Webb [180], M. Langlais [188], [189], R. C. MacCamy [205], [206], P. Marcati [210], P. Marcati and R. Serafini [213], and the author [302], [304], [306], [307].

# 2
# Basic Theory of the Solutions

## 2.1 PRELIMINARIES

The basic assumption we place on the birth function F and the aging function G in the problem (ADP) is that each is a Lipschitz continuous function on bounded sets of $L^1$. This Lipschitz continuity condition is of local type, and is satisfied, for example, in the Gurtin-MacCamy model (1.25), (1.26), (1.27). Specifically, we require that

(2.1) $F: L^1 \to R^n$, there is an increasing function $c_1$: $[0,\infty) \to [0,\infty)$ such that $|F(\phi_1) - F(\phi_2)| \le c_1(r)\|\phi_1 - \phi_2\|_{L^1}$ for all $\phi_1,\phi_2 \in L^1$ such that $\|\phi_1\|_{L^1}, \|\phi_2\|_{L^1} \le r$.

(2.2) $G: L^1 \to L^1$, there is an increasing function $c_2$: $[0,\infty) \to [0,\infty)$ such that $\|G(\phi_1) - G(\phi_2)\|_{L^1} \le c_2(r)\|\phi_1 - \phi_2\|_{L^1}$ for all $\phi_1,\phi_2 \in L^1$ such that $\|\phi_1\|_{L^1}, \|\phi_2\|_{L^1} \le r$.

We now give three technical lemmas. The first lemma allows us to view an element in $L_T = C([0,T];L^1)$ as an element in $L^1((0,\infty) \times (0,T);R^n)$. The second lemma assures us that whenever $\ell$ belongs to $L_T$, then the integrals in (1.49) exist. The third lemma characterizes compact sets in $L^1$.

LEMMA 2.1 Let $T > 0$ and let $\ell \in L_T$. There is a unique element in $L^1((0,\infty) \times (0,T);R^n)$ (which we also denote by $\ell$) such that

(2.3) For each $t \in [0,T]$, $\ell(a,t) = \ell(t)(a)$ for almost everywhere $a > 0$.

$$\int_0^T \|\ell(t)\|_{L^1}\, dt = \int_0^T \left[\int_0^\infty |\ell(a,t)|\, da\right] dt = \int_0^\infty \left[\int_0^T |\ell(a,t)|\, dt\right] da = \int_0^\infty \int_0^T |\ell(a,t)|\, dt\, da \tag{2.4}$$

*Proof.* By Theorem 17, p. 198 in [94] there is a measurable function k on the product space $[0,\infty) \times [0,T]$, which is uniquely determined except for a set of measure zero in $(0,\infty) \times (0,T)$, such that $k(\cdot,t) = \ell(t)$ for almost all $t \in (0,T)$, say $t \in (0,T) - E$, where E has measure zero. For $t \in [0,T]$ let $\ell(t)(a)$ be any representative of the equivalence class $\ell(t) \in L^1$. Define a function $\ell$ from $[0,\infty) \times [0,T]$ to $R^n$ by

$$\ell(a,t) = \begin{cases} k(a,t) & \text{if } t \in (0,T) - E,\ a \geq 0 \\ \ell(t)(a) & \text{if } t \in E,\ a \geq 0 \end{cases} \tag{2.5}$$

Since $m((0,\infty) \times E) = m(0,\infty)m(E) = 0$ (see [94], Corollary 6, p. 187), the function $\ell(a,t)$ defined by (2.5) agrees with $k(a,t)$ except on a set of measure zero in the product space. Since $k(a,t)$ is measurable in the product space, so is $\ell(a,t)$. By the Fubini-Tonelli theorem (see [94], Theorem 13, p. 193 and Corollary 15, p. 194) a measurable function on the product space $(0,\infty) \times (0,T)$ is integrable on the product space if and only if one of its iterated integrals exists, in which case the other iterated integral exists and all three of these integrals are equal. Then, (2.4) follows immediately, since the function $t \to \ell(t)$ is continuous from $[0,T]$ to $L^1$, so that the iterated integral

$$\int_0^T \left[\int_0^\infty |\ell(a,t)|\, da\right] dt = \int_0^T \|\ell(t)\|_{L^1}\, dt$$

exists. □

LEMMA 2.2 Let (2.2) hold, let $T > 0$, let $\Gamma_T \overset{\text{def}}{=} \{(c,s): 0 < s < T, -s < c < \infty\}$, and let $\ell \in L_T$. The following hold:

(2.6) The function $t \to G(\ell(\cdot,t))$ from $[0,T]$ to $L^1$ belongs to $L_T$.

(2.7) There exists $h \in L^1((0,\infty) \times (0,T);R^n)$ such that for each $t \in [0,T]$, $h(a,t) = G(\ell(\cdot,t))(a)$ for almost everywhere $a > 0$.

(2.8) There exists $k \in L^1(\Gamma_T;R^n)$ such that $k(c,s) = h(s + c, s)$ for almost everywhere $(c,s) \in \Gamma_T$, and

$$\int_0^T \left[\int_{-s}^{\infty} k(c,s)\, dc\right] ds = \int_{-T}^{\infty} \left[\int_{\max\{0,-c\}}^{T} k(c,s)\, ds\right] dc$$

*Proof.* Since G is continuous from $L^1$ to $L^1$, (2.6) must hold. Since (2.6) holds, (2.7) follows by Lemma 2.1. Define $k: \Gamma_T \to R^n$ by $k(c,s) \overset{\text{def}}{=} h(s + c, s)$ for almost all $(c,s) \in \Gamma_T$. Since h is measurable on $(0,\infty) \times (0,T)$, k is measurable on $\Gamma_T$. Further, $\|G(\ell(\cdot,t))\|_{L^1}$ is continuous, and hence integrable, in t. Thus,

$$\begin{aligned}\int_0^T \left[\int_0^{\infty} G(\ell(\cdot,t))(a)\, da\right] dt &= \int_0^T \left[\int_0^{\infty} h(a,t)\, da\right] dt \\ &= \int_0^T \left[\int_{-s}^{\infty} h(s + c, s)\, dc\right] ds \\ &= \int_0^T \left[\int_{-s}^{\infty} k(c,s)\, dc\right] ds\end{aligned}$$

Since k is measurable on $\Gamma_T$ and one of its iterated integrals exist, we can apply the Fubini-Tonelli theorem to obtain (2.8) (see Figure 2.1 below). □

REMARK 2.1 Lemma 2.1 and Lemma 2.2 provide a justification for the following notation: Let $T > 0$ and let $\ell \in L_T$. The function $t \to G(\ell(\cdot,t))$ from $[0,T]$ to $L^1$ belongs to $L_T$, and by Lemma 2.1 can be identified with an integrable function $(a,t) \to G(\ell(\cdot,t))(a)$ on

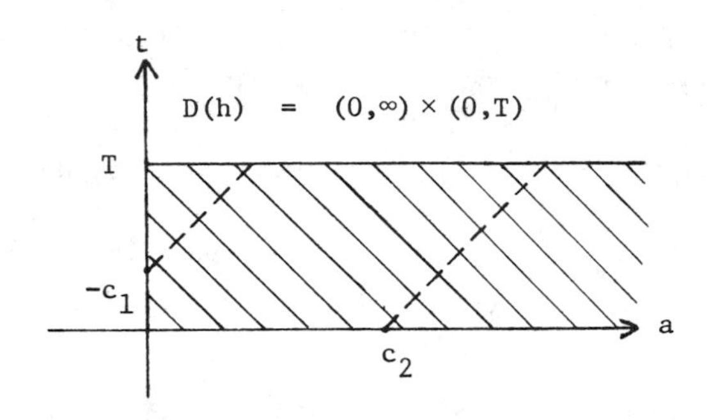


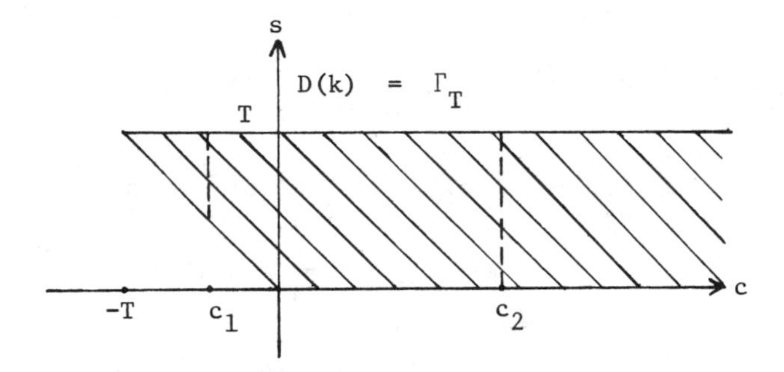


*Figure 2.1*

$(0,\infty) \times (0,T)$. By Lemma 2.2 we can identify this function $(a,t) \to G(\ell(\cdot,t))(a)$ with an integrable function $(c,s) \to G(\ell(\cdot,s))(s + c)$ on $\Gamma_T$. Therefore, for almost all $c > -T$ the function $s \to G(\ell(\cdot,s))(s + c)$ is integrable on $(\max\{0,-c\},T)$, and in this sense the integrals in (1.49) exist. Furthermore, in this notation (2.8) yields

$$(2.9) \qquad \int_0^T \left[ \int_{-s}^{\infty} G(\ell(\cdot,s))(s + c)\, dc \right] ds$$

$$= \int_{-T}^{\infty} \left[ \int_{\max\{0,-c\}}^{T} G(\ell(\cdot,s))(s + c)\, ds \right] dc$$

The proof of the following lemma may be found in [94], Theorem 20, p. 298.

LEMMA 2.3 A closed and bounded subset M of $L^1$ is compact if and only if the following two conditions hold:

(2.10) $\lim_{h\to 0} \int_0^\infty |\phi(a) - \phi(a+h)|\, da = 0$ uniformly for $\phi \in M$ (where $\phi(a+h)$ is taken as 0 if $a + h < 0$).

(2.11) $\lim_{h\to\infty} \int_h^\infty |\phi(a)|\, da = 0$ uniformly for $\phi \in M$.

REMARK 2.2 We remark that translation is a continuous operation in $L^1$ in the following sense: Let $h \in R$ and define $T_h\colon L^1 \to L^1$ by

$$(T_h\phi)(a) = \begin{cases} \phi(a+h) & \text{a.e. } a > \max\{0,-h\} \\ 0 & \text{a.e. } a \in (0,\max\{0,-h\}) \end{cases}$$

for $\phi \in L^1$. If $\phi_n \to \phi$ in $L^1$, then

$$\|T_h\phi_n - T_h\phi\|_{L^1} = \int_{\max\{0,-h\}}^\infty |\phi_n(a+h) - \phi(a+h)|\, da$$
$$\leq \int_0^\infty |\phi_n(a) - \phi(a)|\, da \to 0 \text{ as } n \to \infty$$

Further, if $\phi \in L^1$, then

$$\lim_{h\to 0} \|T_h\phi - \phi\|_{L^1} = 0$$

by (2.10) of Lemma 2.3 (take $M = \{\phi\}$).

## 2.2 LOCAL EXISTENCE AND CONTINUOUS DEPENDENCE ON INITIAL AGE DISTRIBUTIONS

We will first prove that a solution of the integral equation (1.49) is also a solution of (ADP).

PROPOSITION 2.1 Let (2.1), (2.2) hold, let $T > 0$, let $\phi \in L^1$, and let $\ell \in L_T$. If $\ell$ is a solution of (1.49) on $[0,T]$, then $\ell$ is a solution of (ADP) on $[0,T]$.

*Proof.* First, $\ell(\cdot,0) = \phi$, since $\ell(\cdot,t)$ satisfies (1.49) at $t = 0$. Next, let $0 \le t < T$ and let $0 < h < T - t$. From (1.49) and Remark 2.1 we have that

$$\int_0^\infty |h^{-1}[\ell(a + h, t + h) - \ell(a,t)] - G(\ell(\cdot,t))(a)|\, da$$

$$= \int_0^\infty \left| h^{-1} \int_t^{t+h} [G(\ell(\cdot,s))(s + a - t) - G(\ell(\cdot,t))(a)]\, ds \right| da$$

$$\le h^{-1} \int_t^{t+h} \left[ \int_0^\infty |G(\ell(\cdot,s))(s + a - t) - G(\ell(\cdot,t))(a)|\, da \right] ds$$

$$\le h^{-1} \int_t^{t+h} \left[ \int_0^\infty |G(\ell(\cdot,s))(s + a - t) - G(\ell(\cdot,t))(s + a - t)|\, da \right.$$

$$\left. + \int_0^\infty |G(\ell(\cdot,t))(s + a - t) - G(\ell(\cdot,t))(a)|\, da \right] ds$$

$$\le \sup_{t \le s \le t+h} \left[ \|G(\ell(\cdot,s)) - G(\ell(\cdot,t))\|_{L^1} + \int_0^\infty |G(\ell(\cdot,t))(s + a - t) \right.$$

$$\left. - G(\ell(\cdot,t))(a)|\, da \right]$$

This last expression approaches 0 as h approaches 0 by the continuity of the function $s \to G(\ell(\cdot,s))$ from $[0,T]$ to $L^1$ and by the continuity of translation in $L^1$ (see Remark 2.2). Hence, (1.43) holds.

Next, let $0 \le t < T$ and let $0 < h < T - t$. From (1.49) we have that

$$h^{-1} \int_0^h |\ell(a, t + h) - F(\ell(\cdot,t))|\, da$$

$$\le h^{-1} \int_0^h |F(\ell(\cdot, t + h - a)) - F(\ell(\cdot,t))|\, da$$

$$+ h^{-1} \int_0^h \left[ \int_{t+h-a}^{t+h} |G(\ell(\cdot,s))(s + a - t - h)|\, ds \right] da \overset{\text{def}}{=} K_1 + K_2$$

Obviously, $K_1$ approaches 0 as h approaches 0 by the continuity of the function $s \to F(\ell(\cdot,s))$ from $[0,T]$ to $R^n$. Also, we may

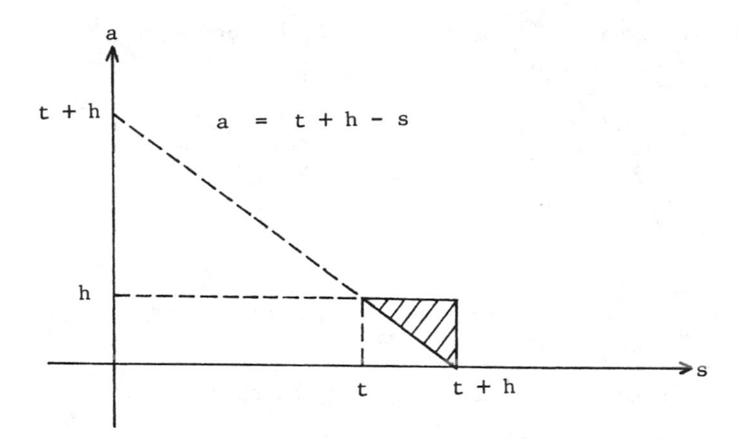


*Figure 2.2*

interchange the order of integration in $K_2$ (see Remark 2.1 and Figure 2.2 above) to obtain

$$K_2 = h^{-1} \int_t^{t+h} \left[ \int_{t+h-s}^{h} |G(\ell(\cdot,s))(s + a - t - h)| \, da \right] ds$$

$$= h^{-1} \int_t^{t+h} \left[ \int_0^{s-t} |G(\ell(\cdot,s))(a)| \, da \right] ds$$

Thus, $K_2$ approaches 0 as h approaches 0, since

$$\lim_{s \to t^+} \int_0^{s-t} |G(\ell(\cdot,s))(a)| \, da \le \lim_{s \to t^+} \left[ \|G(\ell(\cdot,s)) - G(\ell(\cdot,t))\|_{L^1} + \int_0^{s-t} |G(\ell(\cdot,t))(a)| \, da \right] = 0$$

Hence, (1.44) holds. □

We now prove the existence of a unique local solution to the integral equation (1.49).

PROPOSITION 2.2 Let (2.1), (2.2) hold and let $r > 0$. There exists $T > 0$ such that if $\phi \in L^1$ and $\|\phi\|_{L^1} \le r$, then there is a unique function $\ell \in L_T$ such that $\ell$ is a solution of (1.49) on $[0,T]$.

*Proof.* Choose $T > 0$ such that

$$T \frac{c_1(2r) + c_2(2r) + [|F(0)| + \|G(0)\|_{L^1}]}{2r} + \frac{1}{2} \le 1 \tag{2.12}$$

Let $\phi \in L^1$ such that $\|\phi\|_{L^1} \le r$. Define

$$M \overset{\text{def}}{=} \{\ell \in L_T\colon \ \ell(\cdot,0) = \phi \text{ and } \|\ell\|_{L_T} \le 2r\}$$

Obviously, M is a closed subset of $L_T$. Define a mapping K on M as follows: for $\ell \in M$, $t \in [0,T]$,

$$K\ell(a,t) \overset{\text{def}}{=} \begin{cases} F(\ell(\cdot,\ t - a)) + \int_{t-a}^{t} G(\ell(\cdot,s))(s + a - t)\, ds & \text{a.e. } a \in (0,t) \\ \phi(a - t) + \int_0^t G(\ell(\cdot,s))(s + a - t)\, ds & \text{a.e. } a \in (t,\infty) \end{cases} \tag{2.13}$$

We will prove that K is a strict contraction from M into M.

Let $\ell \in M$, $t \in [0,T]$. By using (2.1), (2.2), and by interchanging the order of integration, we obtain

$$\begin{aligned} &\int_0^\infty |K\ell(a,t)|\, da \\ &\le \int_0^t \left[|F(\ell(\cdot,\ t - a))| + \int_{t-a}^t |G(\ell(\cdot,s))(s + a - t)|\, ds\right] da \\ &\quad + \int_t^\infty \left[|\phi(a - t)| + \int_0^t |G(\ell(\cdot,s))(s + a - t)|\, ds\right] da \\ &\le \int_0^t |F(\ell(\cdot,s))|\, ds + \int_0^t \left[\int_{t-s}^t |G(\ell(\cdot,s))(s + a - t)|\, da\right] ds \\ &\quad + \int_0^\infty |\phi(a)|\, da + \int_0^t \left[\int_t^\infty |G(\ell(\cdot,s))(s + a - t)|\, da\right] ds \\ &\le c_1(2r) \int_0^t \|\ell(\cdot,s)\|_{L^1}\, ds + \int_0^t |F(0)|\, ds \end{aligned} \tag{2.14}$$

$$+ \|\phi\|_{L^1} + \int_0^t \|G(\ell(\cdot,s))\|_{L^1} \, ds$$

$$\le (c_1(2r) + c_2(2r)) \int_0^t \|\ell(\cdot,s)\|_{L^1} \, ds + \int_0^t |F(0)| \, ds$$

$$+ r + \int_0^t \|G(0)\|_{L^1} \, ds$$

$$\le \left[ t \, \frac{c_1(2r) + c_2(2r) + [|F(0)| + \|G(0)\|_{L^1}]}{2r} + \frac{1}{2} \right] 2r \le 2r$$

Thus, $K\ell(\cdot,t) \in L^1$ for each $t \in [0,T]$, and $\sup_{0\le t\le T} \|K\ell(\cdot,t)\|_{L^1} \le 2r$. Obviously, $\ell(\cdot,0) = \phi$.

We next show that for $\ell \in M$ the function $t \to K\ell(\cdot,t)$ is continuous from $[0,T]$ to $L^1$. Let $\ell \in M$ and let $0 \le t < \hat{t} \le T$. Then, from (2.13) we have that

$$\|K\ell(\cdot,t) - K\ell(\cdot,\hat{t})\|_{L^1}$$

$$\le \int_0^t \left| F(\ell(\cdot, t - a)) + \int_{t-a}^t G(\ell(\cdot,s))(s + a - t) \, ds \right.$$

$$\left. - F(\ell(\cdot, \hat{t} - a)) - \int_{\hat{t}-a}^{\hat{t}} G(\ell(\cdot,s))(s + a - \hat{t}) \, ds \right| da$$

$$+ \int_t^{\hat{t}} \left| \phi(a - t) + \int_0^t G(\ell(\cdot,s))(s + a - t) \, ds \right.$$

$$\left. - F(\ell(\cdot, \hat{t} - a)) - \int_{\hat{t}-a}^{\hat{t}} G(\ell(\cdot,s))(s + a - \hat{t}) \, ds \right| da$$

$$+ \int_{\hat{t}}^{\infty} \left| \phi(a - t) + \int_0^t G(\ell(\cdot,s))(s + a - t) \, ds \right.$$

$$\left. - \phi(a - \hat{t}) - \int_0^{\hat{t}} G(\ell(\cdot,s))(s + a - \hat{t}) \, ds \right| da$$

$$\stackrel{\text{def}}{=} J_1 + J_2 + J_3$$

Consider first $J_1$. Obviously,

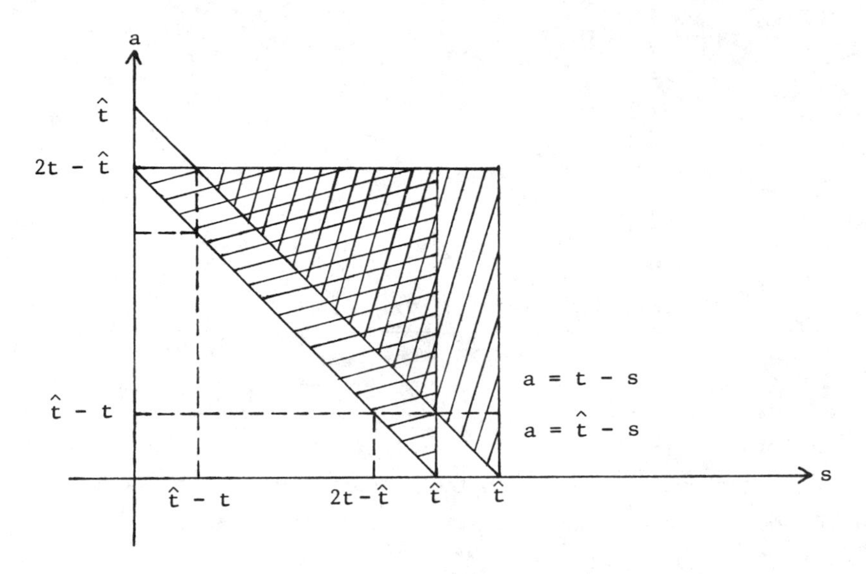


*Figure 2.3*

$$\int_0^t |F(\ell(\cdot, t - a)) - F(\ell(\cdot, \hat{t} - a))| \, da \to 0 \text{ as } |t - \hat{t}| \to 0$$

since $s \to \ell(\cdot,s)$ is continuous from $[0,T]$ to $L^1$ and F is continuous from $L^1$ to $R^n$. Next, if $0 < \hat{t} - t < t$, then (see Figure 2.3)

$$\int_0^t \left| \int_{t-a}^t G(\ell(\cdot,s))(s + a - t) \, ds - \int_{\hat{t}-a}^{\hat{t}} G(\ell(\cdot,s))(s + a - t) \, ds \right| da$$

$$\leq \int_0^{\hat{t}-t} \left[ \int_{t-a}^t |G(\ell(\cdot,s))(s + a - t)| \, ds \right.$$

$$\left. + \int_{\hat{t}-a}^{\hat{t}} |G(\ell(\cdot,s))(s + a - \hat{t})| \, ds \right] da$$

$$+ \int_{\hat{t}-t}^t \left[ \int_{t-a}^{\hat{t}-a} |G(\ell(\cdot,s))(s + a - t)| \, ds \right.$$

$$+ \int_{\hat{t}-a}^t |G(\ell(\cdot,s))(s + a - t) - G(\ell(\cdot,s))(s + a - \hat{t})| \, ds$$

$$\left. + \int_t^{\hat{t}} |G(\Pi(\cdot,s))(s + a - \hat{t})| \, ds \right] da$$

$$= \int_{2t-\hat{t}}^{t} \left[ \int_{t-s}^{\hat{t}-t} |G(\ell(\cdot,s))(s + a - t)| \, da \right] ds$$

$$+ \int_{t}^{\hat{t}} \left[ \int_{\hat{t}-s}^{\hat{t}-t} |G(\ell(\cdot,s))(s + a - \hat{t})| \, da \right] ds$$

$$+ \int_{2t-\hat{t}}^{t} \left[ \int_{\hat{t}-t}^{\hat{t}-s} |G(\ell(\cdot,s))(s + a - t)| \, da \right] ds$$

$$+ \int_{\hat{t}-t}^{2t-\hat{t}} \left[ \int_{t-s}^{\hat{t}-s} |G(\ell(\cdot,s))(s + a - t)| \, da \right] ds$$

$$+ \int_{0}^{\hat{t}-t} \left[ \int_{t-s}^{t} |G(\ell(\cdot,s))(s + a - t)| \, da \right] ds$$

$$+ \int_{\hat{t}-t}^{t} \left[ \int_{\hat{t}-s}^{t} |G(\ell(\cdot,s))(s + a - t) - G(\ell(\cdot,s))(s + a - \hat{t})| \, da \right] ds$$

$$+ \int_{t}^{\hat{t}} \left[ \int_{\hat{t}-t}^{t} |G(\ell(\cdot,s))(s + a - \hat{t})| \, da \right] ds$$

$$= \int_{\hat{t}-t}^{t} \left[ \int_{t-s}^{\hat{t}-s} |G(\ell(\cdot,s))(s + a - t)| \, da \right] ds$$

$$+ \int_{0}^{\hat{t}-t} \left[ \int_{t-s}^{t} |G(\ell(\cdot,s))(s + a - t)| \, da \right] ds$$

$$+ \int_{t}^{\hat{t}} \left[ \int_{\hat{t}-s}^{t} |G(\ell(\cdot,s))(s + a - \hat{t})| \, da \right] ds$$

$$+ \int_{\hat{t}-t}^{t} \left[ \int_{\hat{t}-s}^{t} |G(\ell(\cdot,s))(s + a - t) - G(\ell(\cdot,s))(s + a - \hat{t})| \, da \right] ds$$

$$\leq \int_{\hat{t}-t}^{t} \left[ \int_{0}^{\hat{t}-t} |G(\ell(\cdot,s))(b)| \, db \right] ds + \int_{0}^{\hat{t}-t} \|G(\ell(\cdot,s))\|_{L^1} \, ds$$

$$+ \int_{t}^{\hat{t}} \|G(\ell(\cdot,s))\|_{L^1} \, ds$$

$$+ \int_{\hat{t}-t}^{t} \left[ \int_{\hat{t}-t}^{s} |G(\ell(\cdot,s))(b) - G(\ell(\cdot,s))(b + t - \hat{t})| \, db \right] ds$$

Of these last four integrals, the second and third approach 0 as $|t - \hat{t}|$ approaches 0 by the continuity of the function $s \to G(\ell(\cdot,s))$ from $[0,T]$ to $L^1$; the fourth approaches 0 as $|t - \hat{t}|$ approaches 0 by the uniform continuity of translation on the compact set $\{G(\ell(\cdot,s)) : s \in [0,T]\}$ in $L^1$ (see Lemma 2.3); the first approaches 0 as $|t - \hat{t}|$ approaches 0 by the Lebesgue convergence theorem (see [249], Theorem 15, p. 88), since for each $s \in [0,T]$,

$$\left|\int_0^{\hat{t}-t} G(\ell(\cdot,s))(b)\, db\right| \le \|G(\ell(\cdot,s))\|_{L^1}$$

and

$$\int_0^{\hat{t}-t} G(\ell(\cdot,s)(b)\, db \to 0 \text{ as } |t - \hat{t}| \to 0$$

Thus, we have shown that $J_1$ approaches 0 as $|t - \hat{t}|$ approaches 0 if $0 < \hat{t} - t < t$, and a similar argument holds if $t = 0$.

Next, consider $J_2$. Observe that

$$J_2 \le \int_0^{\hat{t}-t} |\phi(a)|\, da + \int_0^{\hat{t}-t}\left[\int_0^t |G(\ell(\cdot,s))(s + c)|\, ds\right] dc$$
$$+ \int_0^{\hat{t}-t} |F(\ell(\cdot,s))|\, ds + \int_{t-\hat{t}}^0 \left[\int_{-c}^{\hat{t}} |G(\ell(\cdot,s))(s + c)|\, ds\right] dc$$

Now use the fact that

$$\int_0^t |G(\ell(\cdot,s))(s + c)|\, ds \quad \text{and} \quad \int_0^{\hat{t}} |G(\ell(\cdot,s))(s + c)|\, ds$$

are integrable in c (see Remark 2.1), and $|F(\ell(\cdot,s))|$ is integrable in s, to show that $J_2$ approaches 0 as $|t - \hat{t}|$ approaches 0.

Last, consider $J_3$. Observe that

$$J_3 \le \int_{\hat{t}}^{\infty} |\phi(a - t) - \phi(a - \hat{t})|\, da$$
$$+ \int_{\hat{t}}^{\infty}\left[\int_0^t |G(\ell(\cdot,s))(s + a - t) - G(\ell(\cdot,s))(s + a - \hat{t})|\, ds\right] da$$

$$+ \int_{\hat{t}}^{\infty}\left[\int_{t}^{\hat{t}} |G(\ell(\cdot,s))(s + a - \hat{t})| \, ds\right] da$$

$$\leq \int_0^{\infty} |\phi(a + \hat{t} - t) - \phi(a)| \, da$$

$$+ \int_0^{t}\left[\int_{\hat{t}}^{\infty} |G(\ell(\cdot,s))(s + a - t) - G(\ell(\cdot,s))(s + a - \hat{t})| \, da\right] ds$$

$$+ \int_t^{\hat{t}} \|G(\ell(\cdot,s))\|_{L^1} \, ds$$

Now use the fact that $G(\ell(\cdot,s))$ is continuous in s from $[0,T]$ to $L^1$, and the fact that translation is uniformly continuous on the compact set $\{G(\ell(\cdot,s)) : 0 \leq s \leq T\}$ in $L^1$, to show that $J_3$ approaches 0 as $|t - \hat{t}|$ approaches 0.

We have thus shown that $K\ell(\cdot,t)$ is continuous as a function in t from $[0,T]$ to $L^1$ for each $\ell \in M$. Hence, K maps into M. Now use (2.12) and an argument similar to that in (2.14) to show that

$$\|K\ell_1 - K\ell_2\|_{L_T} < \frac{1}{2} \|\ell_1 - \ell_2\|_{L_T} \qquad \ell_1,\ell_2 \in L_T$$

Consequently, K is a strict contraction in M and by the contraction mapping theorem (see [217], Theorem 1.1, p. 114), there is a unique point $\ell \in M$ such that $K\ell = \ell$. This unique fixed point $\ell$ of K in M is the unique solution of (1.49) on $[0,T]$. □

The next proposition shows that the solutions of (ADP) depend continuously on the initial age distributions.

PROPOSITION 2.3 Let (2.1), (2.2) hold, let $\phi$, $\hat{\phi} \in L^1$, let $T > 0$, and let $\ell$, $\hat{\ell} \in L_T$ such that $\ell$, $\hat{\ell}$ is the solution of (ADP) on $[0,T]$ for $\phi$, $\hat{\phi}$, respectively. Let $r > 0$ such that $\|\ell\|_{L_T}$, $\|\hat{\ell}\|_{L_T} \leq r$. Then,

$$\text{(2.15)} \qquad \|\ell(\cdot,t) - \hat{\ell}(\cdot,t)\|_{L^1}$$

$$\leq \exp[(c_1(r) + c_2(r))t]\|\phi - \hat{\phi}\|_{L^1} \qquad \text{for } 0 \leq t \leq T$$

*Proof.* Let $0 \le t \le T$ and define

$$V(t) \overset{\text{def}}{=} \int_0^\infty |\ell(a,t) - \hat{\ell}(a,t)|\, da$$

$$= \int_{-t}^\infty |\ell(t + c, t) - \hat{\ell}(t + c, t)|\, dc$$

It suffices to show that for $0 \le t < T$,

$$\limsup_{h\to 0^+} h^{-1}[V(t + h) - V(t)] \le [c_1(r) + c_2(r)]V(t) \tag{2.16}$$

since (2.15) follows from (2.16) (see [183], Theorem 1.4.1, p. 15). For $0 < h < T - t$,

$$\begin{aligned}
&h^{-1}[V(t + h) - V(t)] \\
&= h^{-1} \int_{-t-h}^{-t} |\ell(t + h + c, t + h) - \hat{\ell}(t + h + c, t + h)|\, dc \\
&\quad + h^{-1} \int_{-t}^\infty [|\ell(t + h + c, t + h) - \hat{\ell}(t + h + c, t + h)| \\
&\quad - |\ell(t + c, t) - \hat{\ell}(t + c, t)|]\, dc \\
&\le h^{-1} \int_0^h [|\ell(a, t + h) - F(\ell(\cdot,t))| + |F(\ell(\cdot,t)) - F(\hat{\ell}(\cdot,t))| \\
&\quad + |F(\hat{\ell}(\cdot,t)) - \hat{\ell}(a, t + h)|]\, da \\
&\quad + \int_0^\infty [|h^{-1}[\ell(a + h, t + h) - \ell(a,t)] - G(\ell(\cdot,t))(a)| \\
&\quad + |G(\ell(\cdot,t))(a) - G(\hat{\ell}(\cdot,t))(a)| \\
&\quad + |h^{-1}[\hat{\ell}(a + h, t + h) - \hat{\ell}(a,t)] - G(\hat{\ell}(\cdot,t))(a)|]\, da
\end{aligned}$$

From (1.43), (1.44) we obtain

$$\begin{aligned}
&\limsup_{h\to 0^+} h^{-1}[V(t + h) - V(t)] \\
&\le |F(\ell(\cdot,t)) - F(\hat{\ell}(\cdot,t))| + \int_0^\infty |G(\ell(\cdot,t))(a) - G(\hat{\ell}(\cdot,t))(a)|\, da
\end{aligned}$$

and from (2.1), (2.2) we then obtain (2.16). □

We now collect the consequences of Propositions 2.1, 2.2, 2.3 into

THEOREM 2.1 Let (2.1), (2.2) hold and let $\phi \in L^1$. There exists $T > 0$ and $\ell \in L_T$ such that $\ell$ is a solution of (ADP) on $[0,T]$. Furthermore, if $T > 0$, then there is at most one solution of (ADP) on $[0,T]$.

*Proof.* The first assertion follows immediately from Propositions 2.1 and 2.2. The second assertion follows immediately from Proposition 2.3 by taking $\phi = \phi$ in (2.15). □

The following three examples of one-species population models provide simple illustrations for Theorem 2.1. The models are treated in [130], [132], [133] by M. Gurtin and R. MacCamy and by M. Gurtin in [127].

EXAMPLE 2.1 Let $R^n = R$. Let $\alpha \geq 0$, let $\beta$: $R \to [0,\infty)$ such that $\beta$ is continuously differentiable, let $P$: $L^1 \to R$ be defined by $P\phi \overset{\text{def}}{=} \int_0^\infty \phi(a)\, da$, and let $F$: $L^1 \to R$ be defined by

$$(2.17) \qquad F(\phi) \overset{\text{def}}{=} \int_0^\infty \beta(P\phi)e^{-\alpha a}\phi(a)\, da \qquad \phi \in L^1$$

In the case that $\alpha > 0$, this birth function corresponds to a concentration of reproductive capacity in the youngest ages. Let $\mu$: $R \to [0,\infty)$ such that $\mu$ is continuously differentiable, and let $G$: $L^1 \to L^1$ be defined by

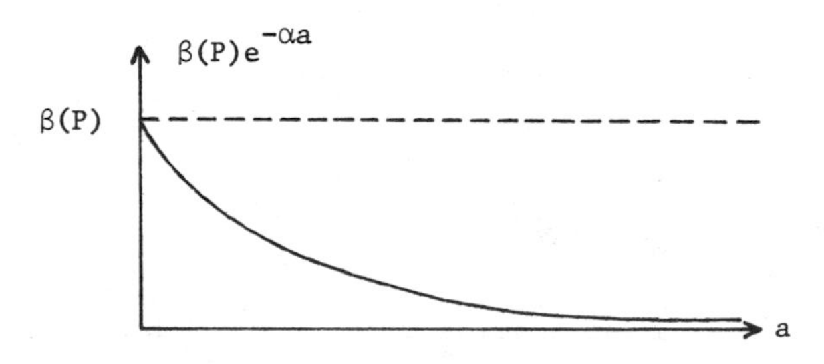


*Figure 2.4*

$$(2.18) \qquad G(\phi)(a) \overset{\text{def}}{=} -\mu(P\phi)\phi(a) \qquad \phi \in L^1, \text{ a.e. } a > 0$$

This aging function corresponds to a mortality process in a harsh environment, in that the mortality modulus $\mu$ depends on population size, but not on age.

We claim that F satisfies (2.1) and G satisfies (2.2). To establish this claim let $r > 0$ and define

$$c_1(r) \overset{\text{def}}{=} r \max_{0 \le |P| \le r} |\beta'(P)| + \max_{0 \le |P| \le r} |\beta(P)|$$

$$c_2(r) \overset{\text{def}}{=} r \max_{0 \le |P| \le r} |\mu'(P)| + \max_{0 \le |P| \le r} |\mu(P)|$$

Then, for $\phi_1, \phi_2 \in L^1$ such that $\|\phi_1\|_{L^1}, \|\phi_2\|_{L^1} \le r$, we have

$$\begin{aligned} |F(\phi_1) - F(\phi_2)| &\le \int_0^\infty |\beta(P\phi_1) - \beta(P\phi_2)| e^{-\alpha a}\phi_1(a)\, da \\ &\quad + \int_0^\infty |\beta(P\phi_2)\ e^{-\alpha a}|\phi_1(a) - \phi_2(a)|\, da \\ &\le \max_{0 \le |P| \le r} |\beta'(P)||P\phi_1 - P\phi_2|\ \|\phi_1\|_{L^1} \\ &\quad + \max_{0 \le |P| \le r} |\beta(P)|\ \|\phi_1 - \phi_2\|_{L^1} \\ &\le c_1(r)\|\phi_1 - \phi_2\|_{L^1} \end{aligned}$$

and

$$\begin{aligned} \|G(\phi_1) - G(\phi_2)\|_{L^1} &\le \int_0^\infty |\mu(P\phi_1) - \mu(P\phi_2)||\phi_1(a)|\, da \\ &\quad + \int_0^\infty |\mu(P\phi_2)||\phi_1(a) - \phi_2(a)|\, da \\ &\le \max_{0 \le |P| \le r} |\mu'(P)||P\phi_1 - P\phi_2|\ \|\phi_1\|_{L^1} \\ &\quad + \max_{0 \le |P| \le r} |\mu(P)|\ \|\phi_1 - \phi_2\|_{L^1} \\ &\le c_2(r)\|\phi_1 - \phi_2\|_{L^1} \end{aligned}$$

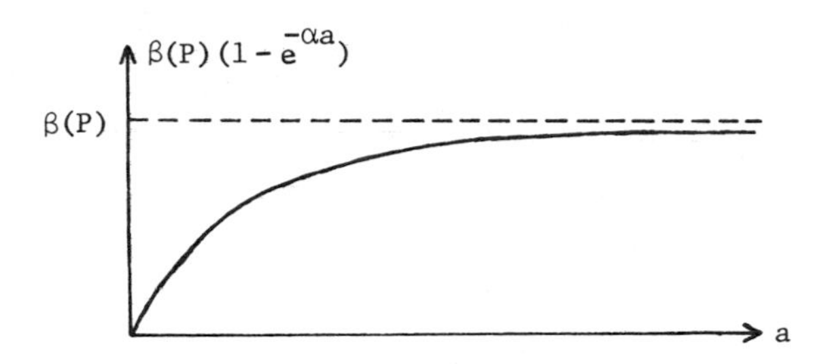


*Figure 2.5*

Thus, we may apply Theorem 2.1 to obtain the existence of a unique local solution to (ADP) for this birth function F, aging function G, and a given initial age distribution $\phi \in L^1$.

EXAMPLE 2.2 Let $R^n = R$. Let $\alpha > 0$ and let F and G be defined as in Example 2.1, except that we replace the formula for F in (2.17) by

$$(2.19) \qquad F(\phi) \overset{\text{def}}{=} \int_0^\infty \beta(P\phi)(1 - e^{-\alpha a})\phi(a)\, da \qquad \phi \in L^1$$

This birth function F corresponds to a reproductive process in which members of the species continue to reproduce as long as they survive. The same argument as in Example 2.1 shows that this birth function F satisfies (2.1).

EXAMPLE 2.3 Let $R^n = R$. Let $\alpha > 0$ and let F and G be defined as in Example 2.1, except that we replace the formula for F in (2.17) by

$$(2.20) \qquad F(\phi) \overset{\text{def}}{=} \int_0^\infty \beta(P\phi)ae^{-\alpha a}\phi(a)\, da \qquad \phi \in L^1$$

This birth function F corresponds to a reproductive process in which reproductive capacity is concentrated in an intermediate age range. A modification of the argument in Example 2.1 shows that F satisfies (2.1), where

$$c_1(r) \overset{\text{def}}{=} \alpha^{-1}e^{-1}[r \max_{0\le|P|\le r} |\beta'(P)| + \max_{0\le|P|\le r} |\beta(P)|]$$

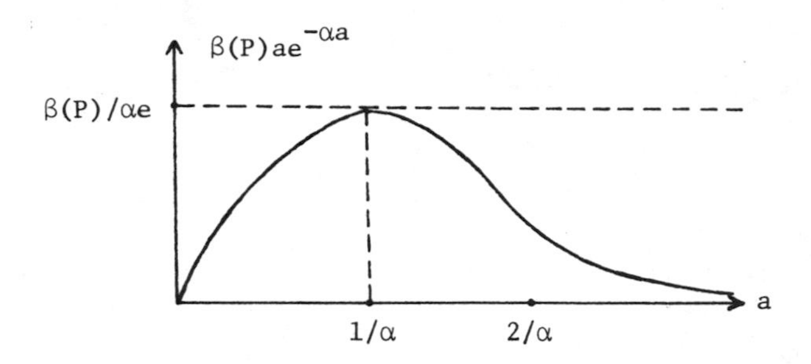


*Figure 2.6*

(since $0 \le ae^{-\alpha a} \le \alpha^{-1}e^{-1}$ for all $a \ge 0$).

## 2.3 THE SEMIGROUP PROPERTY AND CONTINUABILITY OF THE SOLUTIONS

We first prove that the solutions of (1.49) possess the semigroup property.

PROPOSITION 2.4 Let (2.1), (2.2) hold, let $\phi \in L^1$, let $T > 0$, and let $\ell \in L_T$ such that $\ell$ is a solution of (1.49) on $[0,T]$. Let $\hat{T} > 0$ and let $\ell \in L_{\hat{T}}$ such that for $t \in [0,\hat{T}]$

$$(2.21) \qquad \hat{\ell}(a,t) = \begin{cases} F(\hat{\ell}(\cdot, t - a)) + \int_0^a G(\hat{\ell}(\cdot, s + t - a))(s)\, ds & \text{a.e. } a \in (0,t) \\ \ell(a - t, T) + \int_{a-t}^a G(\hat{\ell}(\cdot, s + t - a))(s)\, ds & \text{a.e. } a \in (t,\infty) \end{cases}$$

Define $\ell(\cdot,t) = \hat{\ell}(\cdot,t - T)$ for $T < t \le T + \hat{T}$. Then, $\ell \in L_{T+\hat{T}}$ and $\ell$ is a solution of (1.49) on $[0, T + \hat{T}]$.

*Proof.* It suffices to show that $\ell(\cdot,t)$ satisfies (1.49) for $t \in (T, T + \hat{T}]$. Let $t \in (T, T + \hat{T}]$. For almost all $a \in (0, t - T)$

$$\ell(a,t) = \hat{\ell}(a, t - T)$$

$$= F(\hat{\ell}(\cdot, t - T - a)) + \int_0^a G(\hat{\ell}(\cdot, s + t - T - a))(s)\, ds$$

$$= F(\ell(\cdot, t - a)) + \int_0^a G(\ell(\cdot, s + t - a))(s)\, ds$$

For almost all $a \in (t - T, t)$

$$\ell(a,t) = \hat{\ell}(a, t - T)$$

$$= \ell(a - t + T, T) + \int_{a-t+T}^{a} G(\hat{\ell}(\cdot, s + t - T - a))(s)\, ds$$

$$= F(\ell(\cdot, t - a)) + \int_0^{a-t+T} G(\ell(\cdot, s + t - a))(s)\, ds$$

$$+ \int_{a-t+T}^{a} G(\ell(\cdot, s + t - a))(s)\, ds$$

$$= F(\ell(\cdot, t - a)) + \int_0^a G(\ell(\cdot, s + t - a))(s)\, ds$$

For almost all $a \in (t,\omega)$

$$\ell(a,t) = \hat{\ell}(a, t - T)$$

$$= \ell(a - t + T, T) + \int_{a-t+T}^{a} G(\hat{\ell}(\cdot, s + t - T - a))(s)\, ds$$

$$= \phi(a - t) + \int_{a-t}^{a-t+T} G(\ell(\cdot, s + t - a))(s)\, ds$$

$$+ \int_{a-t+T}^{a} G(\ell(\cdot, s + t - a))(s)\, ds$$

$$= \phi(a - t) + \int_{a-t}^{a} G(\ell(\cdot, s + t - a))(s)\, ds$$

Thus, $\ell$ is a solution of (1.49) on $[0, T + \hat{T}]$. □

As a consequence of Proposition 2.4 we can now establish the equivalence of the problem (ADP) and the problem (1.49).

THEOREM 2.2 Let (2.1), (2.2) hold, let $\phi \in L^1$, let $T > 0$, and let $\ell \in L_T$. Then, $\ell$ is a solution of (ADP) on $[0,T]$ if and only if $\ell$ is a solution of (1.49) on $[0,T]$.

*Proof.* If $\ell$ is a solution of (1.49) on $[0,T]$, then $\ell$ is a solution of (ADP) on $[0,T]$ by Proposition 2.1. Suppose that $\ell$ is a solution of (ADP) on $[0,T]$. By Proposition 2.2 there exists $T_0 \in (0,T]$ and $\ell_0 \in L_{T_0}$ such that $\ell_0$ is a solution of (1.49) on $[0,T_0]$. By Proposition 2.1 $\ell_0$ is a solution of (ADP) on $[0,T_0]$. By Theorem 2.1 $\ell$ restricted to $[0,T_0]$ must agree with $\ell_0$ on $[0,T_0]$, since solutions of (ADP) are unique.

Let $T_1 \overset{\text{def}}{=} \sup\{T_0 \in (0,T]$: there exists $\ell_0 \in L_{T_0}$ such that $\ell_0$ is a solution of (1.49) on $[0,T_0]$ and $\ell$ and $\ell_0$ agree on $[0,T_0]\}$. Let $\ell_1(\cdot,t)$ be defined from $[0,T_1]$ to $L^1$ as follows: if $t \in [0,T_1)$, then $\ell_1(\cdot,t) = \ell_0(\cdot,t)$, where $\ell_0 \in L_{T_0}$, $t < T_0 \leq T_1$, and $\ell_0$ and the restriction of $\ell$ to $[0,T_0]$ agree. Since $\ell$ is continuous from $[0,T_1]$ to $L^1$, so is $\ell_1$. Thus, $\ell_1 \in L_{T_1}$. Further, $\ell_1$ is a solution of (1.49) on $[0,T_1]$, since $\ell_1$ satisfies (1.49) for $t = T_1$ by the continuity of F and G.

Assume that $T_1 < T$. Let $\hat{\phi} = \ell_1(\cdot,T_1)$. By Proposition 2.2 there exists $\hat{T} \in (0, T - T_1]$ and $\hat{\ell} \in L_{\hat{T}}$ such that $\hat{\ell}$ is a solution of (1.49) on $[0,\hat{T}]$ with $\phi$ replaced by $\hat{\phi}$ in (1.45). By Proposition 2.4, $\ell_1$ can be extended to $[0, T_1 + \hat{T}]$ such that $\ell_1 \in L_{T_1+\hat{T}}$ and $\ell_1$ is a solution of (1.49) on $[0, T_1 + \hat{T}]$. By Proposition 2.1 this extended $\ell_1$ is a solution of (ADP) on $[0, T_1 + T]$. By Theorem 2.1 the solution of (ADP) on $[0, T_1 + \hat{T}]$ is unique, and so $\ell_1$ and $\ell$ must agree on $[0, T_1 + \hat{T}]$. The definition of $T_1$ is thus contradicted, whereby $T_1 = T$ and $\ell$ is a solution of (1.49) on $[0,T]$. □

The problem of the continuability of the local solution of (ADP) to a solution defined for all time depends on the existence of an a priori bound. Before investigating this problem we state

DEFINITION 2.1 Let $\phi \in L^1$. The *maximal interval of existence of the solution of* (ADP), denoted by $[0,T_\phi)$, is the interval with the property that if $0 < T < T_\phi$, then there exists $\ell \in L_T$ such that $\ell$ is a solution of (ADP) on $[0,T]$.

We note that by the uniqueness of solutions to (ADP) on $[0,T]$ guaranteed by Theorem 2.1, if $0 < T < \hat{T}$, $\ell \in L_T$, $\hat{\ell} \in L_{\hat{T}}$, $\ell$ is a solution of (ADP) on $[0,T]$, and $\hat{\ell}$ is a solution of (ADP) on $[0,\hat{T}]$, then $\ell$ and $\hat{\ell}$ must agree on $[0,T]$. We also note that in Definition 2.1 we allow the possibility that $T_\phi = \infty$. We now define

DEFINITION 2.2 Let $\phi \in L^1$ and let $\ell$ be a function from $[0,T_\phi)$ to $L^1$. We define $\ell$ to be the *solution of* (ADP) *on* $[0,T_\phi)$ provided that for all $T \in (0,T_\phi)$, $\ell$ restricted to $[0,T]$ is the solution of (ADP) on $[0,T]$.

THEOREM 2.3 Let (2.1), (2.2) hold, let $\phi \in L^1$, and let $\ell$ be the solution of (ADP) on $[0,T_\phi)$. If $T_\phi < \infty$, then $\limsup_{t \to T_\phi^-} \|\ell(\cdot,t)\|_{L^1} = \infty$.

*Proof.* Let $T_\phi < \infty$ and assume that there exists $r > 0$ such that $\|\ell(\cdot,t)\|_{L^1} \le r$ for $t \in [0,T_\phi)$. By Proposition 2.2 there exists $T \in (0,T_\phi)$ such that if $\hat{\phi} \in L^1$ and $\|\hat{\phi}\| \le r$, then there exists a solution of (1.49) on $[0,T]$ with $\phi$ replaced by $\hat{\phi}$ in (1.45). Let $\ell \in L_{T_\phi - T/2}$ such that $\ell$ is a solution of (ADP) (and hence by Theorem 2.2 of (1.49)) on $[0, T_\phi - T/2]$. Since $\|\ell(\cdot, T - T/2)\|_{L^1} \le r$, there exists $\hat{\ell} \in L_T$ such that for $t \in [0,T]$

$$\hat{\ell}(a,t) = \begin{cases} F(\hat{\ell}(\cdot, t - a)) + \int_0^a G(\hat{\ell}(\cdot, s + t - a))(s)\, ds, & \text{a.e. } a \in (0,t) \\ \ell(a - t, T_\phi - T/2) + \int_{a-t}^a G(\hat{\ell}(\cdot, s + t - a))(s)\, ds & \\ & \text{a.e. } a \in (t,\infty) \end{cases}$$

Extend $\ell(\cdot,t)$ from $[0, T_\phi - T/2]$ to $[0, T_\phi + T/2]$ by defining $\ell(\cdot,t) = \hat{\ell}(\cdot, t - T_\phi + T/2)$ for $t \in [T_\phi - T/2, T_\phi + T/2]$. By Proposition 2.4, $\ell \in L_{T_\phi + T/2}$ and $\ell$ is a solution of (1.49) [and hence of (ADP)] on $[0, T_\phi + T/2]$. Therefore, we have contradicted the maximality of $T_\phi$ and so $\limsup_{t \to T_\phi^-} \|\ell(\cdot,t)\|_{L^1} = \infty$. □

## 2.4 POSITIVITY OF SOLUTIONS

In physical applications it is the positive solutions of (ADP) which have significance. It is therefore important to have criteria on the birth and aging functions which assure that positive-valued initial age distributions will yield positive-valued solutions. In this section we establish such criteria. We first introduce some notation.

Let $R^n_+$ denote the positive cone in $R^n$:

$$R^n_+ = \{x = (x_1,\ldots,x_n)^T \in R^n: \ x_i \geq 0 \text{ for } i = 1, \ldots, n\}$$

Let $L^1_+$ denote the positive cone in $L^1$:

$$L^1_+ = \{\phi \in L^1: \ \phi(a) \in R^n_+ \text{ for almost all } a > 0\}$$

Let $L_{T,+}$ denote the positive cone in $L_T$:

$$L_{T,+} = \{\ell \in L_T: \ \ell(\cdot,T) \in L^1_+ \text{ for } t \in [0,T]\}$$

We will place the following hypotheses on the birth function F and the aging function G:

(2.22) $\quad F(L^1_+) \subset R^n_+$

(2.23) $\quad$ There is an increasing function $c_3$: $[0,\infty) \to [0,\infty)$ such that if $r > 0$ and $\phi \in L^1_+$ with $\|\phi\|_{L^1} \leq r$, then $G(\phi) + c_3(r)\phi \in L^1_+$.

PROPOSITION 2.5 Let (2.1), (2.2), (2.22), (2.23) hold, let $\phi \in L^1_+$, let $r = \|\phi\|_{L^1}$, and let $\alpha = c_3(2r)$. There exists $T > 0$ and a unique function $\ell \in L_{T,+}$ such that for $t \in [0,T]$,

$$(2.24)\qquad \ell(a,t) = \begin{cases} e^{-\alpha a}F(\ell(\cdot,\ t - a)) + \int_{t-a}^{t} e^{-(t-s)\alpha} \\ \times\ (G + \alpha I)(\ell(\cdot,s))(s + a - t)\ ds & \text{a.e. } a \in (0,t) \\ e^{-t\alpha}\phi(a - t) + \int_0^t e^{-(t-s)\alpha}(G + \alpha I)(\ell(\cdot,s))(s + a - t)\ ds \\ & \text{a.e. } a \in (t,\infty) \end{cases}$$

*Proof.* Choose $T > 0$ such that

$$(2.25)\qquad T\,\frac{c_1(2r) + c_2(2r) + c_3(2r) + [|F(0)| + \|G(0)\|_{L^1}]}{2r} + \frac{1}{2} \le 1$$

Let M be the closed subset of $L_{T,+}$ defined by

$$M \overset{\text{def}}{=} \{\ell \in L_{T,+}\colon\ \ell(\cdot,0) = \phi \quad \text{and} \quad \|\ell\|_{L_T} \le 2r\}$$

Define a mapping K on M as follows: if $\ell \in M$, then $K\ell(a,t)$ is given by the right-hand side of (2.24). An argument very similar to that of Proposition 2.2 may now be used to show that K is a mapping from M into M, and by virtue of (2.25), a strict contraction. Thus, K has a unique fixed point in M, which is also the unique solution of (2.24) in $L_{T,+}$. □

PROPOSITION 2.6 Let (2.1), (2.2) hold, let $\phi \in L^1$, let $\alpha \in R$, let $T > 0$, and let $\ell \in L_T$ such that

$$(2.26)\qquad \ell(a,t) = \begin{cases} e^{-a\alpha}F(\ell(\cdot,\ t - a)) + \int_{t-a}^{t} e^{-(t-s)\alpha} \\ \times\ G(\ell(\cdot,s))(s + a - t)\ ds & \text{a.e. } a \in (0,t) \\ e^{-t\alpha}\phi(a - t) + \int_0^t e^{-(t-s)\alpha}G(\ell(\cdot,s))(s + a - t)\ ds \\ & \text{a.e. } a \in (t,\infty) \end{cases}$$

Then, $\ell$ satisfies (1.44), (1.45), and for $t \in [0,T)$

$$(2.27)\qquad \lim_{h\to 0^+} \int_0^\infty \Big|h^{-1}[\ell(a + h,\ t + h) - \ell(a,t)] \\ - G(\ell(\cdot,t))(a) + \alpha\ell(a,t)\Big|\ da = 0$$

*Proof.* Let $t \in [0,T)$, $0 < h < T - t$, and define

$$H_1 \stackrel{\text{def}}{=} \int_0^t |h^{-1}[\ell(a + h, t + h) - \ell(a,t)] - G(\ell(\cdot,t))(a) + \alpha\ell(a,t)| \, da$$

$$H_2 \stackrel{\text{def}}{=} \int_t^\infty |h^{-1}[\ell(a + h, t + h) - \ell(a,t)] - G(\ell(\cdot,t))(a) + \alpha\ell(a,t)| \, da$$

From (2.26) we obtain

$$\begin{aligned} H_1 \stackrel{\text{def}}{=} \int_0^t \Bigg| & [h^{-1}(e^{-h\alpha} - 1) + \alpha]e^{-a\alpha}F(\ell(\cdot, t - a)) \\ & + [h^{-1}(e^{-h\alpha} - 1) + \alpha] \int_{t-a}^t e^{-(t-s)\alpha}G(\ell(\cdot,s))(s + a - t) \, ds \\ & + h^{-1} \int_t^{t+h} e^{-(t+h-s)\alpha}G(\ell(\cdot,s))(s + a - t) \, ds \\ & - G(\ell(\cdot,t))(a) \Bigg| \, da \\ \le |h^{-1}(e^{-h\alpha} - 1) + \alpha| & \left\{ \int_0^t e^{-a\alpha}|F(\ell(\cdot, t - a))| \, da \right. \\ & \left. + \int_0^t \left[ \int_{t-a}^t e^{-(t-s)\alpha}|G(\ell(\cdot,s))(s + a - t)| \, ds \right] da \right\} \\ & + h^{-1} \int_t^{t+h} \left[ \int_0^t |e^{-(t+h-s)\alpha}G(\ell(\cdot,s))(s + a - t) \right. \\ & \left. - G(\ell(\cdot,t))(a)| \, da \right] ds \stackrel{\text{def}}{=} K_1 + K_2 \end{aligned}$$

Obviously, $K_1$ converges to 0 as h converges to 0. Further,

$$\begin{aligned} K_2 \le \sup_{t \le s \le t+h} \Bigg[ & \int_0^t e^{-(t+h-s)\alpha}|G(\ell(\cdot,s))(s + a - t) \\ & - G(\ell(\cdot,t))(s + a - t)| \, da \\ & + \int_0^t e^{-(t+h-s)\alpha}|G(\ell(\cdot,t))(s + a - t) - G(\ell(\cdot,t))(a)| \, da \\ & + \int_0^t |e^{-(t+h-s)\alpha} - 1||G(\ell(\cdot,t))(a)| \, da \Bigg] \end{aligned}$$

Then, $K_2$ converges to 0 as h converges to 0 by the continuity of the function $s \to G(\ell(\cdot,s))$ from $[0,T]$ to $L^1$ and the continuity of translation in $L^1$ (see Remark 2.2). A similar argument may be used to show that $H_2$ converges to 0 as h converges to 0. Thus, (2.27) is established. Last, the argument of Proposition 2.1 may be used to show that (1.44) holds, and it is obvious that (1.45) holds. □

PROPOSITION 2.7 Let (2.1), (2.2), (2.22), (2.23) hold and let $\phi \in L^1_+$. There exists $T > 0$ and a function $\ell \in L_{T,+}$ satisfying (1.43), (1.44), (1.45).

*Proof.* By Proposition 2.5 there exists $T > 0$ and a function $\ell \in L_{T,+}$ satisfying (2.24) for $0 \le t \le T$. Set $G = G + \alpha I$ in (2.26), and Proposition 2.6 yields that $\ell$ satisfies (1.44), (1.45), and (2.27) with G replaced by $G + \alpha I$. Thus, $\ell$ satisfies (1.43) as well. □

THEOREM 2.4 Let (2.1), (2.2), (2.22), (2.23) hold and let $\phi \in L^1_+$. The solution $\ell$ of (ADP) on $[0,T_\phi)$ has the property that $\ell(\cdot,t) \in L^1_+$ for $0 \le t < T_\phi$.

*Proof.* Assume for contradiction that there exists $t_0 \in (0,T_\phi)$ such that $\ell(\cdot,t_0) \notin L^1_+$. Let $T \overset{\text{def}}{=} \inf\{t_1 \in (0,T_\phi) : \ell(\cdot,t_1) \notin L^1_+\}$. By the continuity of $\ell(\cdot,t)$ in t as a function from $[0,T_\phi)$ to $L^1$ and the closedness of $L^1_+$ we have that $\ell(\cdot,T) \in L^1_+$. By Proposition 2.4, Proposition 2.7, and Theorem 2.1 there exists $\hat{T} > 0$ such that $\ell \in L_{T+\hat{T},+}$. But then the definition of T is contradicted, and so the conclusion of the theorem must hold. □

THEOREM 2.5 Let (2.1), (2.21), (2.22), (2.23) hold, let $\phi \in L^1_+$, let $\ell$ be the solution of (ADP) on $[0,T_\phi)$, and let there exist $\omega \in R$ such that

$$(2.28) \qquad \sum_{i=1}^{n} F(\ell(\cdot,t))_i + \int_0^\infty G(\ell(\cdot,t))_i(a)\, da$$

$$\le \omega \sum_{i=1}^{n} \int_0^\infty \ell_i(a,t)\, da \qquad t \in [0,T_\phi)$$

where the subscript i denotes the i-th component in $R^n$. Then, $T_\phi = \infty$ and

$$(2.29) \qquad \|\ell(\cdot,t)\|_{L^1} \le e^{\omega t}\|\phi\|_{L^1} \qquad 0 \le t < T_\phi$$

*Proof.* Define $V(t) \overset{\text{def}}{=} \|\ell(\cdot,t)\|_{L^1}$ for $0 \le t < T_\phi$. By Theorem 2.4

$$V(t) = \sum_{i=1}^{n} \int_0^\infty \ell_i(a,t)\, da \qquad 0 \le t < T_\phi$$

For $0 \le t < T_\phi$, $0 < h < T_\phi - t$ we have that

$$\begin{aligned}
&h^{-1}[V(t+h) - V(t)] \\
&= h^{-1} \sum_{i=1}^{n} \left[\int_{-t-h}^{\infty} \ell_i(t+h+c,\, t+h)\, dc - \int_{-t}^{\infty} \ell_i(t+c,\, t)\, dc\right] \\
&= \sum_{i=1}^{n} \left\{h^{-1} \int_{-t-h}^{-t} \ell_i(t+h+c,\, t+h)\, dc \right. \\
&\quad \left. + \int_{-t}^{\infty} h^{-1}[\ell_i(t+h+c,\, t+h) - \ell_i(t+c,\, t)]\, dc\right\} \\
&\le \sum_{i=1}^{n} \left\{h^{-1} \int_0^h |\ell_i(a,\, t+h) - F(\ell(\cdot,t))_i|\, da \right. \\
&\quad + F(\ell(\cdot,t))_i + \int_0^\infty G(\ell(\cdot,t))_i(a)\, da \\
&\quad \left. + \int_0^\infty |h^{-1}[\ell_i(a+h,\, t+h) - \ell_i(a,t)] - G(\ell(\cdot,t))_i(a)|\, da\right\}
\end{aligned}$$

From (1.43), (1.44), and (2.28) we obtain

$$\begin{aligned}
&\limsup_{h\to 0^+} h^{-1}[V(t+h) - V(t)] \\
&\le \sum_{i=1}^{n} F(\ell(\cdot,t))_i + \int_0^\infty G(\ell(\cdot,t))_i(a)\, da \le \omega V(t)
\end{aligned}$$

Now solve this differential inequality to obtain (2.29) (see Theorem 1.4.1, p. 15 in [183]). By Theorem 2.3 we must have that $T_\phi = \infty$. □

The following examples provide illustrations for Theorems 2.4 and 2.5.

EXAMPLE 2.4 Let $R^n = R$. Let F and G be defined as in Example 2.1 [that is, F has the form (2.17) and G has the form (2.18)]. In addition, let there exist a constant $\omega \in R$ such that

$$(2.30) \qquad \beta(P) - \mu(P) \le \omega \qquad \text{for all } P \ge 0$$

Obviously, F satisfies (2.22), since $\beta(P) \ge 0$ for all $P \ge 0$. Also, G satisfies (2.23) with $c_3(r) \overset{\text{def}}{=} \max_{0 \le P \le r} \mu(P)$, since for $\phi \in L^1_+$, $\|\phi\|_{L^1} \le r$, we have that

$$\begin{aligned} G(\phi)(a) + c_3(r)\phi(a) &= -\mu(P\phi)\phi(a) + c_3(r)\phi(a) \\ &\ge -c_3(r)\phi(a) + c_3(r)\phi(a) = 0 \qquad \text{a.e. } a > 0 \end{aligned}$$

The hypothesis of Theorem 2.4 is therefore satisfied, and so for $\phi \in L^1_+$ the solution $\ell$ of (ADP) on $[0,T_\phi)$ satisfies $\ell(\cdot,t) \in L^1_+$ for $0 \le t < T_\phi$. Furthermore, the hypothesis of Theorem 2.5 is satisfied, since for $\phi \in L^1_+$,

$$\begin{aligned} F(\phi) + \int_0^\infty G(\phi)(a)\, da &= \int_0^\infty [\beta(P\phi)e^{-\alpha a} - \mu(P\phi)]\phi(a)\, da \\ &\le \omega \int_0^\infty \phi(a)\, da \end{aligned}$$

Thus, $T_\phi = \infty$ for all $\phi \in L^1_+$ and $\|\ell(\cdot,t)\|_{L^1} \le e^{\omega t}\|\phi\|_{L^1}$ for all $t \ge 0$.

EXAMPLE 2.5 Let $R^n = R$. Let F and G be defined as in Example 2.2 [that is, F has the form (2.19) and G has the form (2.18)] and let (2.30) hold. As in Example 2.4 the hypotheses of Theorems 2.4 and 2.5 are satisfied. Thus, for this birth function F and aging function G the solution of (ADP) corresponding to the initial age distribution $\phi \in L^1_+$ is defined for all $t \ge 0$, lies in $L^1_+$ for all $t \ge 0$, and satisfies (2.29).

EXAMPLE 2.6 Let $R^n = R$. Let F and G be defined as in Example 2.3 [that is, F has the form (2.20) and G has the form (2.18)]. Let there exist a constant $\omega \in R$ such that

(2.31) $$\beta(P)\alpha^{-1}e^{-1} - \mu(P) \leq \omega \qquad \text{for all } P \geq 0$$

Again, the hypotheses of Theorems 2.4 and 2.5 are satisfied, and for any $\phi \in L^1_+$ the solution of (ADP) corresponding to this birth function F and aging function G is defined for all $t \geq 0$, lies in $L^1_+$ for all $t \geq 0$, and satisfies (2.29).

## 2.5 CONTINUOUS DEPENDENCE ON THE BIRTH AND AGING FUNCTIONS

In physical applications it is important to know that a small change in the birth and aging functions will produce only a small change in the corresponding solutions of (ADP). To formulate this problem mathematically we state the following hypotheses:

(2.32) For $F_k$: $L^1 \to R^n$, $k = 0, 1, 2, \ldots$, there is an increasing function $c_1$: $[0,\infty) \to [0,\infty)$ such that $|F_k(\phi_1) - F_k(\phi_2)| \leq c_1(r)\|\phi_1 - \phi_2\|_{L^1}$ for all $\phi_1, \phi_2 \in L^1$ such that $\|\phi_1\|_{L^1}$, $\|\phi_2\|_{L^1} \leq r$ (where $c_1$ is independent of $k = 0, 1, 2, \ldots$).

(2.33) For $G_k$: $L^1 \to L^1$, $k = 0, 1, 2, \ldots$, there is an increasing function $c_2$: $[0,\infty) \to [0,\infty)$ such that $\|G_k(\phi_1) - G_k(\phi_2)\|_{L^1} \leq c_2(r)\|\phi_1 - \phi_2\|_{L^1}$ for all $\phi_1, \phi_2 \in L^1$ such that $\|\phi_1\|_{L^1}$, $\|\phi_2\|_{L^1} \leq r$ (where $c_2$ is independent of $k = 0, 1, 2, \ldots$).

(2.34) For each $\phi \in L^1$ $\lim_{k\to\infty} F_k(\phi) = F_0(\phi)$.

(2.35) For each $\phi \in L^1$ $\lim_{k\to\infty} G_k(\phi) = G_0(\phi)$.

Let $\{\phi_k\}_{k=0}^{\infty}$ be a sequence in $L^1$ such that $\phi_k \to \phi_0$ and observe that under the hypotheses (2.32), (2.33), Theorem 2.1 guarantees the existence of a positive number $T_k$ and a function $\ell_k \in L_{T_k}$ such that $\ell_k$ is a solution of the problem

(2.36) $$\lim_{h\to 0^+} \int_0^\infty |h^{-1}[\ell_k(a + h, t + h) - \ell_k(a,t)] - G_k(\ell_k(\cdot,t))(a)| \, da = 0 \qquad 0 \le t < T_k$$

(2.37) $$\lim_{h\to 0^+} h^{-1} \int_0^h |\ell_k(a, t + h) - F_k(\ell_k(\cdot,t))| \, da = 0 \qquad 0 \le t < T_k$$

(2.38) $$\ell_k(\cdot,0) = \phi_k$$

[Note that $T_k$ can, in fact, be chosen independently of k by virtue of (2.12) and hypotheses (2.32), (2.33), (2.34), (2.35)]. Equivalently, by Theorem 2.2, $\ell_k$ is a solution of the problem

(2.39) $$\ell_k(a,t) = \begin{cases} F_k(\ell_k(\cdot, t - a)) + \int_{t-a}^t G_k(\ell_k(\cdot,s))(s + a - t) \, ds & \text{a.e. } a \in (0,t), \ 0 \le t \le T_k \\ \phi_k(a - t) + \int_0^t G_k(\ell_k(\cdot,s))(s + a - t) \, ds & \text{a.e. } a \in (t,\infty), \ 0 \le t \le T_k \end{cases}$$

THEOREM 2.6 Let (2.32), (2.33), (2.34), (2.35) hold, let $\{\phi_k\}_{k=0}^\infty$ be a sequence in $L^1$ such that $\lim_{k\to\infty} \phi_k = \phi_0$, and for each integer $k = 0, 1, 2, \ldots$, let $T_k > 0$ and $\ell_k \in L_{T_k}$ such that $\ell_k$ is a solution of (2.36), (2.37), (2.38). Suppose that $0 < T \le \inf_{k=0,1,2,\ldots} T_k$ and that r is a positive constant such that $\|\ell_k\|_{L_T} \le r$ for $k = 0, 1, 2, \ldots$. Then,

(2.40) $$\lim_{k\to\infty} \|\ell_k - \ell_0\|_{L_T} = 0$$

*Proof.* We first claim that if $\varepsilon > 0$, there exists a positive integer K such that if $k \ge K$, then the following hold:

(2.41) $$\sup_{0\le t\le T} |F_k(\ell_0(\cdot,t)) - F_0(\ell_0(\cdot,t))| < \varepsilon$$

(2.42) $$\sup_{0\le t\le T} \|G_k(\ell_0(\cdot,t)) - G_0(\ell_0(\cdot,t))\|_{L^1} < \varepsilon$$

To prove (2.41) observe that since $\ell_0 \in L^1_T$, the mapping $t \to \ell_0(\cdot,t)$ is uniformly continuous from $[0,T]$ to $L^1$. Let $\varepsilon > 0$ and let $\{t_1, \ldots, t_m\}$ be a finite set in $[0,T]$ such that if $t \in [0,T]$, then there exists some $i$, $1 \le i \le m$, such that

$$\|\ell_0(\cdot,t) - \ell_0(\cdot,t_i)\|_{L^1} < \varepsilon \tag{2.43}$$

By (2.34) there exists a positive integer $K$ such that if $k \ge K$ and $i = 1, \ldots, m$, then

$$|F_k(\ell_0(\cdot,t_i)) - F_0(\ell_0(\cdot,t_i))| < \varepsilon$$

Let $t \in [0,T]$, let $i$ be as in (2.43), and let $k \ge K$. Then,

$$\begin{aligned}
&|F_k(\ell_0(\cdot,t)) - F_0(\ell_0(\cdot,t))| \\
&\le |F_k(\ell_0(\cdot,t)) - F_k(\ell_0(\cdot,t_i))| + |F_k(\ell_0(\cdot,t_i)) \\
&\quad - F_0(\ell_0(\cdot,t_i))| + |F_0(\ell_0(\cdot,t_i)) - F_0(\ell_0(\cdot,t))| \\
&\le c_1(r)\|\ell_0(\cdot,t) - \ell_0(\cdot,t_i)\|_{L^1} + \varepsilon + c_1(r)\|\ell_0(\cdot,t) - \ell_0(\cdot,t_i)\|_{L^1} \\
&\le 2c_1(r)\varepsilon + \varepsilon
\end{aligned}$$

Then, (2.41) follows and a similar argument proves (2.42).

Now let $\varepsilon > 0$, let $t \in [0,T]$, let $K$ be a positive integer such that (2.41), (2.42) hold for $k \ge K$, and let $k \ge K$. From (2.39) we obtain that for $0 \le t \le T$

$$\begin{aligned}
&\|\ell_k(\cdot,t) - \ell_0(\cdot,t)\|_{L^1} \\
&\le \int_0^t |F_k(\ell_k(\cdot, t - a)) - F_0(\ell_0(\cdot, t - a))|\, da \\
&\quad + \int_0^t \left[ \int_{t-a}^t |G_k(\ell_k(\cdot,s))(s + a - t) \right. \\
&\quad \left. - G_0(\ell_0(\cdot,s))(s + a - t)|\, ds \right] da \\
&\quad + \int_t^\infty |\phi_k(a - t) - \phi_0(a - t)|\, da \\
&\quad + \int_t^\infty \left[ \int_0^t |G_k(\ell_k(\cdot,s))(s + a - t) - G_0(\ell_0(\cdot,s))(s + a - t)|\, ds \right] da
\end{aligned}$$

$$\le \int_0^t [|F_k(\ell_k(\cdot,s)) - F_k(\ell_0(\cdot,s))| + |F_k(\ell_0(\cdot,s))$$

$$- F_0(\ell_0(\cdot,s))|]\, ds + \int_0^t \left\{ \int_0^\infty \left[ |G_k(\ell_k(\cdot,s))(a) - G_k(\ell_0(\cdot,s))(a)| \right.\right.$$

$$\left.\left. + |G_k(\ell_0(\cdot,s))(a) - G_0(\ell_0(\cdot,s))(a)| \right] da \right\} ds + \|\phi_k - \phi_0\|_{L^1}$$

$$\le \int_0^t [c_1(r) + c_2(r)] \|\ell_k(\cdot,s) - \ell_0(\cdot,s)\|_{L^1}\, ds$$

$$+ 2\varepsilon T + \|\phi_k - \phi_0\|_{L^1}$$

From Gronwall's inequality (see [217], Proposition 1.4, p. 204) we obtain

$$\|\ell_k(\cdot,t) - \ell_0(\cdot,t)\|_{L^1}$$

$$\le (2\varepsilon T + \|\phi_k - \phi_0\|_{L^1}) \exp[(c_1(r) + c_2(r))T]$$

and thus (2.40) is proved. □

## 2.6 REGULARITY OF THE SOLUTIONS

In this section we establish some continuity and differentiability properties of the solutions of (ADP). Such regularity properties of the solutions usually depend on additional assumptions on the birth function F, the aging function G, or the initial age distribution $\phi$. In our first result, however, no such additional assumptions are required.

THEOREM 2.7 Let (2.1), (2.2) hold, let $\phi \in L^1$, and let $\ell$ be the solution of (ADP) on $[0,T_\phi)$. For $c > -T_\phi$ define

(2.44) $$w_c(t) \overset{\text{def}}{=} \ell(t + c, t) \qquad t \in [t_c, T_\phi),\ t_c \overset{\text{def}}{=} \max\{0,-c\}$$

(see Figure 1.1). The following hold:

(2.45) For a.e. $c > -T_\phi$, $w_c(t)$ is continuous in t from $[t_c, T_\phi)$ to $R^n$.

(2.46) For a.e. $c > -T_\phi$, $w_c(t)$ is differentiable a.e. in t on $[t_c, T_\phi)$, and $d/dt\, w_c(t) = G(\ell(\cdot,t))(t + c)$ for a.e. $t \in [t_c, T_\phi)$.

*Proof.* We first claim that for almost all $c > -T_\phi$, $w_c(t)$ is integrable in t on each interval $(t_c, T)$, $T < T_\phi$. This claim follows from (2.8), since we may take G = I in the statement of Lemma 2.2. We next claim that for almost all $c > -T_\phi$ and $0 < T < T_\phi$, the equivalence class of $w_c$ in $L^1((t_c,T);R^n)$ contains a continuous representative of $w_c$.

To establish this claim let $\phi_0$ be a representative of the equivalence class of $\phi$. From (2.8) we see that for $0 < T < T_\phi$

$$\int_{-T}^{0}\left[\int_{-c}^{T}\left|\ell(t + c, t) - F(\ell(\cdot,-c)) - \int_{-c}^{t} G(\ell(\cdot,s))(s + c)\, ds\right| dt\right] dc$$

$$+ \int_{0}^{\infty}\left[\int_{0}^{T}\left|\ell(t + c, t) - \phi_0(c) - \int_{0}^{t} G(\ell(\cdot,s))(s + c)\, ds\right| dt\right] dc$$

$$= \int_{0}^{T}\left[\int_{-t}^{0}\left|\ell(t + c, t) - F(\ell(\cdot,-c)) - \int_{-c}^{t} G(\ell(\cdot,s))(s + c)\, ds\right| dc\right] dt$$

$$+ \int_{0}^{T}\left[\int_{0}^{\infty}\left|\ell(t + c, t) - \phi_0(c) - \int_{0}^{t} G(\ell(\cdot,s))(s + c)\, ds\right| dc\right] dt$$

$$= \int_{0}^{T}\left[\int_{0}^{t}\left|\ell(a,t) - F(\ell(\cdot, t - a)) - \int_{t-a}^{t} G(\ell(\cdot,s))(s + a - t)\, ds\right| da\right] dt$$

$$+ \int_{0}^{T}\left[\int_{t}^{\infty}\left|\ell(a,t) - \phi_0(a - t) - \int_{0}^{t} G(\ell(\cdot,s))(s + a - t)\, ds\right| da\right] dt$$

$$= 0$$

since $\ell$ is a solution of (ADP).

Thus, for almost all $c \in (-T,0)$

$$\int_{-c}^{T}\left|\ell(t + c, t) - F(\ell(\cdot,-c)) - \int_{-c}^{t} G(\ell(\cdot,s))(s + c)\, ds\right| dt = 0$$

and for almost all $c \in (0,\infty)$

$$\int_0^T \left| \ell(t + c, t) - \phi_0(c) - \int_0^t G(\ell(\cdot,s))(s + c)\, ds \right| dt = 0$$

Then, for almost all $c \in (-T,0)$

$$w_c(t) = F(\ell(\cdot,-c)) + \int_{-c}^t G(\ell(\cdot,s))(s + c)\, ds \qquad \text{a.e. } t \in (-c,T) \tag{2.47}$$

and for almost all $c \in (0,\infty)$

$$w_c(t) = \phi_0(c) + \int_0^t G(\ell(\cdot,s))(s + c)\, ds \qquad \text{a.e. } t \in (0,T) \tag{2.48}$$

The claims (2.45) and (2.46) now follow from the integrability of the function $s \to G(\ell(\cdot,s))(s + c)$, as assured by (2.8) of Lemma 2.2. □

In our next result we will suppose that the aging function G is a pure mortality function having the form (1.48). We are then able to show that the solutions of (ADP) satisfy additional continuity and differentiability properties. We require that the aging function G satisfies

(2.49) For $G: L^1 \to L^1$, there exists $\mu: [0,\infty) \times L^1 \to B(R^n,R^n)$ such that if $\phi \in L^1$, then $G(\phi)(a) = -\mu(a,\phi)\phi(a)$ for almost all $a > 0$, where $\mu$ satisfies: (i) there is an increasing function $c_4: [0,\infty) \to [0,\infty)$ such that $|\mu(a,\phi) - \mu(\hat{a},\phi)| \le c_4(\|\phi\|_{L^1})|a - \hat{a}|$ for all $\phi \in L^1$, $a, \hat{a} \ge 0$; (ii) there is an increasing function $c_5: [0,\infty) \to [0,\infty)$ such that $|\mu(a,\phi)| \le c_5(\|\phi\|_{L^1})$ for all $\phi \in L^1$, $a \ge 0$; and (iii) there is an increasing function $c_6: [0,\infty) \to [0,\infty)$ such that $|\mu(a,\phi) - \mu(a,\hat{\phi})| \le c_6(r)\|\phi - \hat{\phi}\|_{L^1}$ for all $a \ge 0$ and $\phi, \hat{\phi} \in L^1$ such that $\|\phi\|_{L^1}, \|\hat{\phi}\|_{L^1} \le r$.

REMARK 2.3 We observe that if G satisfies (2.49), then G automatically satisfies (2.2). First, if $\phi \in L^1$, then $G(\phi) \in L^1$, since

$\mu(a,\phi)$ is continuous and bounded in a by (i) and (ii) of (2.49). Second, if $\phi$, $\hat{\phi} \in L^1$ such that $\|\phi\|_{L^1}$, $\|\hat{\phi}\|_{L^1} \leq r$, then

$$\begin{aligned}\|G(\phi) - G(\hat{\phi})\|_{L^1} &= \int_0^\infty |\mu(a,\phi)\phi(a) - \mu(a,\hat{\phi})\hat{\phi}(a)|\, da \\ &\leq \int_0^\infty |\mu(a,\phi)|\,|\phi(a) - \hat{\phi}(a)|\, da \\ &\quad + \int_0^\infty |\mu(a,\phi) - \mu(a,\hat{\phi})|\,|\hat{\phi}(a)|\, da \\ &\leq c_5(r)\|\phi - \hat{\phi}\|_{L^1} + c_6(r)\|\phi - \hat{\phi}\|_{L^1} r\end{aligned}$$

Therefore, we may take $c_2(r) = c_5(r) + c_6(r)r$ in (2.2).

Our next objective will be to obtain a representation for the solutions of (ADP) in the case that the aging function G satisfies (2.49). This representation is given in terms of an evolution operator associated with the mortality modulus $\mu$. In the case of the Gurtin-MacCamy scalar model of Section 1.3, it reduces to the formula in (1.30).

DEFINITION 2.3 Let (2.49) hold, let $T > 0$, let $\ell \in L_T$, let $c \geq -T$, and let $t \in [t_c, T]$, where $t_c \overset{\text{def}}{=} \max\{0,-c\}$. Define the operator $A(t;c,\ell) \in B(R^n,R^n)$ by $A(t;c,\ell) = -\mu(t + c, \ell(\cdot,t))$.

In the scalar case, that is, when $R^n = R$, the family of operators $\{A(t;c,\ell) : t_c \leq t \leq T\}$ is associated with the evolution operator

$$U(t,s;c,\ell) = \exp\left[-\int_s^t \mu(\tau + c, \ell(\cdot,\tau))\, d\tau\right] \qquad t_c \leq s \leq t \leq T$$

In the general case of $R^n$ we have

PROPOSITION 2.8 Let (2.49) hold, let $T > 0$, let $\ell \in L_T$, and let $c \geq -T$. There is a family of operators $\{U(t,s;c,\ell) : t_c \leq s \leq t \leq T\}$

in $B(R^n,R^n)$ with the following properties:

(2.50) $|U(t,s;c,\ell)| \le \exp[c_5(\|\ell\|_{L_T})(t - s)]$ for $t_c \le s \le t \le T$

(2.51) $U(t,\tau;c,\ell)U(\tau,s;c,\ell) = U(t,s;c,\ell)$ for $t_c \le s \le \tau \le t \le T$

(2.52) $U(t,t;c,\ell) = I$ for $t_c \le t \le T$

(2.53)
$$\frac{\partial}{\partial t} U(t,s;c,\ell) = A(t;c,\ell)U(t,s;c,\ell) = -\mu(t + c, \ell(\cdot,t))U(t,s;c,\ell)$$
for $t_c \le s \le t \le T$

(2.54)
$$\frac{\partial}{\partial s} U(t,s;c,\ell) = -U(t,s;c,\ell)A(s;c,\ell) = U(t,s;c,\ell)\mu(s + c, \ell(\cdot,s))$$
for $t_c \le s \le t \le T$

(2.55) If $x \in R^n$, $t_c \le s \le T$, and $u$ is a function from $[s,T]$ to $R^n$ such that $u(s) = x$ and $d/dt\, u(t) = A(t;c,\ell)u(t)$, $s \le t \le T$, then $u(t) = U(t,s;c,\ell)x$, $s \le t \le T$.

*Proof.* We observe that the function $t \to A(t;c,\ell)$ is continuous from $[t_c,T]$ to $B(R^n,R^n)$ by virtue of (2.49). The existence of the family of operators $\{U(t,s;c,\ell) : t_c \le s \le t \le T\}$ in $B(R^n,R^n)$ satisfying the properties (2.50)-(2.55) follows directly from the general results proved in [169], pp. 188-190. □

PROPOSITION 2.9 Let the hypothesis of Proposition 2.8 hold. The family of operators $\{U(t,s;c,\ell) : t_c \le s \le t \le T\}$ satisfies the following:

(2.56) $|U(t,s;c,\ell) - U(\hat{t},s;c,\ell)| \le |t - \hat{t}|c_5(\|\ell\|_{L_T})$
$\exp[c_5(\|\ell\|_{L_T})T]$ for $t_c \le s \le t \le \hat{t} \le T$

(2.57) $|U(t,s;c,\ell) - U(t,\hat{s};c,\ell)| \le |s - \hat{s}|c_5(\|\ell\|_{L_T})$
$\exp[c_5(\|\ell\|_{L_T})T]$ for $t_c \le s \le \hat{s} \le t \le T$

(2.58) $|U(t,s;c,\ell) - U(t,s;\hat{c},\ell)| \le |c - \hat{c}|c_4(\|\ell\|_{L_T})$
$\exp[c_5(\|\ell\|_{L_T})(t - s)]T$ for $\hat{c} > c \ge -T$, $t_c \le s \le t \le T$

(2.59) $\quad |U(t,s;c,\ell) - U(t,s;c,\hat{\ell})| \le \|\ell - \hat{\ell}\|_{L_T} c_6(r)$
$\exp[c_5(r)(t - s)]T$ for $\ell, \hat{\ell} \in L_T$, $t_c^T \le s \le t \le T$, and
$\|\ell\|_{L_T}, \|\hat{\ell}\|_{L_T} \le r$

*Proof.* The properties (2.56) and (2.57) follow from (2.49), (2.50), (2.52), since by integration of (2.53) and (2.54) we obtain

$$U(t,s;c,\ell) = I - \int_s^t \mu(\tau + c, \ell(\cdot,\tau))U(\tau,s;c,\ell)\,d\tau \qquad (2.60)$$
$$t_c \le s \le t \le T$$

$$U(t,s;c,\ell) = I + \int_s^t U(t,\tau;c,\ell)\mu(\tau + c, \ell(\cdot,\tau))\,d\tau \qquad (2.61)$$
$$t_c \le s \le t \le T$$

The properties (2.58) and (2.59) follow from (2.49) and the fact that

$$\frac{\partial}{\partial\tau}[U(t,\tau;c,\ell)U(\tau,s;\hat{c},\hat{\ell})] = -U(t,\tau;c,\ell)[A(\tau;c,\ell) - A(\tau;\hat{c},\hat{\ell})]U(\tau,s;\hat{c},\hat{\ell})$$

which by (2.50) implies that

$$|U(t,s;c,\ell) - U(t,s;\hat{c},\hat{\ell})|$$
$$= \left|\int_s^t U(t,\tau;c,\ell)[A(\tau;c,\ell) - A(\tau;\hat{c},\hat{\ell})]U(\tau,s;\hat{c},\hat{\ell})\,d\tau\right|$$
$$\le \int_s^t \exp[c_5(\|\ell\|_{L_T})(t - \tau)]\ |\mu(\tau + c, \ell(\cdot,\tau)) - \mu(\tau + \hat{c}, \hat{\ell}(\cdot,\tau))|\ \exp[c_5(\|\hat{\ell}\|_{L_T})(\tau - s)\,d\tau$$

(see [190], p. 190). □

THEOREM 2.8 Let (2.1), (2.49) hold, let $\phi \in L^1$, and let $\ell$ be the solution of (ADP) on $[0,T_\phi)$. Then, for $0 \le t < T_\phi$

$$\ell(a,t) = \begin{cases} U(t, t - a; a - t, \ell)F(\ell(\cdot, t - a)) & 0 \le a \le t \\ U(t,0; a - t, \ell)\phi(a - t) & \text{a.e. } a \in (t,\infty) \end{cases} \qquad (2.62)$$

*Proof.* From (2.45) and (2.47) we have that for almost all $c \in (-T_\phi,0)$, $w_c(t)$ is continuous in $t$ from $[-c,T_\phi)$ to $R^n$, and

$$(2.63)\qquad w_c(t) = F(\ell(\cdot,-c)) - \int_{-c}^{t} \mu(s + c, \ell(\cdot,s))w_c(s)\,ds$$

for almost all $t \in (-c,T_\phi)$. By the continuity of $w_c$ and $\mu$, (2.63) holds for all $t \in [-c,T_\phi)$. Thus

$$\frac{d}{dt} w_c(t) = -\mu(t + c, \ell(\cdot,t))w_c(t) \qquad t \in [-c,T]$$
$$w_c(-c) = F(\ell(\cdot,-c))$$

By (2.55) we must have

$$(2.64)\qquad w_c(t) = U(t,-c;c,\ell)F(\ell(\cdot,-c)) \qquad t \in [-c,T_\phi)$$

We have now shown that for almost all $c \in (-T_\phi,0)$ (2.64) holds for all $t \in [-c,T_\phi)$. Thus, for all $t \in [0,T_\phi)$ and almost all $a \in (0,t)$

$$(2.65)\qquad w_{a-t}(t) = \ell(a,t) = U(t,\ t - a;\ a - t,\ \ell)F(\ell(\cdot,\ t - a))$$

Since the right-hand side of (2.65) is continuous in $a$ from $[0,t]$ to $R^n$ by (2.57) and (2.58), we conclude that the first line in (2.62) holds. A similar argument uses (2.45) and (2.48) to establish the second line in (2.62). □

THEOREM 2.9 Let (2.1), (2.49) hold, let $\phi \in L^1$, and let $\ell$ be the solution of (ADP) on $[0,T_\phi)$. The following hold:

(2.66) $\ell(a,t)$ is continuous on the triangle $0 \le a \le t < T_\phi$, $D\ell(a,t) = -\mu(a,\ell(\cdot,t))\ell(a,t)$ on this triangle [where $D$ is as in (1.6)], and $\ell(0,t) = F(\ell(\cdot,t))$ for $0 \le t < T_\phi$.

(2.67) For $c \in (-T_\phi,0]$ $w_c$ [defined as in (2.44)] is continuously differentiable on $[-c,T_\phi)$ and $d/dt\ w_c(t) = -\mu(t + c, \ell(\cdot,t))w_c(t)$ for all $t \in [-c,T_\phi)$.

(2.68) For almost all $c \in (0,\infty)$ $w_c$ [defined as in (2.44)] is continuously differentiable on $[0,T_\phi)$ and $d/dt\ w_c(t) =$

$-\mu(t + c, \ell(\cdot,t))w_c(t)$ for all $t \in [0,T_\phi)$.

(2.69) If $\phi$ is continuous on $[0,\infty)$ and $\phi(0) = F(\phi)$, then $\ell(a,t)$ is continuous on the strip $[0,\infty) \times [0,T_\phi)$, $D\ell(a,t) = -\mu(a,\ell(\cdot,t))\ell(a,t)$ on this strip, and $\ell(0,t) = F(\ell(\cdot,t))$ for $0 \leq t < T_\phi$.

(2.70) If there exists $a_0 > 0$ such that $\phi(a) = 0$ for almost all $a \in (a_0,\infty)$, then for each $t \in [0,T_\phi)$, $\ell(a,t) = 0$ for almost all $a \in (a_0 + t, \infty)$.

*Proof.* The claim (2.66) follows from the representation (2.62) and Propositions 2.8 and 2.9, as does the claim (2.67). The claim (2.68) follows from (2.46), since if

$$\frac{d}{dt} w_c(t) = G(\ell(\cdot,t))(t + c) = -\mu(t + c, \ell(\cdot,t))w_c(t) \tag{2.71}$$

for almost all $t \in [0,T_\phi)$, then the continuity of $w_c$ assured by (2.45) implies that $w_c$ is continuously differentiable on $[0,T_\phi)$ and satisfies (2.71) for all $t \in [0,T_\phi)$. To prove the claim (2.69) let $\phi$ be continuous on $[0,\infty)$ and let $t \in (0,T_\phi)$. From (2.62) we see that $\ell(a,t)$ can be redefined for $a \geq t$ such that $\ell(a,t)$ is continuous in a for $a \geq t$. Further, if $\phi(0) = F(\phi)$, then $F(\phi) = \lim_{a\to t^-} \ell(a,t) = \lim_{a\to t^+} \ell(a,t) = \phi(0)$. Thus, $\ell(a,t)$ is continuous at $a = t$, and hence everywhere on the strip $[0,\infty) \times [0,T_\phi)$. The claim (2.69) now follows from (2.62) and Propositions 2.8 and 2.9. Last, the claim (2.70) follows immediately from (2.62). □

From Theorem 2.9 we see that if the aging function G is a pure mortality function of the form (2.49), then the solution of (ADP) exhibits a "smoothing action" on the initial age distribution $\phi \in L^1$. Although the initial age distribution $\phi$ may be only integrable, the solution $\ell(a,t)$, regarded as a function of age a for a fixed time $t > 0$, is not only integrable on $(0,\infty)$, but continuous on $[0,t]$.

If, instead of making assumptions on the form of the aging function, we assume differentiability conditions on the birth and aging functions, we can obtain still further regularity of the solutions of (ADP). We require the following

DEFINITION 2.4 Let X and Y be Banach spaces and let K be a mapping from X to Y. If $\hat{x} \in D(K)$, then K is ***Fréchet differentiable*** at $\hat{x}$ provided that $K(x) = K(\hat{x}) + K'(\hat{x})(x - \hat{x}) + o(x - \hat{x})$ for all $x \in D(K)$, where $K'(\hat{x})$ is a bounded linear operator from X to Y, o is a function from X to Y, and b is a continuous increasing function from $[0,\infty)$ to $[0,\infty)$ such that $b(0) = 0$ and $\|o(x)\| \le b(r)\|x\|$ for all $x \in X$ such that $\|x\| \le r$. If K is Fréchet differentiable at each $\hat{x} \in D(K)$, then K is ***continuously Fréchet differentiable*** on D(K) provided that $|K'(x_1) - K'(x_2)| \le d(r)\|x_1 - x_2\|$ for all $x_1, x_2 \in D(K)$ such that $\|x_1\|, \|x_2\| \le r$, where d is a continuous increasing function from $[0,\infty)$ to $[0,\infty)$.

THEOREM 2.10 Let (2.1), (2.2) hold, let F be continuously Fréchet differentiable from $L^1$ to $R^n$ and let G be continuously Fréchet differentiable from $L^1$ to $L^1$. Let $\phi \in L^1$ such that $\phi$ is absolutely continuous on $[0,\infty)$, $\phi' \in L^1$, $\phi(0) = F(\phi)$, and let $\ell$ be the solution of (ADP) on $[0,T_\phi)$. The following hold:

(2.72) The mapping $t \to \ell(\cdot,t)$ is continuously differentiable from $[0,T_\phi)$ to $L^1$.

(2.73) For $0 \le t \le T < T_\phi$, $\|\frac{d}{dt}\ell(\cdot,t)\|_{L^1} \le \|\phi' - G(\phi)\|_{L^1}$

$$\exp[t(\sup_{0\le s\le t}|F'(\ell(\cdot,s))| + \sup_{0\le s\le t}|G'(\ell(\cdot,s))|)].$$

*Proof.* As in Definition 2.4 let

(2.74) $d_1$: $[0,\infty) \to [0,\infty)$ such that $d_1$ is continuous, increasing, and satisfies $|F'(\phi_1) - F'(\phi_2)| \le d_1(r)\|\phi_1 - \phi_2\|_{L^1}$ for $\|\phi_1\|_{L^1}, \|\phi_2\|_{L^1} \le r$.

(2.75) $d_2$: $[0,\infty) \to [0,\infty)$ such that $d_2$ is continuous, increasing, and satisfies $|G'(\phi_1) - G'(\phi_2)| \le d_2(r)\|\phi_1 - \phi_2\|_{L^1}$ for $\|\phi_1\|_{L^1}, \|\phi_2\|_{L^1} \le r$.

Consider the integral equation in two independent variables

(2.76)

$$k(a,t) = \begin{cases} F'(\ell(\cdot,t-a))k(\cdot,t-a) + \int_{t-a}^{t} (G'(\ell(\cdot,s))k(\cdot,s))(s+a-t)\,ds & \text{a.e. } a \in (0,t) \\ -\phi'(a-t) + G(\phi)(a-t) + \int_{0}^{t} (G'(\ell(\cdot,s))k(\cdot,s))(s+a-t)\,ds & \text{a.e. } a \in (t,\infty) \end{cases}$$

We will use the method of successive integrations to show that there exists a solution k of (2.76) such that $k \in L_T$ for each $T \in [0,T_\phi)$. We will then show that for $t \in [0,T_\phi)$

$$(2.77) \qquad \ell(\cdot,t) = \phi + \int_0^t k(\cdot,s)\,ds$$

For notational convenience define $K(t) \overset{\text{def}}{=} F'(\ell(\cdot,t))$, $0 \le t < T_\phi$, $H(t) \overset{\text{def}}{=} G'(\ell(\cdot,t))$, $0 \le t < T_\phi$, $\psi \overset{\text{def}}{=} -\phi' + G(\phi)$, and for each $T \in [0,T_\phi)$ define $h_1(T) \overset{\text{def}}{=} \sup_{0\le t\le T}|K(t)|$, $h_2(T) \overset{\text{def}}{=} \sup_{0\le t\le T}|H(t)|$. Fix $T \in [0,T_\phi)$. Let $k_0 \overset{\text{def}}{=} 0$ in $L^1$ and for n = 1, 2, ... and $0 \le t \le T$ let

$$k_n(a,t) \overset{\text{def}}{=} \begin{cases} K(t-a)k_{n-1}(\cdot,t-a) + \int_{t-a}^{t} (H(s)k_{n-1}(\cdot,s))(s+a-t)\,ds & \text{a.e. } a \in (0,t) \\ \psi(a-t) + \int_{0}^{t} (H(s)k_{n-1}(\cdot,s))(s+a-t)\,ds & \text{a.e. } a \in (t,\infty) \end{cases}$$

Using an argument similar to the one of Proposition 2.2, one shows that $k_n(\cdot,t) \in L^1$ for $0 \le t \le T$, n = 1, 2, .... Further, for $t \in [0,T]$

$$\|k_1(\cdot,t) - k_0(\cdot,t)\|_{L^1} = \|\psi\|_{L^1}$$

$$\|k_2(\cdot,t) - k_1(\cdot,t)\|_{L^1}$$

$$\le \int_0^t |K(t-a)[k_1(\cdot, t-a) - k_0(\cdot, t-a)]|\,da$$

$$+ \int_0^t \left[ \int_{t-a}^t |(H(s)k_1(\cdot,s))(s+a-t)\right.$$

$$- (H(s)k_0(\cdot,s))(s + a - t)|\ ds\Big]\ da$$

$$+ \int_t^\infty \Big[\int_0^t |(H(s)k_1(\cdot,s))(s + a - t)$$

$$- (H(s)k_0(\cdot,s))(s + a - t)|\ ds\Big]\ da$$

$$\le \int_0^t |K(s)[k_1(\cdot,s) - k_0(\cdot,s)]|\ ds$$

$$+ \int_0^t \Big[\int_{t-s}^\infty |(H(s)k_1(\cdot,s))(s + a - t)$$

$$- (H(s)k_0(\cdot,s))(s + a - t)|\ da\Big]\ ds$$

$$\le h_1(T) \int_0^t \|\psi\|_{L^1}\ ds + \int_0^t \|H(s)[k_1(\cdot,s) - k_0(\cdot,s)]\|_{L^1}\ ds$$

$$\le [h_1(T) + h_2(T)]t\|\psi\|_{L^1}$$

and for $n = 2, 3, \ldots,$

$$\|k_{n+1}(\cdot,t) - k_n(\cdot,t)\|_{L^1}$$

$$\le [h_1(T) + h_2(T)] \int_0^t \|k_n(\cdot,s) - k_{n-1}(\cdot,s)\|_{L^1}\ ds$$

$$\le [h_1(T) + h_2(T)]^n \frac{t^n}{n!} \|\psi\|_{L^1}$$

Using standard arguments, one now shows that $\lim_{n\to\infty} k_n(\cdot,t) \overset{\text{def}}{=} k(\cdot,t)$ exists for $t \in [0,T]$ and $k$ is the unique solution of (2.76). Further, an argument similar to that of Proposition 2.2 shows that $k \in L_T$.

To establish (2.77) define

$$\hat{\ell}(\cdot,t) \overset{\text{def}}{=} \phi + \int_0^t k(\cdot,s)\ ds \qquad 0 \le t \le T$$

and let $r > 0$ such that $\|\ell(\cdot,t)\|_{L^1}$, $\|\hat{\ell}(\cdot,t)\|_{L^1}$, $\|k(\cdot,t)\|_{L^1}$ are all $\le r$ for $0 \le t \le T$. Observe that for $0 < a < t < T$

$$(2.78)\quad \phi(0) = F(\phi) = F(\hat{\ell}(\cdot,\ t - a)) - \int_a^t \frac{d}{ds} F(\hat{\ell}(\cdot,\ s - a))\ ds$$

$$= F(\hat{\ell}(\cdot,\ t - a)) - \int_a^t F'(\hat{\ell}(\cdot,\ s - a))k(\cdot,\ s - a)\ ds$$

$$(2.79)\quad \int_0^a G(\hat{\phi})(\tau)\ d\tau = \int_0^a G(\hat{\ell}(\cdot,0))(\tau)\ d\tau = \int_0^a G(\hat{\ell}(\cdot,\tau))(\tau)\ d\tau$$

$$- \int_0^a \left[ \int_{a-t}^a \frac{d}{ds} G(\hat{\ell}(\cdot,\ \tau + s - a))(\tau)\ ds \right] d\tau$$

$$= \int_0^a G(\hat{\ell}(\cdot,\tau))(\tau)\ d\tau$$

$$- \int_0^a \left[ \int_{a-s}^a (G'(\hat{\ell}(\cdot,\ \tau + s - a))k(\cdot,\ \tau + s - a))(\tau)\ d\tau \right] ds$$

$$(2.80)\quad \int_0^a G(\hat{\ell}(\cdot,\tau))(\tau)\ d\tau = \int_0^a G(\hat{\ell}(\cdot,\ \tau + t - a))(\tau)\ d\tau$$

$$- \int_0^a \left[ \int_a^t \frac{d}{ds} G(\hat{\ell}(\cdot,\ \tau + s - a))(\tau)\ ds \right] d\tau$$

$$= \int_0^a G(\hat{\ell}(\cdot,\ \tau + t - a))(\tau)\ d\tau$$

$$- \int_a^t \left[ \int_0^a (G'(\hat{\ell}(\cdot,\ \tau + s - a))k(\cdot,\ \tau + s - a))(\tau)\ d\tau \right] ds$$

For $t \in [0,T]$ (2.78), (2.79), (2.80) imply that

$$(2.81)\quad \int_0^t |\ell(a,t) - \hat{\ell}(a,t)|\ da = \int_0^t \Big| F(\ell(\cdot,\ t - a))$$

$$+ \int_0^a G(\ell(\cdot,\ \tau + t - a))(\tau)\ d\tau - \phi(a) - \int_0^a k(a,s)\ ds$$

$$- \int_a^t k(a,s)\ ds \Big|\ da$$

$$= \int_0^t \Big| F(\ell(\cdot,\ t - a)) + \int_0^a G(\ell(\cdot,\ \tau + t - s))(\tau)\ d\tau - \phi(a)$$

$$- \int_0^a \Big[ -\phi'(a - s) + G(\phi)(a - s)$$

$$+ \int_{a-s}^a G'(\ell(\cdot, \tau + s - a))k(\cdot, \tau + s - a))(\tau)\, d\tau \Big]\, ds$$

$$- \int_a^t \Big[ F'(\ell(\cdot, s - a))k(\cdot, s - a)$$

$$+ \int_0^a (G'(\ell(\cdot, \tau + s - a))k(\cdot, \tau + s - a))(\tau)\, d\tau \Big]\, ds \Big|\, da$$

$$= \int_0^t \Big| F(\ell(\cdot, t - a)) + \int_0^a G(\ell(\cdot, \tau + t - a))(\tau)\, d\tau - F(\hat{\ell}(\cdot, t - a))$$

$$+ \int_a^t F'(\hat{\ell}(\cdot, s - a))k(\cdot, s - a)\, ds$$

$$+ \int_0^a \Big[ \int_{a-s}^a G'(\hat{\ell}(\cdot, \tau + s - a))k(\cdot, \tau + s - a))(\tau)\, d\tau \Big]\, ds$$

$$- \int_0^a G(\hat{\ell}(\cdot, \tau + t - a))(\tau)\, d\tau$$

$$+ \int_a^t \Big[ \int_0^a (G'(\hat{\ell}(\cdot, \tau + s - a))k(\cdot, \tau + s - a))(\tau)\, d\tau \Big]\, ds$$

$$- \int_0^a \Big[ \int_{a-s}^a G'(\ell(\cdot, \tau + s - a))k(\cdot, \tau + s - a))(\tau)\, d\tau \Big]\, ds$$

$$- \int_a^t F'(\ell(\cdot, s - a))k(\cdot, s - a)\, ds$$

$$- \int_a^t \Big[ \int_0^a (G'(\ell(\cdot, \tau + s - a))k(\cdot, \tau + s - a))(\tau)\, d\tau \Big]\, ds \Big|\, da$$

$$\leq \int_0^t |F(\ell(\cdot, t - a)) - F(\hat{\ell}(\cdot, t - a))|\, da$$

$$+ \int_0^t \Big[ \int_{t-a}^t |G(\ell(\cdot,s))(s + a - t) - G(\hat{\ell}(\cdot,s))(s + a - t)|\, ds \Big]\, da$$

$$+ \int_0^t \Big[ \int_0^{t-a} |[F'(\hat{\ell}(\cdot,\sigma)) - F'(\ell(\cdot,\sigma))]k(\cdot,\sigma)|\, d\sigma \Big]\, da$$

$$+ \int_0^t \left( \int_0^a \left\{ \int_0^s |([G'(\hat{\ell}(\cdot,\sigma)) - G'(\ell(\cdot,\sigma))]k(\cdot,\sigma))(\sigma + a - s)| \, d\sigma \right\} ds \right) da$$

$$+ \int_0^t \left( \int_a^t \left\{ \int_{s-a}^s |([G'(\hat{\ell}(\cdot,\sigma)) - G'(\ell(\cdot,\sigma))]k(\cdot,\sigma))(\sigma + a - s)| \, d\sigma \right\} ds \right) da$$

$$\leq \int_0^t |F(\ell(\cdot,s)) - F(\hat{\ell}(\cdot,s))| \, ds$$

$$+ \int_0^t \left[ \int_{t-s}^t |G(\ell(\cdot,s))(s + a - t) - G(\hat{\ell}(\cdot,s))(s + a - t)| \, da \right] ds$$

$$+ \int_0^t \left[ \int_0^{t-\sigma} |[F'(\hat{\ell}(\cdot,\sigma)) - F'(\ell(\cdot,\sigma))]k(\cdot,\sigma)| \, da \right] d\sigma$$

$$+ \int_0^t \left( \int_0^a \left\{ \int_0^\sigma |([G'(\hat{\ell}(\cdot,\sigma)) - G'(\ell(\cdot,\sigma))]k(\cdot,\sigma))(\sigma + a - s)| \, ds \right\} d\sigma \right) da$$

$$+ \int_0^t \left( \int_a^t \left\{ \int_{\sigma-a}^\sigma |([G'(\hat{\ell}(\cdot,\sigma)) - G'(\ell(\cdot,\sigma))]k(\cdot,\sigma))(\sigma + a - s)| \, ds \right\} d\sigma \right) da$$

$$\leq \int_0^t c_1(r) \|\ell(\cdot,s) - \hat{\ell}(\cdot,s)\|_{L^1} \, ds + \int_0^t c_2(r) \|\ell(\cdot,s) - \hat{\ell}(\cdot,s)\|_{L^1} \, ds$$

$$+ \int_0^t T d_1(r) \|\hat{\ell}(\cdot,\sigma) - \ell(\cdot,\sigma)\|_{L^1} \|k(\cdot,\sigma)\|_{L^1} \, d\sigma$$

$$+ \int_0^t \left[ \int_0^a d_2(r) \|\hat{\ell}(\cdot,\sigma) - \ell(\cdot,\sigma)\|_{L^1} \|k(\cdot,\sigma)\|_{L^1} \, d\sigma \right] da$$

$$+ \int_0^t \left[ \int_a^t d_2(r) \|\hat{\ell}(\cdot,\sigma) - \ell(\cdot,\sigma)\|_{L^1} \|k(\cdot,\sigma)\|_{L^1} \, d\sigma \right] da$$

$$\leq \int_0^t \{c_1(r) + c_2(r) + T[d_1(r) + d_2(r)]r\} \|\ell(\cdot,s) - \hat{\ell}(\cdot,s)\|_{L^1} \, ds$$

Next, observe that for $0 < t < T$ and $a > t$ (see Figure 2.7 below)

$$(2.82) \qquad \int_{a-t}^a G(\phi)(\tau) \, d\tau = \int_{a-t}^a G(\hat{\ell}(\cdot,0))(\tau) \, d\tau$$

$$= \int_{a-t}^a G(\hat{\ell}(\cdot, \tau + t - a))(\tau) \, d\tau$$

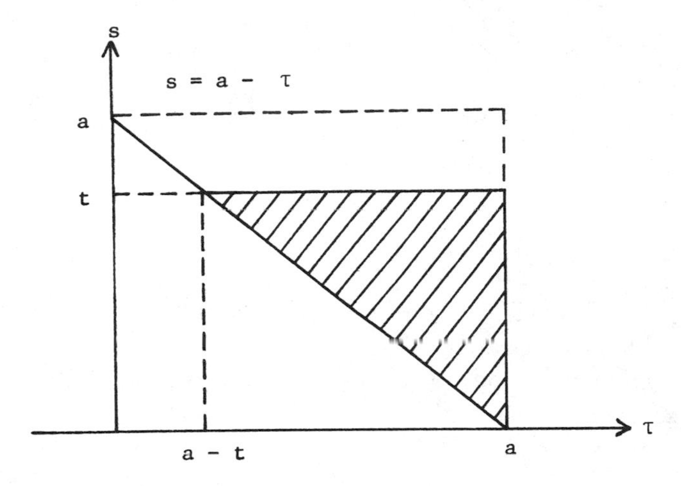


*Figure 2.7*

$$- \int_{a-t}^{a} \left[ \int_{a-\tau}^{t} \frac{d}{ds} G(\hat{\ell}(\cdot, \tau + s - a))(\tau)\, ds \right] d\tau$$

$$= \int_{a-t}^{a} G(\hat{\ell}(\cdot, \tau + t - a))(\tau)\, d\tau$$

$$- \int_{0}^{t} \left[ \int_{a-s}^{a} (G'(\hat{\ell}(\cdot, \tau + s - a))k(\cdot, \tau + s - a))(\tau)\, d\tau \right] da$$

For $t \in [0,T]$ (2.82) implies that

$$\text{(2.83)} \qquad \int_{t}^{\infty} |\ell(a,t) - \hat{\ell}(a,t)|\, da$$

$$= \int_{t}^{\infty} \Big| \phi(a - t) + \int_{a-t}^{a} G(\ell(\cdot, \tau + t - a))(\tau)\, d\tau - \phi(a)$$

$$- \int_{0}^{t} k(a,s)\, ds \Big|\, da$$

$$= \int_{t}^{\infty} \Big| \phi(a - t) + \int_{a-t}^{a} G(\ell(\cdot, \tau + t - a))(\tau)\, d\tau - \phi(a)$$

$$- \int_{0}^{t} \Big[ -\phi'(a - s) + G(\phi)(a - s)$$

$$+ \int_{a-s}^{a} (G'(\ell(\cdot, \tau + s - a))k(\cdot, \tau + s - a))(\tau)\, d\tau \Big]\, ds \Big|\, da$$

$$= \int_{t}^{\infty} \Big| \int_{a-t}^{a} G(\ell(\cdot, \tau + t - a))(\tau)\, d\tau$$

$$- \int_{a-t}^{a} G(\hat{\ell}(\cdot, \tau + t - a))(\tau)\, d\tau$$

$$+ \int_{0}^{t} \Big[ \int_{a-s}^{a} ([G'(\hat{\ell}(\cdot, \tau + s - a))$$

$$- G'(\ell(\cdot, \tau + s - a))]k(\cdot, \tau + s - a))(\tau)\, d\tau \Big]\, ds \Big|\, da$$

$$= \int_{t}^{\infty} \Big| \int_{0}^{t} [G(\ell(\cdot,s))(s + a - t) - G(\hat{\ell}(\cdot,s))(s + a - t)]\, ds$$

$$+ \int_{0}^{t} \Big[ \int_{0}^{s} ([G'(\hat{\ell}(\cdot,\sigma))$$

$$- G'(\ell(\cdot,\sigma))]k(\cdot,\sigma))(\sigma + a - s)\, d\sigma \Big]\, ds \Big|\, da$$

$$\le \int_{0}^{t} \Big[ \int_{t}^{\infty} |G(\ell(\cdot,s))(s + a - t) - G(\hat{\ell}(\cdot,s))(s + a - t)|\, da \Big]\, ds$$

$$+ \int_{0}^{t} \Big[ \int_{0}^{\sigma} \Big( \int_{t}^{\infty} |([G'(\hat{\ell}(\cdot,\sigma))$$

$$- G'(\ell(\cdot,\sigma))]k(\cdot,\sigma))(\sigma + a - s)\, d\sigma \Big]\, ds \Big|\, da$$

$$\le \int_{0}^{t} \|G(\ell(\cdot,s)) - G(\hat{\ell}(\cdot,s))\|_{L^1}\, ds$$

$$+ \int_{0}^{t} \Big[ \int_{0}^{\infty} \|[G'(\hat{\ell}(\cdot,\sigma)) - G'(\ell(\cdot,\sigma))]k(\cdot,\sigma)\|_{L^1}\, ds \Big]\, d\sigma$$

$$\le \int_{0}^{t} c_2(r) \|\ell(\cdot,s) - \hat{\ell}(\cdot,s)\|_{L^1}\, ds$$

$$+ \int_{0}^{t} \Big[ \int_{0}^{\sigma} d_2(r) \|\ell(\cdot,\sigma) - \hat{\ell}(\cdot,\sigma)\|_{L^1} \|k(\cdot,\sigma)\|_{L^1}\, ds \Big]\, d\sigma$$

$$\le [c_2(r) + Td_2(r)r] \int_{0}^{t} \|\ell(\cdot,s) - \hat{\ell}(\cdot,s)\|_{L^1}\, ds$$

Now use (2.81), (2.83), and Gronwall's inequality (see [217], Proposition 1.4, p. 204) to show that $\hat{\ell}(\cdot,t) = \ell(\cdot,t)$ for $t \in [0,T]$. The claim (2.72) then follows immediately. The claim (2.73) follows from another application of Gronwall's inequality and the estimate

$$\|k(\cdot,t)\|_{L^1} = \int_0^t |k(a,t)|\, da + \int_t^\infty |k(a,t)|\, da$$

$$\le \int_0^t |F'(\ell(\cdot, t - a))k(\cdot, t - a)|\, da$$

$$+ \int_0^t \left[ \int_{t-a}^t |(G'(\ell(\cdot,s))k(\cdot,s))(s + a - t)|\, ds \right] da$$

$$+ \int_t^\infty |\phi'(a - t) - G(\phi)(a - t)|\, da$$

$$+ \int_t^\infty \left[ \int_0^t |(G'(\ell(\cdot,s))k(\cdot,s))(s + a - t)|\, ds \right] da$$

$$\le [h_1(T) + h_2(T)] \int_0^t \|k(\cdot,s)\|_{L^1}\, ds + \|\phi' - G(\phi)\|_{L^1}$$

for $t \in [0,T]$. □

REMARK 2.4 We note that Examples 2.1, 2.2, and 2.3 may be used to illustrate Theorems 2.8, 2.9, and 2.10. If we require that $\beta\colon R \to [0,\infty)$ and $\mu\colon R \to [0,\infty)$ are both twice continuously differentiable, then F defined in (2.17), (2.19), or (2.20) and G defined in (2.18) satisfy the hypothesis of Theorems 2.8, 2.9, and 2.10. It is obvious that this G satisfies (2.49). Also, F and G are continuously Fréchet differentiable in the sense of Definition 2.4, since

(2.84) $$F'(\hat{\phi})\phi = \beta'(P\hat{\phi})P\phi \int_0^\infty f(a)\phi(a)\, da + \beta(P\hat{\phi}) \int_0^\infty f(a)\phi(a)\, da, \quad \phi,$$

$\hat{\phi} \in L^1$ [where $f(a) = e^{-\alpha a}$, $1 - e^{-\alpha a}$, or $ae^{-\alpha a}$].

(2.85) $$(G'(\hat{\phi})\phi)(a) = -\mu'(P\hat{\phi})P\phi\ \phi(a) - \mu(P\hat{\phi})\phi(a), \quad \hat{\phi}, \phi \in L^1,$$

a.e. $a > 0$.

There are additional regularity properties of the solutions of (ADP) which can be established by analyzing the infinitesimal generator of the strongly continuous semigroup of nonlinear operators associated with the problem. These results will be established in the study of the connection between strongly continuous operator semigroups and age-dependent population dynamics, a study whose development is the objective of the next chapter.

## 2.7 NOTES

Many of the results in this chapter first appeared in [308]. In a series of papers M. Marcus and V. Mizel have studied general classes of integral equations with two independent variables. Their theory is applicable to many types of functional differential equations, including the equations of nonlinear age-dependent population dynamics (see [214], [215], and [216]).

It is sometimes useful to choose as a setting for the densities Banach spaces other than $L^1(0,\infty;R^n)$. In [37] M. Chipot has developed basic theory for nonlinear nonautonomous age-dependent population dynamics in the setting of $L^1(0,\infty;R^n)$. In [303] and [305] Banach spaces of continuous functions were chosen as the setting for the density functions.

The continuability in time of the local solutions of (ADP) depends on a priori considerations, since the hypothesis (2.1) on the birth function F and the hypothesis (2.2) on the aging function G involve only local Lipschitz continuity conditions. In [130] M. Gurtin and R. C. MacCamy assume that the birth modulus is bounded in order to assure global existence of the solutions, and give an example to show that if this boundedness condition is not satisfied, then the solutions may become infinite in finite time. In the case of the Gurtin and MacCamy model (1.25), (1.26), (1.27) the positivity of the solutions and their regularity properties can be discerned directly from the representation of the solutions in formulas (1.30), (1.33), and (1.34).

# 3
# The Nonlinear Semigroup Associated with the Solutions

## 3.1 THE NONLINEAR SEMIGROUP

The solutions of (ADP) evolve with time in the state space $L^1$. For this reason it is possible to view the problem (ADP) as an abstract evolution equation in the Banach space $L^1$ and to associate with its solutions a strongly continuous semigroup of nonlinear operators in this space. The theory of abstract nonlinear evolution equations and nonlinear semigroups in Banach spaces has been extensively developed in recent years. In this section we will state some of the basic results in this theory, and we will show that the solutions of (ADP) give rise to such a nonlinear semigroup in the Banach space $L^1$. We first define

DEFINITION 3.1 Let X be a Banach space and let C be a closed set in X. A *strongly continuous nonlinear semigroup* in C is a family of mappings $S(t)$, $t \geq 0$, satisfying the following:

(3.1) $S(t)$ is a continuous mapping from C into C for each $t \geq 0$.

(3.2) $S(0) = I$ (where I denotes the identity mapping in X restricted to C).

(3.3) $S(t_1 + t_2)x = S(t_1)S(t_2)x$ for all $t_1, t_2 \geq 0$, $x \in C$.

(3.4) $S(t)x$ is continuous in t as a function from $[0,\infty)$ to C for each fixed $x \in C$.

We remark that the product notation in (3.3) means composition of mappings. The condition (3.3) is called the *semigroup property* and the condition (3.4) is called the *strong continuity property*. The functions $t \to S(t)x$, for fixed $x \in C$, are called the *trajectories* (or *orbits*) of the nonlinear semigroup.

THEOREM 3.1 Let (2.1), (2.2), (2.22), (2.23) hold and for each $\phi \in L^1_+$ let the maximal interval of existence $[0,T_\phi)$ of the solution of (ADP) be $[0,\infty)$. Let $S(t)$, $t \geq 0$, be the family of mappings in $L^1_+$ defined as follows: for $t \geq 0$, $\phi \in L^1_+$, $S(t)\phi \overset{\text{def}}{=} \ell(\cdot,t)$, where $\ell$ is the solution of (ADP) on $[0,\infty)$. Then, $S(t)$, $t \geq 0$, is a strongly continuous nonlinear semigroup in $L^1_+$.

*Proof.* The properties (3.2) and (3.4) follow immediately from the definition of solution of (ADP) on $[0,T_\phi)$. To prove (3.1) let $t > 0$, let $\phi \in L^1_+$, let $r = 1 + \sup_{0\leq s\leq t}\|S(s)\phi\|_{L^1}$, let $c = c_1(r) + c_2(r)$, and let $\delta = e^{-ct}$. We will show that if $\hat{\phi} \in L^1_+$ and $\|\phi - \hat{\phi}\|_{L^1} < \delta$, then $\sup_{0\leq s\leq t}\|S(s)\hat{\phi}\|_{L^1} \leq r$. Let $\hat{\phi} \in L^1_+$ such that $\|\phi - \hat{\phi}\|_{L^1} < \delta$ and observe that $\|S(0)\hat{\phi}\|_{L^1} = \|\hat{\phi}\|_{L^1} < \|\phi\|_{L^1} + \delta = \|\phi\|_{L^1} + e^{-ct} \leq r - 1 + e^{-ct} < r$. Assume that there exists $t_1 \in (0,t]$ such that $\|S(t_1)\hat{\phi}\|_{L^1} > r$. Then, there exists a smallest number $t_0 \in (0,t_1)$ such that $\|S(t_0)\hat{\phi}\|_{L^1} = r$. From (2.15) we have that $\|S(t_0)\phi - S(t_0)\hat{\phi}\|_{L^1} \leq \|\phi - \hat{\phi}\|_{L^1}e^{ct_0} < \delta e^{ct_0} \leq 1$. Hence, $\|S(t_0)\hat{\phi}\|_{L^1} < \|S(t_0)\phi\|_{L^1} + 1 \leq \sup_{0\leq s\leq t}\|S(s)\phi\|_{L^1} + 1 = r$, and we have a contradiction. Therefore, if $\|\phi - \hat{\phi}\|_{L^1} < \delta$, (2.15) yields that $\|S(t)\phi - S(t)\hat{\phi}\|_{L^1} \leq \|\phi - \hat{\phi}\|_{L^1}e^{ct}$, and so $S(t)$ is continuous at $\phi$. From Theorem 2.4 we have that for each $t > 0$, $S(t)$ maps $L^1_+$ into $L^1_+$. Thus, (3.1) is proved.

To prove (3.3) let $\phi \in L^1_+$, let $t_1 > 0$, let $\ell(\cdot,t) = S(t)\phi$ for $t \geq 0$, let $\hat{\phi} = S(t_1)\phi$, and let $\hat{\ell}(\cdot,t) = S(t)\hat{\phi}$ for $t \geq 0$. By Proposition 2.4 we see that $\ell(\cdot, t_1 + t_2) = \hat{\ell}(\cdot,t_2)$ for $t \geq 0$. Thus, $S(t_1 + t_2)\phi = S(t_2)S(t_1)\phi$, and so (3.3) is proved. □

REMARK 3.1 We remark that the continuity of $S(t)$, for a fixed $t$, implies that the solutions of (ADP) depend continuously on the initial age distributions in the norm of $L^1$. Notice that in the proof of Theorem 3.1 we have actually shown that for each $t > 0$ the mapping $S(t)$ is locally Lipschitz continuous on $L^1_+$ in the following sense: if $\phi \in L^1_+$, there exists a neighborhood $N_\phi$ about $\phi$ in $L^1$ and a constant $M_\phi$ such that if $\phi_1, \phi_2 \in L^1_+ \cap N_\phi$, then $\|S(t)\phi_1 - S(t)\phi_2\|_{L^1} \leq M_\phi \|\phi_1 - \phi_2\|_{L^1}$. That is, it was proved that if $\phi \in L^1$ and $t > 0$, there exists $\delta > 0$ and $r > 0$ such that if $\hat{\phi} \in L^1_+$ and $\|\phi - \hat{\phi}\|_{L^1} < \delta$, then $\sup_{0 \leq s \leq t} \|S(s)\hat{\phi}\|_{L^1} \leq r$. The claimed local Lipschitz continuity of $S(t)$ at $\phi$ then follows from Proposition 2.3. In general, the mappings $S(t)$ will not be globally Lipschitz continuous on $L^1_+$, since we have hypothesized only that $F$ and $G$ are Lipschitz continuous on bounded sets of $L^1$. If, however, $F$ and $G$ are globally Lipschitz continuous on $L^1$ in the sense that $\sup_{r \geq 0} c_1(r) < \infty$ and $\sup_{r \geq 0} c_2(r) < \infty$, then we can prove

PROPOSITION 3.1 Let (2.1), (2.2), (2.22), (2.23) hold, let $\sup_{r>0} c_1(r) \overset{\text{def}}{=} c_1 < \infty$, let $\sup_{r \geq 0} c_2(r) \overset{\text{def}}{=} c_2 < \infty$, and let $\omega \overset{\text{def}}{=} c_1 + c_2$. Then, for all $\phi \in L^1_+$ the maximal interval of existence of the solution of (ADP) is $[0, T_\phi) = [0, \infty)$. Further, if $S(t)$, $t \geq 0$ is the strongly continuous nonlinear semigroup in $L^1_+$ as in Theorem 3.1, then

(3.5) $$\|S(t)\phi_1 - S(t)\phi_2\|_{L^1} \leq e^{\omega t} \|\phi_1 - \phi_2\|_{L^1} \text{ for all } \phi_1, \phi_2 \in L^1_+, \quad t \geq 0.$$

*Proof.* Let $\phi \in L^1_+$ and let $0 \leq t < T_\phi$. From Theorem 2.2 we have that the solutions of (ADP) are solutions of (1.49), and so

$$\begin{aligned}
\|S(t)\phi\|_{L^1} \leq & \int_0^t |F(S(t-a)\phi) - F(0)|\, da + \int_0^t |F(0)|\, da \\
& + \int_0^t \left[ \int_{t-a}^t |G(S(s)\phi)(s+a-t) - G(0)(s+a-t)|\, ds \right] da \\
& + \int_0^t \left[ \int_{t-a}^t |G(0)(s+a-t)|\, ds \right] da + \int_t^\infty |\phi(a-t)|\, da
\end{aligned}$$

$$+ \int_t^\infty \left[ \int_0^t |G(S(s)\phi)(s + a - t) - G(0)(s + a - t)| \, ds \right] da$$

$$+ \int_t^\infty \left[ \int_0^t |G(0)(s + a - t)| \, ds \right] da$$

$$\le c_1 \int_0^t \|S(s)\phi\|_{L^1} \, ds + t|F(0)|$$

$$+ \int_0^t \left[ \int_0^\infty |G(S(s)\phi)(a) - G(0)(a)| \, da \right] ds$$

$$+ \int_0^t \left[ \int_0^\infty |G(0)(a)| \, da \right] ds + \|\phi\|_{L^1}$$

$$\le (c_1 + c_2) \int_0^t \|S(s)\phi\|_{L^1} \, ds + \|\phi\|_{L^1} + t[|F(0)| + \|G(0)\|_{L^1}]$$

By Gronwall's lemma ([217], Proposition 1.4, p. 204) we obtain

$$\|S(t)\phi\|_{L^1} \le e^{\omega t}[\|\phi\|_{L^1} + t(|F(0)| + \|G(0)\|_{L^1})] \tag{3.6}$$

and by Theorem 2.3 we then obtain $T_\phi = \infty$. An application of Proposition 2.3 now establishes (3.5). □

If F and G are bounded linear operators, then the solutions of (ADP) may be associated with a strongly continuous semigroup of bounded linear operators in $L^1$. We state this fact as

PROPOSITION 3.2 Let F be a bounded linear operator from $L^1$ to $R^n$ and let G be a bounded linear operator from $L^1$ to $L^1$. If $\phi \in L^1$, then the solution of (ADP) is defined on $[0,\infty)$. Further, the family of mappings $S(t)$, $t \ge 0$, in $L^1$ defined by $S(t) = \ell(\cdot,t)$, where $\ell(\cdot,t)$ is the solution of (ADP) on $[0,\infty)$, is a strongly continuous semigroup of bounded linear operators in $L^1$ satisfying

$$|S(t)| \le e^{\omega t} \qquad t \ge 0 \qquad \text{where } \omega \overset{\text{def}}{=} |F| + |G| \tag{3.7}$$

*Proof.* The arguments of Theorem 3.1 and Proposition 3.1 may be used to establish that $T_\phi = \infty$ for each $\phi \in L^1$, and that the family of mappings $S(t)$, $t \geq 0$, form a strongly continuous semigroup defined on all of $L^1$. The linearity of each mapping $S(t)$ follows from (1.49) by virtue of the linearity of F and G. The estimate (3.7) follows as in the proof of (3.6). □

We now state some definitions and theorems from nonlinear semigroup theory. For an introduction to this theory, we refer the reader to the monographs of V. Barbu [13], J. Goldstein [116], R. Martin [217], and J. Walker [290]. One of the basic concepts in the theory of both linear and nonlinear semigroups is the concept of the infinitesimal generator. Information concerning the regularity, asymptotic behavior, and numerical approximation of the trajectories of the semigroup can be ascertained from the properties of its infinitesimal generator.

DEFINITION 3.2 Let C be a closed subset of the Banach space X and let $S(t)$, $t \geq 0$, be a strongly continuous nonlinear semigroup in C. The *infinitesimal generator* of $S(t)$, $t \geq 0$, is the mapping B from a subset of C to X such that

$$(3.8) \qquad \lim_{t \to 0^+} t^{-1}[S(t)x - x] \overset{\text{def}}{=} Bx$$

with domain $D(B)$ the set of all $x \in C$ for which the limit in (3.8) exists.

The special class of strongly continuous nonlinear semigroups which satisfy global Lipschitz continuity conditions of the form (3.5) are associated with the theory of nonlinear accretive operators in Banach spaces. We will apply this theory to the problem (ADP), first in the special case that F and G are globally Lipschitz continuous, and then, by means of a truncation device, to the case that F and G are locally Lipschitz continuous in the sense of (2.1) and (2.2). We now state

DEFINITION 3.3 Let A be a mapping from a subset of a Banach space X to X. A is said to be *accretive* in X provided that if $x_1$, $x_2$ belong to the domain $D(A)$ of A and $\lambda > 0$, then

(3.9) $\|(I + \lambda A)x_1 - (I + \lambda A)x_2\| \geq \|x_1 - x_2\|$

Notice from (3.9) that if A is accretive in X and $\lambda > 0$, then $I + \lambda A$ is one-to-one on D(A), and for $y_1$, $y_2$ in the range $R(I + \lambda A)$ of $I + \lambda A$,

(3.10) $\|(I + \lambda A)^{-1}y_1 - (I + \lambda A)^{-1}y_2\| \leq \|y_1 - y_2\|$

We state next one of the fundamental results in the theory of nonlinear semigroups, a proof of which may be found in [58], Theorem I.

PROPOSITION 3.3 (M. Crandall and T. Liggett) Let A be a mapping from a subset of a Banach space X to X and let there exist $\omega \in R$ such that $A + \omega I$ is accretive in X. Let there exist $\lambda_1 > 0$ such that if $0 < \lambda < \lambda_1$, then $R(I + \lambda A) \supset \overline{D(A)}$. Then, for each $x \in \overline{D(A)}$

(3.11) $\lim_{n\to\infty}(I + t/nA)^{-n}x \overset{def}{=} T(t)x$ exists uniformly in bounded intervals of $t \geq 0$.

Moreover, the family of mappings T(t), $t \geq 0$, so defined is a strongly continuous nonlinear semigroup in $\overline{D(A)}$ satisfying

(3.12) $\|T(t)x_1 - T(t)x_2\| \leq e^{\omega t}\|x_1 - x_2\|$ for all $t \geq 0$ and $x_1$, $x_2 \in \overline{D(A)}$.

REMARK 3.2 In Proposition 3.3 one says that the strongly continuous nonlinear semigroup T(t), $t \geq 0$, is generated by A. It is well known that if T(t), $t \geq 0$, is a strongly continuous semigroup of linear operators in a Banach space, then T(t), $t \geq 0$, has a densely defined infinitesimal generator B, and T(t), $t \geq 0$, is generated by -B. The same result holds true for a strongly continuous nonlinear semigroup of contraction mappings defined in a closed convex subset of a Hilbert space. For the case of a strongly continuous nonlinear semigroup in a general Banach space this result may not be true, and in fact, the infinitesimal generator may have empty domain (see [58]). We will show, however, that the nonlinear semigroup S(t), $t \geq 0$, associated with (ADP) as in Theorem 3.1 does have an infinitesimal generator B, which is densely defined in $L^1_+$. We will also show that in the case that F and G are globally Lipschitz continuous as in Proposition 3.1, $S(t)$, $t \geq 0$, is generated by $-B$. To establish this result we will use another fundamental result in nonlinear semigroup theory, the proof of which may be found in [26], Theorem 2.1 and Corollary 2.1 and [27], Corollary 4.3:

PROPOSITION 3.4 (H. Brezis and A. Pazy) Let C be a closed subset of a Banach space X. Let A be a mapping from C to X such that $\overline{D(A)} = C$, let there exist $\omega \in R$ such that $A + \omega I$ is accretive, and let there exist $\lambda_1 > 0$ such that if $0 < \lambda < \lambda_1$, then $R(I + \lambda A) \supset C$. Let $T(t)$, $t \geq 0$, be the strongly continuous nonlinear semigroup in C generated by A as in Proposition 3.3. If $x \in D(A)$ and u is a Lipschitz continuous function from $[0,t_0]$ to X such that $u(0) = x$ and for almost everywhere $t \in (0,t_0)$, u is differentiable at t, $u(t) \in D(A)$, and $d/dt\, u(t) = -Au(t)$, then $u(t) = T(t)x$ for all $t \in [0,t_0]$. Further, if C is convex and $S(t)$, $t \geq 0$, is a family of mappings from C to C such that

(3.13) $\|S(t)x_1 - S(t)x_2\| \leq M(t)\|x_1 - x_2\|$ for all $t \geq 0$, $x_1, x_2 \in C$, where $M(t) = 1 + \omega t + o(t)$ as $t \to \infty$

and

(3.14) $\lim_{t \to 0^+} t^{-1}[S(t)x - x] = -Ax$ for all $x \in D(A)$

then for each $x \in C$

(3.15) $\lim_{n \to \infty} S(t/n)^n x = T(t)x$ uniformly in bounded intervals of $t \geq 0$.

## 3.2 THE INFINITESIMAL GENERATOR

We now identify the infinitesimal generator of the strongly continuous nonlinear semigroup associated with the solutions of (ADP). In Section 3.3 we will show that the domain of the infinitesimal generator is dense in $L^1_+$. We first state

DEFINITION 3.4 Let (2.1), (2.2) hold and define the mapping A from $L^1_+$ to $L^1$ by

(3.16) $A\phi \overset{\text{def}}{=} \phi' - G(\phi)$ for $\phi \in D(A)$, where $D(A) = \{\phi \in L^1_+$: $\phi$ is absolutely continuous on $[0,\infty)$, $\phi' \in L^1$, and $\phi(0) = F(\phi)\}$.

THEOREM 3.2 Let (2.1), (2.2), (2.22), (2.23) hold, let A be defined as in (3.16), and let $S(t)$, $t \geq 0$, be the strongly continuous nonlinear semigroup in $L^1_+$ as in Theorem 3.1. Then, $-A$ is the infinitesimal generator of $S(t)$, $t \geq 0$.

*Proof.* Let B denote the infinitesimal generator of $S(t)$, $t \geq 0$. We first prove that if $\phi \in D(A)$, then $\phi \in D(B)$ and $B\phi = -A\phi$. For $\phi \in D(A)$, $t > 0$, we have

$$\|t^{-1}[S(t)\phi - \phi] + A\phi\|_{L^1}$$

$$= \int_0^\infty |t^{-1}[(S(t)\phi)(a) - \phi(a)] + \phi'(a) - G(\phi)(a)|\ da$$

$$= \int_0^t \left|t^{-1}\left[F(S(t - a)\phi) + \int_{t-a}^t G(S(s)\phi)(s + a - t)\ ds - \phi(a)\right] + \phi'(a) - G(\phi)(a)\right| da$$

$$+ \int_t^\infty \left|t^{-1}\left[\phi(a - t) + \int_0^t G(S(s)\phi)(s + a - t)\ ds - \phi(a)\right] + \phi'(a) - G(\phi)(a)\right| da$$

$$\overset{\text{def}}{=} K_1 + K_2$$

Since $\phi(0) = F(\phi)$,

$$K_1 \leq t^{-1} \int_0^t |F(S(t - a)\phi) - F(\phi)|\ da$$

$$+ \int_0^t |\phi'(a) - t^{-1}[\phi(a) - \phi(0)]|\ da$$

$$+ \int_0^t \left|t^{-1} \int_{t-a}^t G(S(s)\phi)(s + a - t)\ ds - G(\phi)(a)\right| da$$

$$\stackrel{\text{def}}{=} H_1 + H_2 + H_3$$

By the continuity of F,

$$H_1 \le \sup_{0 \le s \le t} |F(S(s)\phi) - F(\phi)| \to 0 \text{ as } t \to 0^+$$

Since $\phi$ is absolutely continuous and $\phi' \in L^1$,

$$\text{(3.17)} \qquad H_2 = \int_0^t \left| \phi'(a) - t^{-1} \int_0^a \phi'(b)\, db \right| da$$

$$\le \int_0^t |\phi'(a)|\, da + \sup_{0 \le a \le t} \int_0^a |\phi'(b)|\, db \to 0 \text{ as } t \to 0^+$$

By the continuity of G and the fact that $G(\phi) \in L^1$,

$$\text{(3.18)} \qquad H_3 \le t^{-1} \int_0^t \left[ \int_{t-s}^t |G(S(s)\phi)(s + a - t)|\, da \right] ds$$

$$+ \int_0^t |G(\phi)(a)|\, da$$

$$\le t^{-1} \int_0^t \left\{ \int_0^s [\,|G(S(s)\phi)(a) - G(\phi)(a)|\right.$$

$$\left. + |G(\phi)(a)|\,]\, da \right\} ds + \int_0^t |G(\phi)(a)|\, da$$

$$\le \sup_{0 \le s \le t} \|G(S(s)\phi) - G(\phi)\|_{L^1} + 2 \int_0^t |G(\phi)(a)|\, da \to 0 \quad \text{as } t \to 0^+$$

Next, observe that

$$K_2 \le \int_t^\infty |t^{-1}[\phi(a - t) - \phi(a)] + \phi'(a)|\, da$$

$$+ \int_t^\infty \left| t^{-1} \int_0^t G(S(s)\phi)(s + a - t)\, ds - G(\phi)(a) \right| da$$

$$\stackrel{\text{def}}{=} J_1 + J_2$$

Since $\phi$ is absolutely continuous and $\phi' \in L^1$,

$$J_1 = \int_t^\infty \left| t^{-1} \int_{a-t}^a \phi'(b)\,db - \phi'(a) \right| da$$

$$= \int_t^\infty \left| t^{-1} \int_0^t [\phi'(a - \tau) - \phi'(a)]\,d\tau \right| da$$

$$\le t^{-1} \int_0^t \left[ \int_t^\infty |\phi'(a - \tau) - \phi'(a)|\,da \right] d\tau$$

$$\le \sup_{0 \le \tau \le t} \int_t^\infty |\phi'(a - \tau) - \phi'(a)|\,da \to 0 \text{ as } t \to 0^+$$

Last, the continuity of G and the fact that $G(\phi) \in L^1$ imply that

$$\text{(3.19)} \qquad J_2 \le t^{-1} \int_0^t \left\{ \int_t^\infty [\,|G(S(s)\phi)(s + a - t) - G(\phi)(s + a - t)| + |G(\phi)(s + a - t) - G(\phi)(a)|\,]\,da \right\} ds$$

$$\le \sup_{0 \le s \le t} \left[ \|G(S(s)\phi) - G(\phi)\|_{L^1} + \int_t^\infty |G(\phi)(s + a - t) - G(\phi)(a)\ \ da \right] \to 0 \text{ as } t \to 0^+$$

Thus, we have that

$$\lim_{t \to 0^+} t^{-1}[S(t)\phi - \phi] = -A\phi$$

and so by the definition of the infinitesimal generator, $\phi \in D(B)$ and $B\phi = -A\phi$.

Now let $\phi \in D(B)$, where

$$B\phi = \lim_{t \to 0^+} t^{-1}[S(t)\phi - \phi]$$

We will prove that $\phi \in D(A)$ and $B\phi = -A\phi$. For $t > 0$, define $\chi_t \in L^1$, $\phi_t \in L^1$ by

$$\chi_t(a) = \begin{cases} 0 & \text{for a.e. } a < t \\ t^{-1} \int_0^t G(S(s)\phi)(s + a - t)\,da & \text{for a.e. } a > t \end{cases}$$

$$\phi_t(a) = \begin{cases} 0 & \text{for a.e. } a < t \\ t^{-1}[\phi(a - t) - \phi(a)] & \text{for a.e. } a > t \end{cases}$$

From (3.19) we have that

$$\|\chi_t - G(\phi)\|_{L^1} = \int_0^t |G(\phi)(a)|\, da + \int_t^\infty \left| t^{-1} \int_0^t G(S(s)\phi)(s + a - t)\, ds - G(\phi)(a) \right| da \to 0 \text{ as } t \to 0^+$$

Next, observe that

$$\|\phi_t + \chi_t - B\phi\|_{L^1} = \int_0^t |B\phi(a)|\, da + \int_t^\infty \left| t^{-1} \left[ \phi(a - t) + \int_0^t G(S(s)\phi)(s + a - t)\, ds - \phi(a) \right] - B\phi(a) \right| da$$
$$\le \int_0^t |B\phi(a)|\, da + \|t^{-1}[S(t)\phi - \phi] - B\phi\|_{L^1} \to 0 \text{ as } t \to 0^+$$

Thus, we must have that $\lim_{t \to 0^+} \phi_t = B\phi - G(\phi)$ in $L^1$. We next use a device used in [151], p. 535. Let $0 < t < u < v$. Then,

$$t^{-1} \int_{u-t}^u \phi(a)\, da - t^{-1} \int_{v-t}^v \phi(a)\, da$$
$$= \int_u^v t^{-1}[\phi(a - t) - \phi(a)]\, da = \int_u^v \phi_t(a)\, da$$

From this last equation we see that there exists a set E of measure zero such that if $0 < u < v$, $u, v \notin E$, then

$$\phi(u) - \phi(v) = \int_u^v [B\phi(a) - G(\phi)(a)]\, da$$

Therefore, $\phi$ is absolutely continuous and $\phi'(a) = -B\phi(a) + G(\phi)(a)$ for almost everywhere $a > 0$. Now observe that

$$\|t^{-1}[S(t)\phi - \phi] - B\phi\|_{L^1}$$

$$\geq \int_0^t \left| t^{-1}\left[F(S(t - a)\phi) + \int_{t-a}^t G(S(s)\phi)(s + a - t)\, ds - \phi(a)\right] - B\phi(a)\right| da$$

$$\geq t^{-1} \int_0^t |F(S(t - a)\phi) - \phi(0)|\, da - \int_0^t |\phi'(a) - t^{-1}[\phi(a) - \phi(0)]|\, da - \int_0^t \left| t^{-1} \int_{t-a}^t G(S(s)\phi)(s + a - t) - G(\phi)(a)\right| da$$

Now let t approach 0 from the right in this last inequality and use (3.17) and (3.18) to conclude that $\phi(0) = F(\phi)$. Thus, $\phi \in D(A)$ and $-A\phi = B\phi$. □

EXAMPLE 3.1 Let $R^n = R^1$. Let F and G be defined as in Example 2.4 (that is, F has the form (2.17) with $\beta$ continuously differentiable, G has the form (2.18) with $\mu$ continuously differentiable, and (2.30) holds). In Example 2.4 it was shown that for each initial age distribution $\phi \in L_+^1$ the solution of (ADP) exists on $[0,\infty)$ and lies in $L_+^1$ for all $t \geq 0$. Thus, Theorem 3.1 applies and there exists a strongly continuous nonlinear semigroup $S(t)$, $t \geq 0$, in $L_+^1$ associated with (ADP) for this birth function F and aging function G. By Theorem 3.2 the infinitesimal generator B of $S(t)$, $t \geq 0$, is given by

(3.20) $B\phi(a) = -\phi'(a) - \mu(P\phi)\phi(a)$, a.e. $a > 0$, $D(B) = \{\phi \in L_+^1$: $\phi$ is absolutely continuous on $[0,\infty)$, $\phi' \in L^1$, and $\phi(0) = \int_0^\infty \beta(P\phi)e^{-\alpha a}\phi(a)\, da\}$.

EXAMPLE 3.2 Let $R^n = R^1$. Let F and G be defined as in Example 2.5 [that is, F has the form (2.19) with $\beta$ continuously differentiable, G has the form (2.18), with $\mu$ continuously differentiable, and (2.30) holds]. The infinitesimal generator of the strongly continuous nonlinear semigroup $S(t)$, $t \geq 0$, in $L^1_+$ associated with (ADP) for this birth function F and aging function G is

(3.21) $B\phi(a) = -\phi'(a) - \mu(P\phi)\phi(a)$, a.e. $a > 0$, $D(B) = \{\phi \in L^1_+$: $\phi$ is absolutely continuous on $[0,\infty)$, $\phi' \in L^1$, and $\phi(0) = \int_0^\infty \beta(P\phi)(1 - e^{-\alpha a})\phi(a)\, da\}$.

EXAMPLE 3.3 Let $R^n = R^1$. Let F and G be defined as in Example 2.6 [that is, F has the form (2.20) with $\beta$ continuously differentiable, G has the form (2.18) with $\mu$ continuously differentiable, and (2.31) holds]. The infinitesimal generator for the strongly continuous nonlinear semigroup $S(t)$, $t \geq 0$, for this (ADP) is

(3.22) $B\phi(a) = -\phi'(a) - \mu(P\phi)\phi(a)$, a.e. $a > 0$, $D(B) = \{\phi \in L^1_+$: $\phi$ is absolutely continuous on $[0,\infty)$, $\phi' \in L^1$, and $\phi(0) = \int_0^\infty \beta(P\phi) a e^{-\alpha a}\phi(a)\, da\}$.

Theorem 3.2 characterizes the infinitesimal generator B of the strongly continuous nonlinear semigroup $S(t)$, $t \geq 0$, associated with the solutions of (ADP), given by Theorem 3.1. Notice that if $\phi$ is in the domain of the infinitesimal generator, then $\phi$ must satisfy the compatibility condition $\phi(0) = F(\phi)$. If $\phi$ is in the domain of the infinitesimal generator, then we can obtain regularity information about the solution $S(t)\phi$ of (ADP) corresponding to $\phi$. In Proposition 3.5 below we prove that if $\phi \in D(B)$, then the trajectory $S(t)\phi$ is Lipschitz continuous in t from bounded intervals of $[0,\infty)$ to $L^1$. In Proposition 3.6 we prove that if G has the form (1.48) and $\phi \in D(B)$, then $S(t)\phi \in D(B)$ for $t > 0$ and the trajectory $S(t)\phi$ is a solution of the abstract evolution equation in $L^1$

(3.23) $$\frac{d^+}{dt} S(t)\phi = B\, S(t)\phi \qquad t \geq 0$$

PROPOSITION 3.5 Let (2.1), (2.2), (2.22), (2.23) hold, let $T_\phi = +\infty$ for all $\phi \in L^1_+$, let $S(t)$, $t \geq 0$, be the strongly continuous nonlinear semigroup in $L^1_+$ as in Theorem 3.1, and let B be the infinitesimal generator of $S(t)$, $t \geq 0$, as in Theorem 3.2. If $\phi \in D(B)$ and $t > 0$, then there exists a constant K such that

$$\|S(t_1)\phi - S(t_2)\phi\|_{L^1} \leq K|t_1 - t_2| \qquad t_1, t_2 \in [0,t] \tag{3.24}$$

*Proof.* Let $\phi \in D(B)$, $t > 0$. From (3.8) we see that there exists $K_1 > 0$ and $h > 0$ such that if $0 < s < h$, then

$$\|S(s)\phi - \phi\|_{L^1} \leq K_1 s$$

Let $0 \leq t_1 < t_2 \leq t$ and let $\{s_i\}_{i=0}^k$ be a sequence such that $s_0 = t_1$, $s_k = t_2$, and $|s_i - s_{i-1}| < h$ for $i = 1, \ldots, k$. Let $r > 0$ such that $\sup_{0 \leq s \leq t} \|S(s)\phi\|_{L^1} \leq r$. From Proposition 2.3 we obtain

$$\begin{aligned}
&\|S(t_1)\phi - S(t_2)\phi\|_{L^1} \\
&\leq \sum_{i=1}^{k} \|S(s_i)\phi - S(s_{i-1})\phi\|_{L^1} \\
&= \sum_{i=1}^{k} \|S(s_{i-1})S(s_i - s_{i-1})\phi - S(s_{i-1})\phi\|_{L^1} \\
&\leq \sum_{i=1}^{k} \exp\{[c_1(r) + c_2(r)]t_{i-1}\}\|S(s_i - s_{i-1})\phi - \phi\|_{L^1} \\
&\leq \exp\{[c_1(r) + c_2(r)]t\} \sum_{i=1}^{k} K_1(s_i - s_{i-1}) \\
&= \exp\{[c_1(r) + c_2(r)]t\}K_1|t_2 - t_1|
\end{aligned}$$

which establishes (3.24). □

If X is a reflexive Banach space, it is known that a Lipschitz continuous function from a closed interval to X must be differentiable almost everywhere (see [13], Theorem 2.1, p. 16). The Banach space $L^1$ is not reflexive, and so we cannot use Proposition 3.5

to automatically conclude differentiability of the trajectory $S(t)\phi$ when $\phi \in D(B)$. If G has the form (1.48), however, we can establish

PROPOSITION 3.6 Let the hypothesis of Proposition 3.5 hold, and in addition, let (2.49) hold. If $t > 0$, then $S(t)[D(B)] \subset D(B)$. Further, if $\phi \in D(B)$, then (3.23) holds.

*Proof.* Let $t > 0$ and let $\phi \in D(A)$, where A is defined as in (3.16). We first use (2.62) to show that $(S(t)\phi)(a)$ is absolutely continuous in a from bounded intervals on $[0,\infty)$ to $R^n$. Let K be a constant as in (3.24) and observe from (2.1) that

$$|F(S(t_1)\phi) - F(S(t_2)\phi)| \le c(\sup_{0 \le s \le t} \|S(s)\phi\|_{L^1})K|t_1 - t_2|$$

for $0 \le t_1 \le t_2 \le t$. Thus, the function $a \to F(S(t - a)\phi)$ is absolutely continuous from $[0,t]$ to $R^n$. Let $\ell(\cdot,s) = S(s)\phi$, $0 \le s \le t$, and observe from (2.57) and (2.58) that the function $a \to |U(t, t - a; a - t, \ell)|$ is absolutely continuous from $[0,t]$ to $[0,\infty)$ and the function $a \to |U(t,0; a - t, \ell)|$ is absolutely continuous in a from bounded intervals of $[t,\infty)$ to $[0,\infty)$. Now use the absolute continuity of $\phi$, formula (2.62), and the fact that the product of absolutely continuous functions is absolutely continuous to argue that $(S(t)\phi)(a)$ is absolutely continuous in a from bounded intervals of $[0,\infty)$ to $R^n$.

We thus have that $(d/da)(S(t)\phi)(a)$ exists for almost all $a > 0$, and $(d/da)(S(t)\phi)(a)$ belongs to $L^1((0,a_1);R^n)$ for all $a_1 > 0$. We must show that $(d/da)(S(t)\phi)(a)$ belongs to $L^1$. From (2.62) and the fact that $\phi \in L^1$, $\phi' \in L^1$, it suffices to show that the function $a \to U(t,0; a - t; \ell)$ is bounded in a and has a bounded derivative in a from $[t,\infty)$ to $B(R^n,R^n)$. The boundedness of this function follows from (2.50), and the boundedness of its derivative follows from (2.58). From (2.62) it is obvious that $(S(t)\phi)(0) = F(S(t)\phi)$, and we have therefore shown that $S(t)\phi \in D(A) = D(B)$. The claim (3.23) now follows immediately from (3.8), since $S(t)\phi \in D(B)$ means that

(3.24) $$\frac{d^+}{dt} S(t)\phi = \lim_{h\to 0^+} h^{-1}[S(t+h)\phi - S(t)\phi] = \lim_{h\to 0^+} h^{-1}[S(h) - I]S(t)\phi = BS(t)\phi \quad \square$$

REMARK 3.3 The calculation (3.24) shows that whenever $(d^+/dt)S(t)\phi$ exists, $S(t)\phi \in D(B)$ and (3.23) holds. Thus, the hypothesis of Theorem 2.10 (namely, (2.1), (2.2) hold and F, G are continuously Fréchet differentiable) implies the same conclusion as Proposition 3.6.

It is well known that if $S(t)$, $t \geq 0$, is a strongly continuous semigroup of linear operators in a Banach space, then for each $t > 0$ the linear operator $S(t)$ maps the domain of the infinitesimal generator into itself. Thus, if F and G are bounded linear operators as in Proposition 3.2, we obtain

PROPOSITION 3.7 Let F be a bounded linear operator from $L^1$ into $R^n$, let G be a bounded linear operator from $L^1$ into $L^1$, and let $S(t)$, $t \geq 0$, be the strongly continuous semigroup of bounded linear operators in $L^1$ as in Proposition 3.2. The infinitesimal generator of $S(t)$, $t \geq 0$, is

(3.25) $B\phi = -\phi' + G(\phi)$, where $D(B) = \{\phi \in L^1\colon\ \phi$ is absolutely continuous on $[0,\infty)$, $\phi' \in L^1$, and $\phi(0) = F(\phi)\}$.

Further,

(3.26) For all $t \geq 0$, $S(t)(D(B)) \subset D(B)$ and $(d/dt)S(t)\phi = BS(t)\phi = S(t)B\phi$ for all $\phi \in D(B)$.

*Proof.* The proof of (3.25) is the same as the proof of Theorem 3.2 except that $\phi$ is not required to lie in $L^1_+$. The proof of (3.26) follows from a general result in the theory of strongly continuous semigroups of linear operators (see [94], Lemma 7, p. 619). $\square$

## 3.3 THE EXPONENTIAL REPRESENTATION

We now demonstrate that the nonlinear semigroup associated with (ADP) can be represented in terms of its infinitesimal generator by means of an exponential formula. We first consider the special case that the birth function F and the aging function G are globally Lipschitz continuous.

PROPOSITION 3.8 Let (2.1), (2.2), (2.22), (2.23) hold, let $\sup_{r\geq 0} c_1(r) \overset{def}{=} c_1 < \infty$ and $\sup_{r\geq 0} c_2(r) \overset{def}{=} c_2 < \infty$, let A be defined as in (3.16), and let $\omega = c_1 + c_2$. The following hold:

(3.27) $\quad R(I + \lambda A) = L^1_+ \qquad$ for $0 < \lambda < \omega^{-1}$

(3.28) $\quad A + \omega I$ is accretive in $L^1$

(3.29) $\quad \overline{D(A)} = L^1_+$

*Proof of (3.27).* Let $0 < \lambda < \omega^{-1}$ and let $\psi \in L^1_+$. We must solve the boundary value problem $\phi + \lambda\phi' - \lambda G(\phi) = \psi$, $\phi(0) = F(\phi)$, with $\phi \in L^1_+$. Let $r = (1 - \lambda\omega)^{-1}[\|\psi\|_{L^1} + \lambda(|F(0)| + \|G(0)\|_{L^1})]$ and let $\alpha = c_3(r)$, where $c_3(r)$ is as in (2.23). Define the closed subset $M \overset{def}{=} \{\phi \in L^1_+ : \|\phi\|_{L^1} \leq r\}$ of $L^1$. Define a mapping K with domain M by

$$K\phi(a) = \exp[-a(1 + \lambda\alpha)\lambda^{-1}]\{F(\phi) + \int_0^a \exp[b(1 + \lambda\alpha)\lambda^{-1}] \times [(G + \alpha I)(\phi)(b) + \lambda^{-1}\psi(b)]\, db\}, \quad \phi \in M,\ a \geq 0. \tag{3.30}$$

Observe from (2.22) and (2.23) that if $\phi \in M$, then $K\phi(a) \in R^n_+$ for all $a \geq 0$. Further, for $\phi \in M$,

$$\|K\phi\|_{L^1} \leq \int_0^\infty \exp[-a(1 + \lambda\alpha)\lambda^{-1}]\,|F(\phi)|\, da + \int_0^\infty \Big\{\exp[-a(1 + \lambda\alpha)\lambda^{-1}] \times \int_0^a \exp[b(1 + \lambda\alpha)\lambda^{-1}][|(G + \alpha I)(\phi)(b)| + \lambda^{-1}|\psi(b)|]\, db\Big\}\, da$$

$$\leq |F(\phi)|\lambda(1 + \lambda\alpha)^{-1} + \int_0^\infty \Big\{\exp[b(1 + \lambda\alpha)\lambda^{-1}]$$

$$\times\ [|(G + \alpha I)(\phi)(b)| + \lambda^{-1}|\psi(b)|] \int_b^\infty \exp[-a(1 + \lambda\alpha)\lambda^{-1}]\ da\Big\}\ db$$

$$\le [c_1\|\phi\|_{L^1} + |F(0)|]\lambda(1 + \lambda\alpha)^{-1}$$

$$+ \int_0^\infty [|(G + \alpha I)(\phi)(b)| + \lambda^{-1}|\psi(b)|]\lambda(1 + \lambda\alpha)^{-1}\ db$$

$$\le [c_1\|\phi\|_{L^1} + |F(0)| + c_2\|\phi\|_{L^1} + \|G(0)\|_{L^1}$$

$$+ \alpha\|\phi\|_{L^1} + \lambda^{-1}\|\psi\|_{L^1}]\lambda(1 + \lambda\alpha)^{-1}$$

$$\le [\lambda(\omega + \alpha)r + \lambda|F(0)| + \lambda\|G(0)\|_{L^1} + \|\psi\|_{L^1}](1 + \lambda\alpha)^{-1}$$

$$= r$$

Therefore, K maps M into M. Further, K is a strict contraction in M, since $\lambda(1 + \lambda\alpha)^{-1}(\omega + \alpha) < 1$ and for $\phi$, $\hat{\phi} \in M$,

$$\|K\phi - K\hat{\phi}\|_{L^1} = \int_0^\infty |K\phi(a) - K\hat{\phi}(a)|\ da$$

$$\le \int_0^\infty \exp[-a(1 + \lambda\alpha)\lambda^{-1}]|F(\phi) - F(\hat{\phi})|\ da$$

$$+ \int_0^\infty \Big\{\exp[-a(1 + \lambda\alpha)\lambda^{-1}] \int_0^a \exp[b(1 + \lambda\alpha)\lambda^{-1}]$$

$$\times\ [|G(\phi)(b) - G(\hat{\phi})(b)| + \alpha|\phi(b) - \hat{\phi}(b)|]\ db\Big\}\ da$$

$$\le |F(\phi) - F(\hat{\phi})|\lambda(1 + \lambda\alpha)^{-1} + \int_0^\infty \Big\{\exp[b(1 + \lambda\alpha)\lambda^{-1}]$$

$$\times\ [|G(\phi)(b) - G(\hat{\phi})(b)| + \alpha|\phi(b) - \hat{\phi}(b)|]$$

$$\times \int_b^\infty \exp[-a(1 + \lambda\alpha)\lambda^{-1}]\ da\Big\}\ db$$

$$\le \lambda(1 + \lambda\alpha)^{-1}\|\phi - \hat{\phi}\|_{L^1}(\omega + \alpha)$$

Thus, by the contraction mapping theorem (see [217], Theorem 1.1, p. 114) K has a unique fixed point $\phi$ in M, that is, $K\phi = \phi$. From

(3.30) we see immediately that this $\phi$ belongs to D(A) and $(I + \lambda A)\phi = \psi$. □

*Proof of (3.28).* Let $\phi$, $\hat{\phi} \in D(A)$ and let $0 < \lambda < \omega^{-1}$. To verify (3.9) it suffices to show that

$$(3.31) \qquad \|(I + \lambda A)\phi - (I + \lambda A)\hat{\phi}\|_{L^1} \geq (1 - \lambda\omega)\|\phi - \hat{\phi}\|_{L^1}$$

since if $\hat{\lambda} = \lambda(1 - \lambda\omega)^{-1}$, then $\hat{\lambda} > 0$, $\lambda = \hat{\lambda}(1 + \hat{\lambda}\omega)^{-1}$, and

$$\|[I + \hat{\lambda}(A + \omega I)]\phi - [I + \hat{\lambda}(A + \omega I)]\hat{\phi}\|_{L^1}$$

$$= (I + \hat{\lambda}\omega)\|(I + \lambda A)\phi - (I + \lambda A)\hat{\phi}\|_{L^1} \geq \|\phi - \hat{\phi}\|_{L^1}$$

Let $\psi = (I + \lambda A)\phi$, $\hat{\psi} = (I + \lambda A)\hat{\phi}$. Then, $\psi = \phi + \lambda\phi' - \lambda G(\phi)$, $\phi(0) = F(\phi)$, and $\hat{\psi} = \hat{\phi} + \lambda\hat{\phi}' - \lambda G(\hat{\phi})$, $\hat{\phi}(0) = F(\hat{\phi})$. By an integration we obtain

$$(3.32) \qquad \phi(a) = \exp[-a\lambda^{-1}]\left\{F(\phi) + \int_0^a \exp[b\lambda^{-1}][G(\phi)(b) + \lambda^{-1}\psi(b)]\, db\right\}$$

and a similar equation for $\hat{\phi}$, $\hat{\psi}$. Then,

$$\begin{aligned}
\|\phi - \hat{\phi}\|_{L^1} &\leq \int_0^\infty \exp[-a\lambda^{-1}]\,|F(\phi) - F(\hat{\phi})|\, da \\
&\quad + \int_0^\infty \left\{\exp[-a\lambda^{-1}] \int_0^a \exp[b\lambda^{-1}][|G(\phi)(b) - G(\hat{\phi})(b)| \right. \\
&\quad \left. + \lambda^{-1}|\psi(b) - \hat{\psi}(b)|]\, db\right\} da \\
&\leq \lambda|F(\phi) - F(\hat{\phi})| + \int_0^\infty \left\{\exp[b\lambda^{-1}][|G(\phi)(b) - G(\hat{\phi})(b)| \right. \\
&\quad \left. + \lambda^{-1}|\psi(b) - \hat{\psi}(b)|] \int_b^\infty \exp[-a\lambda^{-1}]\, da\right\} db \\
&\leq \lambda c_1\|\phi - \hat{\phi}\|_{L^1} + \lambda c_2\|\phi - \hat{\phi}\|_{L^1} + \|\psi - \hat{\psi}\|_{L^1}
\end{aligned}$$

from which (3.31) follows immediately. □

*Proof of (3.29).* Let $\psi \in L^1_+$. We first claim that

$$\lim_{\lambda \to 0^+} \int_0^\infty \left| \int_0^a \exp[-t\lambda^{-1}]\lambda^{-1}\psi(a - t)\, dt - \psi(a) \right| da = 0 \tag{3.33}$$

To prove (3.33) let $t > 0$ and define $\psi^t \in L^1_+$ by

$$\psi^t(a) = \begin{cases} 0 & 0 < a < t \\ \psi(a - t) & \text{a.e. } a > t \end{cases}$$

Since

$$\|\psi^t - \psi\|_{L^1} = \int_0^t |\psi(a)|\, da + \int_0^\infty |\psi(a) - \psi(a + t)|\, da$$

we have that

$$\lim_{t \to 0^+} \|\psi^t - \psi\|_{L^1} = 0 \tag{3.34}$$

Then, (3.33) follows from (3.34), since for $0 < \lambda < \omega^{-1}$ and $\varepsilon > 0$,

$$\int_0^\infty \left| \int_0^a \exp[-t\lambda^{-1}]\lambda^{-1}\psi(a - t)\, dt - \psi(a) \right| da$$

$$= \int_0^\infty \left| \int_0^a \exp[-t\lambda^{-1}]\lambda^{-1}\psi(a - t)\, dt - \int_0^\infty \exp[-t\lambda^{-1}]\lambda^{-1}\psi(a)\, dt \right| da$$

$$= \int_0^\infty \left| \int_0^\infty \exp[-t\lambda^{-1}]\lambda^{-1}[\psi^t(a) - \psi(a)]\, dt \right| da$$

$$\le \int_0^\infty \left\{ \exp[-t\lambda^{-1}]\lambda^{-1} \int_0^\infty |\psi^t(a) - \psi(a)|\, da \right\} dt$$

$$\le \sup_{0 \le t \le \varepsilon} \|\psi^t - \psi\|_{L^1} \int_0^\varepsilon \exp[-t\lambda^{-1}]\lambda^{-1}\, dt$$

$$+ 2\|\psi\|_{L^1} \int_\varepsilon^\infty \exp[-t\lambda^{-1}]\lambda^{-1}\, dt$$

$$\le \sup_{0 \le t \le \varepsilon} \|\psi^t - \psi\|_{L^1} + 2\|\psi\|_{L^1} \exp[-\varepsilon\lambda^{-1}]$$

Next, let $0 < \lambda < \omega^{-1}$ and set $\phi = (I + \lambda A)^{-1}\psi$. From (3.31) we obtain

$$\|\phi\|_{L^1} \le \|(I + \lambda A)^{-1}\psi - (I + \lambda A)^{-1}0\|_{L^1} + \|(I + \lambda A)^{-1}0\|_{L^1} \tag{3.35}$$

$$\le (1 - \lambda\omega)^{-1}\|\psi\|_{L^1} + \|(I + \lambda A)^{-1}0\|_{L^1}$$

and from (3.32) we obtain

$$\|(I + \lambda A)^{-1}0\|_{L^1} \le \int_0^\infty \exp[-a\lambda^{-1}]\,|F((I + \lambda A)^{-1}0)|\,da \tag{3.36}$$

$$+ \int_0^\infty \left\{\exp[-a\lambda^{-1}] \int_0^a \exp[b\lambda^{-1}]\,|G((I + \lambda A)^{-1}0)(b)|\,db\right\} da$$

$$\le \int_0^\infty \exp[-a\lambda^{-1}][c_1\|(I + \lambda A)^{-1}0\|_{L^1} + |F(0)|]\,da$$

$$+ \int_0^\infty \left\{\exp[b\lambda^{-1}][|G((I + \lambda A)^{-1}0)(b) - G(0)(b)|\right.$$

$$\left. + |G(0)(b)|] \int_b^\infty \exp[-a\lambda^{-1}]\,da\right\} db$$

$$\le \lambda[c_1\|(I + \lambda A)^{-1}0\|_{L^1} + |F(0)|]$$

$$+ \lambda[c_2\|(I + \lambda A)^{-1}0\|_{L^1} + \|G(0)\|_{L^1}]$$

$$= \lambda\omega\|(I + \lambda A)^{-1}0\|_{L^1} + \lambda[|F(0)| + \|G(0)\|_{L^1}]$$

Then, (3.35) and (3.36) imply

$$\|\phi\|_{L^1} \le (1 - \lambda\omega)^{-1}\{\|\psi\|_{L^1} + \lambda[|F(0)| + \|G(0)\|_{L^1}]\} \tag{3.37}$$

From (3.32) and (3.37) we then obtain

$$\|(I + \lambda A)^{-1}\psi - \psi\|_{L^1} = \int_0^\infty \left|\exp[-a\lambda^{-1}]\left\{F(\phi)\right.\right.$$

$$\left.\left. + \int_0^a \exp[b\lambda^{-1}][G(\phi)(b) + \lambda^{-1}\psi(b)]\,db\right\} - \psi(a)\right| da$$

$$\leq \lambda|F(\phi)| + \int_0^\infty \left\{\exp[b\lambda^{-1}]|G(\phi)(b)| \int_b^\infty \exp[-a\lambda^{-1}]\, da\right\} db$$

$$+ \int_0^\infty \left|\int_0^a \exp[(b - a)\lambda^{-1}]\lambda^{-1}\psi(b)\, db - \psi(a)\right| da$$

$$\leq \lambda[c_1\|\phi\|_{L^1} + |F(0)|] + \lambda[c_2\|\phi\|_{L^1} + \|G(0)\|_{L^1}]$$

$$+ \int_0^\infty \left|\int_0^a \exp[-t\lambda^{-1}]\lambda^{-1}\psi(a - t)\, dt - \psi(a)\right| da$$

From (3.33) and (3.37) we now see that

$$(3.38) \qquad \lim_{\lambda\to 0^+} \|(I + \lambda A)^{-1}\psi - \psi\|_{L^1} = 0$$

Thus, (3.29) follows from (3.38), since $(I + \lambda A)^{-1}\psi \in D(A)$ for $0 < \lambda < \omega^{-1}$. □

The next proposition establishes the exponential representation for the case that the birth function F and the aging function G are globally Lipschitz continuous.

PROPOSITION 3.9 Let (2.1), (2.2), (2.22), (2.23) hold, let $\sup_{r\geq 0} c_1(r) \overset{\text{def}}{=} c_1 < \infty$, $\sup_{r\geq 0} c_2(r) \overset{\text{def}}{=} c_2 < \infty$, let $S(t)$, $t \geq 0$, be the strongly continuous nonlinear semigroup as in Proposition 3.1, and let A be defined as in (3.16). If $\phi \in L^1_+$, then

$$(3.39) \qquad \lim_{n\to\infty} \left(I + \frac{t}{n} A\right)^{-n} \phi = S(t)\phi \text{ uniformly in bounded intervals of } t \geq 0.$$

*Proof.* By Theorem 3.2 we know that -A is the infinitesimal generator of $S(t)$, $t \geq 0$. By Proposition 3.8 we know that A satisfies the hypothesis of Proposition 3.3 with $\overline{D(A)} = L^1_+$ and $\omega = c_1 + c_2$. Thus, there exists a strongly continuous nonlinear semigroup $T(t)$, $t \geq 0$, in $L^1_+$ such that for $\phi \in L^1_+$, $t \geq 0$,

$$(3.40) \qquad \lim_{n\to\infty} \left(I + \frac{t}{n} A\right)^{-n} \phi = T(t)\phi$$

where the convergence in (3.40) is uniform in bounded intervals of $t$ in $[0,\infty)$. We claim that the hypothesis of Proposition 3.4 is fulfilled. From Proposition 3.1 we see that we may take $M(t) = \exp((c_1 + c_2)t)$ in (3.13). By the definition of infinitesimal generator we see that (3.14) is satisfied. Further, $L^1_+$ is convex. Therefore, from (3.3) and (3.15) we obtain $S(t) = T(t)$ for $t \geq 0$. Hence, (3.39) follows immediately from (3.40). □

We next consider the case that the birth function $F$ and the aging function $G$ are locally Lipschitz continuous in the sense of (2.1) and (2.2). We first show that the radial truncations of $F$ and $G$ satisfy the hypothesis of Proposition 3.9.

PROPOSITION 3.10 Let (2.1), (2.21), (2.22), (2.23) hold and let $r > 0$. Define

$$(3.41) \qquad F_r(\phi) \stackrel{\text{def}}{=} \begin{cases} F(\phi) & \text{if } \phi \in L^1 \text{ and } \|\phi\|_{L^1} \leq r \\ F\left(\dfrac{r\phi}{\|\phi\|_{L^1}}\right) & \text{if } \phi \in L^1 \text{ and } \|\phi\|_{L^1} > r \end{cases}$$

$$(3.42) \qquad G_r(\phi) \stackrel{\text{def}}{=} \begin{cases} G(\phi) & \text{if } \phi \in L^1 \text{ and } \|\phi\|_{L^1} \leq r \\ G\left(\dfrac{r\phi}{\|\phi\|_{L^1}}\right) & \text{if } \phi \in L^1 \text{ and } \|\phi\|_{L^1} > r \end{cases}$$

Then, $F_r$ and $G_r$ satisfy the following:

$$(3.43) \qquad |F_r(\phi) - F_r(\hat{\phi})| \leq 2c_1(r)\|\phi - \hat{\phi}\|_{L^1} \qquad \phi, \hat{\phi} \in L^1$$

$$(3.44) \qquad \|G_r(\phi) - G_r(\hat{\phi})\|_{L^1} \leq 2c_2(r)\|\phi - \hat{\phi}\|_{L^1} \qquad \phi, \hat{\phi} \in L^1$$

$$(3.45) \qquad F_r(L^1_+) \subset R^n_+$$

(3.46) $\quad G_r(\phi) + c_3(r_1)\phi \in L^1_+$ for all $\phi \in L^1_+$ such that $\|\phi\|_{L^1} \leq r_1$ [where $c_3(r_1)$ is as in (2.23)].

*Proof.* The claim (3.45) follows immediately from (2.22). The claim (3.46) follows immediately from (2.23), since if $\phi \in L^1_+$ and $r < \|\phi\|_{L^1} \le r_1$, then $\|r\phi/\|\phi\|_{L^1}\|_{L^1} \le r_1$, which implies that

$$G\left(\frac{r\phi}{\|\phi\|_{L^1}}\right) + c_3(r_1)\frac{r\phi}{\|\phi\|_{L^1}} \in L^1_+$$

which in turn implies that

$$G_r(\phi) + c_3(r_1)\phi = G\left(\frac{r\phi}{\|\phi\|_{L^1}}\right) + c_3(r_1)\frac{r\phi}{\|\phi\|_{L^1}} + c_3(r_1)\left(1 - \frac{r}{\|\phi\|_{L^1}}\right)\phi \in L^1_+$$

To prove (3.43) let $\phi, \hat{\phi} \in L^1$ such that $\|\phi\|_{L^1}, \|\hat{\phi}\|_{L^1} \ge r$. Then,

$$\begin{aligned}|F_r(\phi) - F_r(\hat{\phi})| &\le c_1(r)\left\|\frac{r\phi}{\|\phi\|_{L^1}} - \frac{r\hat{\phi}}{\|\hat{\phi}\|_{L^1}}\right\|_{L^1} \\ &= \left(\frac{c_1(r)r}{\|\phi\|_{L^1}\|\hat{\phi}\|_{L^1}}\right)\|\|\hat{\phi}\|_{L^1}\phi - \|\phi\|_{L^1}\hat{\phi}\|_{L^1} \\ &\le \left(\frac{c_1(r)}{\|\phi\|_{L^1}}\right)[\|\phi\|_{L^1}|\|\phi\|_{L^1} - \|\hat{\phi}\|_{L^1}| + \|\phi\|_{L^1}\|\phi - \hat{\phi}\|_{L^1}] \\ &\le 2c_1(r)\|\phi - \hat{\phi}\|_{L^1}\end{aligned}$$

If $\phi, \hat{\phi} \in L^1$ such that $\|\phi\|_{L^1} > r$ and $\|\hat{\phi}\|_{L^1} \le r$, then

$$\begin{aligned}|F_r(\phi) - F_r(\hat{\phi})| &\le c_1(r)\left\|\frac{r\phi}{\|\phi\|_{L^1}} - \hat{\phi}\right\|_{L^1} \\ &= \left(\frac{c_1(r)}{\|\phi\|_{L^1}}\right)\|r\phi - \|\phi\|_{L^1}\hat{\phi}\|_{L^1} \\ &\le \left(\frac{c_1(r)}{\|\phi\|_{L^1}}\right)[r\|\phi - \hat{\phi}\|_{L^1} + \|\hat{\phi}\|_{L^1}(\|\phi\|_{L^1} - r)]\end{aligned}$$

$$\leq \left(\frac{c_1(r)}{\|\phi\|_{L^1}}\right)[r\|\phi - \hat{\phi}\|_{L^1} + \|\hat{\phi}\|_{L^1}(\|\phi\|_{L^1} - \|\hat{\phi}\|_{L^1})]$$

$$\leq 2c_1(r)\|\phi - \hat{\phi}\|_{L^1}$$

If $\phi$, $\hat{\phi} \in L^1$ such that $\|\phi\|_{L^1}$, $\|\hat{\phi}\|_{L^1} \leq r$, then

$$|F_r(\phi) - F_r(\hat{\phi})| = |F(\phi) - F(\hat{\phi})| \leq c_1(r)\|\phi - \hat{\phi}\|_{L^1}$$

Thus, (3.43) is established and a similar argument proves (3.44). □

DEFINITION 3.5 Let (2.1), (2.2), (2.22), (2.23) hold, let $r > 0$, let $F_r$ and $G_r$ be defined as in (3.41) and (3.42), respectively, and let the mapping $A_r$ from $L^1_+$ to $L^1$ be defined by

(3.47) $\quad A_r\phi \overset{\text{def}}{=} \phi' - G_r(\phi)$ for $\phi \in D(A_r)$, where $D(A_r) = \{\phi \in L^1_+ :$ $\phi$ is absolutely continuous on $[0,\infty)$, $\phi' \in L^1$, and $\phi(0) = F_r(\phi)\}$.

THEOREM 3.3 Let (2.1), (2.2), (2.22), (2.23) hold, let $T_\phi = \infty$ for each $\phi \in L^1_+$, and let $S(t)$, $t \geq 0$, be the strongly continuous nonlinear semigroup in $L^1_+$ as in Theorem 3.1. If $\phi \in L^1_+$, $t > 0$, $r \geq \sup_{0\leq s\leq t}\|S(s)\phi\|_{L^1}$, and $A_r$ is defined as in (3.47), then

(3.48) $$\lim_{n\to\infty}\left(I + \frac{s}{n}A_r\right)^{-n}\phi = S(s)\phi \text{ uniformly for } s \in [0,t]$$

*Proof.* From Proposition 3.1 and Proposition 3.10 we see that there exists a strongly continuous nonlinear semigroup $S_r(t)$, $t \geq 0$, in $L^1$ such that for $\phi \in L^1_+$, $S_r(t)\phi$ is the solution of

(3.49) $$(S_r(t)\phi)(a) = \begin{cases} F_r(S_r(t - a)\phi) + \int_{t-a}^{t} G_r(S_r(s)\phi)(s + a - t)\, ds & \text{a.e. } a \in (0,t) \\ \phi(a - t) + \int_0^t G_r(S_r(s))(s + a - t)\, ds & \text{a.e. } a \in (t,\infty) \end{cases}$$

From Theorem 2.1 and Theorem 2.2 we know that $S_r(\cdot)\phi$ is the unique solution of (3.49). From Proposition 3.9 we have that

$$(3.50) \qquad \lim_{n\to\infty} \left(I + \frac{s}{n} A_r\right)^{-n} \phi = S_r(s)\phi \text{ uniformly for } s \in [0,t]$$

Since $\sup_{0\le s\le t}\|S(s)\phi\|_{L^1} \le r$, $F(S(s)\phi) = F_r(S(s)\phi)$ and $G(S(s)\phi) = G_r(S(s)\phi)$ for $0 \le s \le t$. Since $S(\cdot)\phi$ is the solution of (1.49) and $S_r(\cdot)\phi$ is the unique solution of (3.49), we must have that $S_r(s)\phi = S(s)\phi$ for $0 \le s \le t$. Consequently, (3.48) follows from (3.50). □

THEOREM 3.4 Let (2.1), (2.2), (2.22), (2.23) hold and let A be defined as in (3.16). Then, $\overline{D(A)} = L^1_+$. Further, let $T_\phi = \infty$ for all $\phi \in L^1_+$, let $S(t)$, $t \ge 0$, be the strongly continuous nonlinear semigroup in $L^1_+$ as in Theorem 3.1, let $\phi \in D(A)$, and let u be a Lipschitz continuous function from $[0,t]$ to $L^1$ such that $u(0) = \phi$ and for almost all $s \in (0,t)$, u is differentiable at s, $u(s) \in D(A)$, and $(d/ds)u(s) = -Au(s)$. Then, $u(s) = S(s)\phi$ for $s \in [0,t]$.

*Proof.* Let $\phi \in L^1_+$, let $r > \|\phi\|_{L^1}$, and let $A_r$ be as in (3.47). By Proposition 3.10 and by (3.29) of Proposition 3.8 there is a sequence $\{\phi_k\} \subset D(A_r)$ such that $\lim_{k\to\infty} \phi_k = \phi$. For k sufficiently large $F_r(\phi_k) = F(\phi_k)$. Thus, $\phi_k \in D(A)$ for k sufficiently large by (3.16). Consequently, $\phi$ is a limit point of D(A).

Now let $S(t)$, $t \ge 0$, and u be as in the statement of the theorem, let $r > 0$ such that $\sup_{0\le s\le t}\|u(s)\|_{L^1} \le r$, $\sup_{0\le s\le t}\|S(s)\phi\|_{L^1} \le r$, and let $A_r$ be as in (3.47). Then, $(d/ds)u(s) = -A_r u(s) = -Au(s)$ for almost all $s \in (0,t)$. By Proposition 3.4 $u(s) = \lim_{n\to\infty}(I + s/nA_r)^{-n}\phi$ for $s \in [0,t]$. Then, $u(s) = S(s)\phi$ for $s \in [0,t]$ by (3.48) of Theorem 3.3. □

We conclude this section by collecting some results for the strongly continuous semigroup of bounded linear operators associated with (ADP) in the case that F and G are bounded linear operators. These results follow from the general theory of strongly continuous semigroups of bounded linear operators in Banach spaces (see [94], Chapter VIII, Section 1).

PROPOSITION 3.11 Let F be a bounded linear operator from $L^1$ into $R^n$, let G be a bounded linear operator from $L^1$ to $L^1$, let S(t), $t \geq 0$, be the strongly continuous semigroup of bounded linear operators in $L^1$ as in Proposition 3.2, let B be the infinitesimal generator of S(t), $t \geq 0$, as in Proposition 3.7, and let $\omega = |F| + |G|$ as in (3.7). The following hold:

(3.51) $\overline{D(B)} = L^1$

(3.52) $-B + \omega I$ is accretive in $L^1$.

(3.53) $(I - \lambda B)^{-1}$ is a bounded everywhere defined linear operator in $L^1$ for all $0 < \lambda < \omega^{-1}$.

(3.54) For each $\phi \in L^1$, $\lim_{n\to\infty}(I - t/nB)^{-n}\phi = S(t)\phi$ uniformly in bounded intervals of t.

## 3.4 COMPACTNESS OF THE TRAJECTORIES

The property of compactness of the trajectories of a nonlinear semigroup can be very useful in analyzing their ultimate behavior as time evolves. Since the nonlinear semigroup S(t), $t \geq 0$, associated with (ADP) is defined in the infinite dimensional Banach space $L^1$, it is not automatic that a bounded trajectory lies in a compact set. For this reason it is advantageous to treat the questions of boundedness and precompactness of the trajectories separately. In Section 4.2 we will consider the question of boundedness of the trajectories of S(t), $t \geq 0$. In this section we consider the question of their precompactness. To investigate this question we will use the measure of noncompactness due to C. Kuratowski (see [181]).

DEFINITION 3.6 Let X be a complete metric space with metric $\rho$ and let M be a bounded subset of X. The *diameter* of M is the infimum of $\varepsilon > 0$ such that if $x,y \in M$, then $\rho(x,y) \leq \varepsilon$. The *measure of noncompactness* of M, denoted by $\alpha[M]$, is the infimum of $\varepsilon > 0$ such that M can be covered by a finite number of subsets of X each with diameter no larger than $\varepsilon$.

A proof of the following proposition may be found in [217], Lemma 5.2, p. 17.

PROPOSITION 3.12 Let $M_1$ and $M_2$ be bounded subsets of the Banach space X. The following hold:

(3.55) $\alpha[M_1] \leq \alpha[M_2]$ if $M_1 \subset M_2$.

(3.56) $\alpha[M_1] = 0$ if and only if $\bar{M}_1$ is compact.

(3.57) $\alpha[M_1 \cup M_2] = \max\{\alpha[M_1], \alpha[M_2]\}$.

(3.58) $\alpha[M_1 + M_2] \leq \alpha[M_1] + \alpha[M_2]$.

(3.59) $\alpha[\overline{co}\, M_1] = \alpha[M_1]$, where $\overline{co}\, M_1$ denotes the closed convex hull of $M_1$.

We will use the following proposition to show that under suitable hypotheses on the birth function F and the aging function G, the bounded trajectories of the nonlinear semigroup associated with (ADP) have compact closure in $L^1$.

PROPOSITION 3.13 Let X be a Banach space, let C be a closed subset of X, let $S(t)$, $t \geq 0$, be a strongly continuous nonlinear semigroup in C, and for each $x \in C$, $t \geq 0$, let $S(t)x = U(t)x + W(t)x$, where $U(t)$, $t \geq 0$, and $W(t)$, $t \geq 0$, are families of mappings from C to X satisfying

(3.60) There exists a function $\delta$: $[0,\infty) \times [0,\infty) \to [0,\infty)$ such that for $r > 0$, $\lim_{t\to\infty} \delta(t,r) = 0$, and if $x \in C$ with $\|x\| \leq r$ and $t \geq 0$, then $\|U(t)x\| \leq \delta(t,r)$.

(3.61) There exists $t_0 \geq 0$ such that if $t > t_0$, then $W(t)$ maps bounded sets of C into sets with compact closure in X.

Let $x \in C$ such that for some $r > 0$, $\sup_{t\geq 0}\|S(t)x\| \leq r$. Then, $\{S(t)x\colon\ t \geq 0\}$ has compact closure in X.

*Proof.* Let $\varepsilon > 0$ and let $t_1 > t_0$ such that $\delta(t_1,r) < \varepsilon$. Observe from (3.4) that the set $\{S(t)x\colon\ 0 \leq t \leq t_1\}$ has compact closure. Using Proposition 3.12 we obtain

$$\begin{aligned}
&\alpha[\{S(t)x : t \geq 0\}] \\
&= \alpha[\{S(t_1 + t)x : t \geq 0\} \cup \{S(t)x : 0 \leq t \leq t_1\}] \\
&= \alpha[\{S(t_1)S(t)x : t \geq 0\} \cup \{S(t)x : 0 \leq t \leq t_1\}] \\
&= \max\{\alpha[S(t_1)S(t)x : t \geq 0\}],0\} \\
&= \alpha[\{U(t_1)S(t)x + W(t_1)S(t)x : t \geq 0\}] \\
&\leq \alpha[\{U(t_1)S(t)x : t \geq 0\}] + \alpha[\{W(t_1)S(t)x : t \geq 0\}] \\
&\leq \delta(t_1,r) + 0 \\
&< \varepsilon
\end{aligned}$$

Thus, $\alpha[\{S(t)x : t \geq 0\}] = 0$ and the conclusion follows by (3.56). □

In order to apply Proposition 3.13 to the strongly continuous nonlinear semigroup $S(t)$, $t \geq 0$, associated with (ADP), we must find a decomposition $S(t) = U(t) + W(t)$ such that $U(t)$ is ultimately small in the sense of (3.60) and $W(t)$ is ultimately compact in the sense of (3.61). We will use a decomposition similar to the one used by J. Hale in [145] and by J. Prüss in [240].

DEFINITION 3.8 Let $S(t)$, $t \geq 0$, be the strongly continuous nonlinear semigroup associated with (ADP) as in Theorem 3.1. For $t \geq 0$ define the mappings $U(t)$, $W(t)$ in $L^1_+$ as follows: for $\phi \in L^1_+$,

$$(3.62) \qquad (U(t)\phi)(a) = \begin{cases} 0 & \text{a.e. } a \in (0,t) \\ (S(t)\phi)(a) & \text{a.e. } a \in (t,\infty) \end{cases}$$

$$(3.63) \qquad (W(t)\phi)(a) = \begin{cases} (S(t)\phi)(a) & \text{a.e. } a \in (0,t) \\ 0 & \text{a.e. } a \in (t,\infty) \end{cases}$$

We may interpret $U(t)\phi$ as the density at time $t$ of that part of the population which was existent at the initial time 0 (these members of the population have age $a$ greater than time $t$), and $W(t)\phi$ as the density at time $t$ of that part of the population born after the initial time 0 (these members of the population have age $a$ less than time $t$). It is reasonable to expect that $\lim_{t\to\infty} \|U(t)\phi\|_{L^1} = 0$, and we now provide a sufficient condition on the aging function G such that the mapping U(t) satisfies (3.60). We require that

(3.64) $G: L^1 \to L^1$ and there exists a constant $\underline{\mu} > 0$ such that $\Sigma_{i=1}^n \int_t^\infty G(\phi)_i(a)\, da \le -\underline{\mu}\, \Sigma_{i=1}^n \int_t^\infty \phi_i(a)\, da$ for all $\phi \in L_+^1$ and $t \ge 0$ (where the subscript i denotes the i-th component).

PROPOSITION 3.14 Let (2.1), (2.2), (2.22), (2.23), (3.64) hold, let $T_\phi = \infty$ for all $\phi \in L_+^1$, let S(t), $t \ge 0$, be the strongly continuous nonlinear semigroup in $L_+^1$ as in Theorem 3.1, and for each $t \ge 0$, let U(t) be defined as in (3.62). Then, for all $\phi \in L_+^1$, $t \ge 0$,

(3.65) $$\|U(t)\phi\|_{L^1} \le e^{-\underline{\mu} t}\|\phi\|_{L^1}$$

*Proof.* For $\phi \in L_+^1$ and $t \ge 0$ define

$$V(t) = \|U(t)\phi\|_{L^1}$$

$$= \sum_{i=1}^n \int_t^\infty (S(t)\phi)_i(a)\, da$$

$$= \sum_{i=1}^n \int_0^\infty (S(t)\phi)_i(t + c)\, dc$$

Then, using (1.49), we have that for $h > 0$,

$$h^{-1}[V(t + h) - V(t)]$$

$$= \sum_{i=1}^n \int_0^\infty h^{-1}[(S(t + h)\phi)_i(t + h + c) - (S(t)\phi)_i(t + c)]\, dc$$

$$= \sum_{i=1}^n \int_0^\infty \left[h^{-1} \int_t^{t+h} G(S(s)\phi)_i(s + c)\, ds\right] dc$$

$$= \sum_{i=1}^n h^{-1} \int_t^{t+h} \left[\int_s^\infty G(S(s)\phi)_i(a)\, da\right] ds$$

$$\le \sum_{i=1}^n h^{-1} \int_t^{t+h} \left[-\underline{\mu} \int_s^\infty (S(s)\phi)_i(a)\, da\right] ds$$

$$= -\underline{\mu} \sum_{i=1}^{n} h^{-1} \int_t^{t+h} \left[ \int_0^\infty (S(s)\phi)_i(s + c)\ dc \right] ds$$

$$= -\underline{\mu}\ h^{-1} \int_t^{t+h} V(s)\ ds$$

Thus, $\limsup_{h\to 0^+} h^{-1}[V(t + h) - V(t)] \leq -\underline{\mu}V(t)$ for $t \geq 0$, and so (3.65) follows from Theorem 1.41, p. 15, in [183]. □

REMARK 3.4 If the aging function G has the form $G(\phi)(a) = -\mu(a,\phi)\phi(a)$ as in (2.49), then the following condition on $\mu$: $[0,\infty) \times L^1 \to B(R^n,R^n)$ guarantees that G satisfies (3.64):

(3.66) $\mu_{ij}(a,\phi) \geq 0$ for all $\phi \in L^1_+$, $a \geq 0$, $i, j = 1, \ldots, n$, and there exists a constant $\underline{\mu} > 0$ such that $\mu_{ii}(a,\phi) \geq \underline{\mu}$ for all $\phi \in L^1_+$, $a \geq 0$, $i = 1, \ldots, n$ [where $\mu_{ij}(a,\phi)$ denotes the ij-th entry of the matrix representation of $\mu(a,\phi) \in B(R^n,R^n)$].

To see that (3.66) implies (3.64) observe that for $\phi \in L^1_+$, $i = 1, \ldots, n$, almost all $a \geq 0$,

$$\begin{aligned} G(\phi)_i(a) &= -[\mu(a,\phi)\phi(a)]_i \\ &= \sum_{j=1}^{n} \mu_{ij}(a,\phi)\phi_j(a) \\ &\leq -\mu_{ii}(a,\phi)\phi_i(a) \\ &\leq -\underline{\mu}\phi_i(a) \end{aligned}$$

In the case that F and G are bounded linear operators it is useful to have a decay estimate on $U(t)\phi$ for all $\phi \in L^1$ rather than just $\phi \in L^1_+$. For this purpose we require that

(3.67) $G\colon\ L^1 \to L^1$ and there exists a constant $\underline{\mu} > 0$ such that $\Sigma_{i=1}^n [\operatorname{sgn} \phi_i(a)][G(\phi)_i(a)] \leq -\underline{\mu}|\phi(a)|$ for all $\phi \in L^1$ and almost all $a > 0$ (where $\operatorname{sgn} z \overset{\text{def}}{=} z/|z|$ if $z \neq 0$ and $\operatorname{sgn} 0 \overset{\text{def}}{=} 0$).

PROPOSITION 3.15 Let F be a bounded linear operator from $L^1$ into $R^n$, let G be a bounded linear operator from $L^1$ into $L^1$, let $S(t)$, $t \geq 0$, be the strongly continuous semigroup of bounded linear operators in $L^1$ as in Proposition 3.2, let G satisfy (2.49) and (3.67), and for each $\phi \in L^1$, $t \geq 0$, let $U(t)\phi$ be defined as in (3.62). Then, for all $\phi \in L^1$, $t \geq 0$,

$$\|U(t)\phi\|_{L^1} \leq e^{-\underline{\mu}t}\|\phi\|_{L^1} \tag{3.68}$$

*Proof.* We will use the following fact, whose proof is easily established:

(3.69) If $u$: $[0,\infty) \to R$ and $d^-/dt\, u(t)$ exists, then $d^-/dt|u(t)|$ exists and is $[\operatorname{sgn} u(t)][d^-/dt\, u(t)]$ if $u(t) \neq 0$, and $-|d^-/dt\, u(t)|$ if $u(t) = 0$.

Let $\phi \in L^1$ and let $i = 1, \ldots, n$. From (2.68) we have that there is a set E of measure zero such that if $c > 0$, $c \notin E$, then for all $t \geq 0$,

$$\frac{d}{dt}(S(t)\phi)_i(t + c) = G(S(t)\phi)_i(t + c) \tag{3.70}$$

Let $c > 0$, $c \notin E$, $t > 0$, $h < 0$, $t + h > 0$, and consider the difference quotients

$$h^{-1}[|(S(t + h)\phi)_i(t + h + c)| - |(S(t)\phi)_i(t + c)|]$$

From (3.69) and (3.70) we have that these difference quotients converge to 0 from the left. Also, from (1.49) and (2.49) we have that

$$\begin{aligned} &|h^{-1}[|(S(t + h)\phi)_i(t + h + c)| - |(S(t)\phi)_i(t + c)|]| \\ &\leq (-h)^{-1}|(S(t + h)\phi)_i(t + h + c) - (S(t)\phi)_i(t + c)| \\ &= (-h)^{-1}\left|\int_t^{t+h} G(S(s)\phi)_i(s + c)\, ds\right| \\ &\leq (-h)^{-1}\int_{t+h}^{t} c_5(\|S(s)\phi\|_{L^1})|(S(s)\phi)(s + c)|\, ds \end{aligned} \tag{3.71}$$

$$\le (-h)^{-1}[\sup_{0\le s\le t} c_5(\|S(s)\phi\|_{L^1})] \int_{t+h}^{t} \Big|\phi(c) + \int_0^s G(S(\tau)\phi)(\tau + c)\, d\tau\Big|\, ds$$

$$\le [\sup_{0\le s\le t} c_5(\|S(s)\phi\|_{L^1})]\Big[|\phi(c)| + \int_0^t |G(S(\tau)\phi)(\tau + c)|\, d\tau\Big]$$

Further, we have that

$$\int_0^\infty \Big[|\phi(c)| + \int_0^t |G(S(\tau)\phi)(\tau + c)|\, d\tau\Big]\, dc \le \|\phi\|_{L^1} + \int_0^t \|G(S(s)\phi)\|_{L^1}\, ds \tag{3.72}$$

Now define

$$V(t) = \|U(t)\phi\|_{L^1} = \sum_{i=1}^{n} \int_t^\infty |(S(t)\phi)_i(a)|\, da = \sum_{i=1}^{n} \int_0^\infty |(S(t)\phi)_i(t + c)|\, dc$$

From (3.67), (3.69), (3.70), (3.71), (3.72), and the Lebesgue convergence theorem ([249], Theorem 15, p. 88) we obtain

$$\frac{d^-}{dt} V(t) = \frac{d^-}{dt} \sum_{i=1}^{n} \int_0^\infty |(S(t)\phi)_i(t + c)|\, dc$$

$$= \int_0^\infty \sum_{i=1}^{n} \frac{d^-}{dt} |(S(t)\phi)_i(t + c)|\, dc$$

$$\le \int_0^\infty \sum_{i=1}^{n} G(S(t)\phi)_i(t + c)\, \mathrm{sgn}(S(t)\phi)_i(t + c)\, dc$$

$$\le \int_0^\infty -\underline{\mu}|(S(t)\phi)(t + c)|\, dc$$

$$= -\underline{\mu}V(t)$$

Thus, $\limsup_{h \to 0^-} h^{-1}[V(t + h) - V(t)] \le -\mu V(t)$ for $t > 0$, and then (3.68) follows from Theorem 1.4.1, p. 15, in [183]. □

We next provide sufficient conditions on the birth function $F$ and the aging function $G$ such that the mapping $W(t)$ defined in (3.63) satisfies (3.61). We will require that $G$ satisfies (2.49) and $F$ satisfies

(3.73) $F$: $L^1 \to R^n$ and there exists $\beta$: $[0,\infty) \times L^1 \to B(R^n,R^n)$ such that if $\phi \in L^1$, then $F(\phi) = \int_0^\infty \beta(a,\phi)\phi(a)\, da$, where $\beta$ satisfies: (i) there is an increasing function $c_7$: $[0,\infty) \to [0,\infty)$ such that if $\phi \in L^1$ and $a, \hat{a} \ge 0$, then $|\beta(a,\phi) - \beta(\hat{a},\phi)| \le c_7(\|\phi\|_{L^1})|a - \hat{a}|$; (ii) there is an increasing function $c_8$: $[0,\infty) \to [0,\infty)$ such that if $\phi \in L^1$ and $a \ge 0$, then $|\beta(a,\phi)| \le c_8(\|\phi\|_{L^1})$; and (iii) there is an increasing function $c_9$: $[0,\infty) \to [0,\infty)$ such that if $a \ge 0$ and $\phi, \hat{\phi} \in L^1_+$ with $\|\phi\|_{L^1}, \|\hat{\phi}\|_{L^1} \le r$, then $|\beta(a,\phi) - \beta(a,\hat{\phi})| \le c_9(r) \Sigma_{i=1}^n |\|\phi_i\|_{L^1} - \|\hat{\phi}\|_{L^1}|$ [where $\phi_i$, $\hat{\phi}_i$ denote the i-th components of $\phi$, $\hat{\phi}$, and $\|\phi_i\|_{L^1}$, $\|\hat{\phi}_i\|_{L^1}$ denote the norms of $\phi_i$, $\hat{\phi}_i$ in $L^1((0,\infty);R)$].

PROPOSITION 3.16 Let (2.1), (2.2), (2.22), (2.23), (2.49), (3.73) hold, let $T_\phi = \infty$ for all $\phi \in L^1_+$, let $S(t)$, $t \ge 0$, be the strongly continuous nonlinear semigroup in $L^1_+$ as in Theorem 3.1, and for each $t \ge 0$ let $W(t)$ be defined as in (3.63). Let $S(t)$, $t \ge 0$, have the property that if $t > 0$ and $M$ is a bounded subset of $L^1_+$, then there exists $r > 0$ such that $\|S(s)\phi\|_{L^1} \le r$ for all $\phi \in M$, $s \in [0,t]$. Then, for each $t > 0$, $W(t)$ maps bounded sets of $L^1_+$ into sets with compact closure in $L^1$.

*Proof.* Let $t > 0$, let $M$ be a bounded set in $L^1_+$, and let $r > 0$ such that $\|S(s)\phi\|_{L^1} \le r$ for all $\phi \in M$, $s \in [0,t]$. For notational convenience we define $r_0 \overset{\text{def}}{=} c_1(r)r + |F(0)|$, $r_1 \overset{\text{def}}{=} \exp[c_5(r)t]$, and $r_2 \overset{\text{def}}{=} c_5(r)$ [where $c_1$ is as in (2.1) and $c_5$ is as in (2.49)]. By Lemma 2.3 it suffices to show that

(3.74) $\lim_{h\to 0} \int_0^t |(W(t)\phi)(a + h) - (W(t)\phi)(a)|\, da = 0$ uniformly for $\phi \in M$ (where $(W(t)\phi)(a)$ is taken as 0 for $a < 0$).

Fix $\phi \in M$ and let $\ell(a,s) \stackrel{\text{def}}{=} (S(s)\phi)(a)$ for $0 \le s \le t$, almost all $a > 0$, and let $B(s) \stackrel{\text{def}}{=} F(\ell(\cdot,s))$ for $0 \le s \le t$. From (2.62) we have that for $0 \le s \le t$, $0 \le a \le s$,

$$\ell(a,s) = U(s, s - a; a - s, \ell)B(s - a) = (W(s)\phi)(a)$$

Observe from (2.1) and (2.50) that

$$(3.75) \qquad |B(s)| \le r_0 \qquad \text{for } 0 \le s \le t$$

$$(3.76) \qquad |\ell(a,s)| \le r_1 r_0 \qquad \text{for } 0 \le a \le s \le t$$

We first claim that

(3.77) $\lim_{h\to 0} \left| \|\ell_i(\cdot, s + h)\|_{L^1} - \|\ell_i(\cdot,s)\|_{L^1} \right| = 0$ uniformly for $\phi \in M$, $s \in [0,t]$, and $i = 1, \ldots, n$ (where $\ell_i$ denotes the i-th component of $\ell$).

To prove (3.77) observe from (2.62) that for $s \in [0,t]$, $s + h \in [0,t]$, $i = 1, \ldots, n$,

$$\left| \|\ell_i(\cdot, s + h)\|_{L^1} - \|\ell_i(\cdot,s)\|_{L^1} \right|$$

$$= \left| \int_0^\infty |\ell_i(a, s + h)|\, da - \int_0^\infty |\ell_i(a,s)|\, da \right|$$

$$= \left| \int_0^{s+h} |[U(s + h, s + h - a; a - s - h, \ell)B(s + h - a)]_i|\, da \right.$$

$$+ \int_{s+h}^\infty |[U(s + h, 0; a - s - h, \ell)\phi(a - s - h)]_i|\, da$$

$$- \int_0^s |[U(s, s - a; a - s, \ell)B(s - a)]_i|\, da$$

$$- \int_s^\infty |[U(s,0; a - s, \ell)\phi(a - s)]_i|\, da$$

$$= \left| \int_0^{s+h} |[U(s + h, a; -a,\ell)B(a)]_i| \, da \right.$$

$$+ \int_0^\infty |[U(s + h, 0; a,\ell)\phi(a)]_i| \, da$$

$$- \int_0^s |[U(s,a;-a,\ell)B(a)]_i| \, da$$

$$\left. - \int_0^\infty |[U(s,0;a,\ell)\phi(a)]_i| \, da \right|$$

$$\leq \left| \int_s^{s+h} |[U(s + h, a; -a,\ell)B(a)]_i| \, da \right|$$

$$+ \int_0^s |[U(s + h, a; -a,\ell)B(a)]_i - [U(s,a;-a,\ell)B(a)]_i| \, da$$

$$+ \int_0^\infty |[U(s + h, 0; a,\ell)\phi(a)]_i - [U(s,0;a,\ell)\phi(a)]_i| \, da$$

$$\leq |h| \sup_{0 \leq a \leq s+h} |U(s + h, a; -a,\ell)| r_0$$

$$+ t \sup_{0 \leq a \leq s} |U(s + h, a; -a,\ell) - U(s,a;-a,\ell)| r_0$$

$$+ \sup_{0 \leq a < \infty} |U(s + h, 0; a,\ell) - U(s,0;-a,\ell)| \, \|\phi\|_{L^1}$$

Now use (2.50) and (2.56) to conclude that this last expression is $\leq |h| r_1 r_0 + t|h| r_2 r_1 r_0 + |h| r_2 r_1 \|\phi\|_{L^1}$. The claim (3.77) now follows immediately.

We next claim that

$$\lim_{h \to 0} |B(s + h) - B(s)| = 0 \text{ uniformly for } \phi \in M, \; s \in [0,t] \tag{3.78}$$

To prove (3.78) observe from (2.62) that for $s \in [0,t]$, $s + h \in [0,t]$,

$$B(s + h) - B(s)$$

$$= \int_0^\infty \beta(a,\ell(\cdot, s + h))\ell(a, s + h) \, da - \int_0^\infty \beta(a,\ell(\cdot,s))\ell(a,s) \, da$$

$$= \int_0^{s+h} \beta(a,\ell(\cdot, s+h))U(s+h, s+h-a; a-s-h, \ell)B(s+h-a)\, da$$

$$+ \int_{s+h}^{\infty} \beta(a,\ell(\cdot, s+h))U(s+h, 0; a-s-h, \ell)\phi(a-s-h)\, da$$

$$- \int_0^{s} \beta(a,\ell(\cdot,s))U(s, s-a; a-s, \ell)B(s-a)\, da$$

$$- \int_s^{\infty} \beta(a,\ell(\cdot,s))U(s,0; a-s, \ell)\phi(a-s)\, da$$

$$= \int_0^{s+h} \beta(s+h-a, \ell(\cdot, s+h))U(s+h, a; -a,\ell)B(a)\, da$$

$$+ \int_0^{\infty} \beta(s+h-a, \ell(\cdot, s+h))U(s+h, 0; a,\ell)\phi(a)\, da$$

$$- \int_0^{s} \beta(s-a, \ell(\cdot,s))U(s,a;-a,\ell)B(a)\, da$$

$$- \int_0^{\infty} \beta(s+a, \ell(\cdot,s))U(s,0;a,\ell)\phi(a)\, da$$

$$= \int_s^{s+h} \beta(s+h-a, \ell(\cdot, s+h))U(s+h, a; -a,\ell)B(a)\, da$$

$$+ \int_0^{s} \{\beta(s+h-a, \ell(\cdot, s+h))U(s+h, a; -a,\ell)$$

$$- \beta(s-a, \ell(\cdot,s))U(s,a;-a,\ell)\}B(a)\, da$$

$$+ \int_0^{\infty} \{\beta(s+h+a, \ell(\cdot, s+h))U(s+h, 0; a,\ell)$$

$$- \beta(s+a, \ell(\cdot,s))U(s,0;a,\ell)\}\phi(a)\, da$$

$$\stackrel{\text{def}}{=} L_1 + L_2 + L_3$$

From (2.50), (3.73), (3.75) we see that

$$|L_1| \le |h| c_8(r) r_1 r_0$$

From (2.50), (2.56), (3.73), (3.75) we see that

$$
\begin{aligned}
|L_2| \le \sup_{0\le a\le s} \{ & |[\beta(s + h - a, \ell(\cdot, s + h)) \\
& - \beta(s - a, \ell(\cdot, s + h))]U(s + h, a; -a,\ell)B(a)| \\
& + |[\beta(s - a, \ell(\cdot, s + h)) - \beta(s - a, \ell(\cdot, s))] \\
& \times U(s + h, a; -a,\ell)B(a)| \\
& + |\beta(s - a, \ell(\cdot,s))[U(s + h, a; -a,\ell) - U(s,a;-a,\ell)]B(a)|\} \\
\le\ & c_7(r)|h|r_1 r_0 \\
& + c_9(r) \sum_{i=1}^{n} |\|\ell_i(\cdot, s + h)\|_{L^1} - \|\ell_i(\cdot,s)\|_{L^1}| r_1 r_0 \\
& + c_8(r)|h|r_2 r_1 r_0
\end{aligned}
$$

From (2.50), (2.56), (3.73) we see that

$$
\begin{aligned}
|L_3| \le \int_0^\infty & |[\beta(s + h + a, \ell(\cdot, s + h)) - \beta(s + a, \ell(\cdot, s + h))] \\
& \times U(s + h, 0; a,\ell)\phi(a) \\
& + [\beta(s + a, \ell(\cdot, s + h)) - \beta(s + a, \ell(\cdot,s))] \\
& \times U(s + h, 0; a,\ell)\phi(a) \\
& + \beta(s + a, \ell(\cdot,s))[U(s + h, 0; a,\ell) \\
& - U(s,0;a,\ell)]\phi(a)|\ da \\
& c_7(r)|h|r_1\|\phi\|_{L^1} \\
& + c_9(r) \sum_{i=1}^{n} |\|\ell_i(\cdot, s + h)\|_{L^1} - \|\ell_i(\cdot,s)\|_{L^1}| r_1 \|\phi\|_{L^1} \\
& + c_8(r)|h|r_2 r_1
\end{aligned}
$$

To claim (3.78) now follows immediately from (3.77).

Last, we prove (3.74). Observe from (3.76) that for $0 \le a \le t$, $0 < h < t$,

$$
\begin{aligned}
& \int_0^t |(W(t)\phi)(a + h) - (W(t)\phi)(a)|\ da \\
& = \int_0^{t-h} |\ell(a + h, t) - \ell(a,t)|\ da + \int_{t-h}^t |\ell(a,t)|\ da \\
& \le \int_0^{t-h} |\ell(a + h, t) - \ell(a,t)|\ da + hr_1 r_0
\end{aligned}
$$

and for $0 \le a \le t$, $-t < h < 0$,

$$\int_0^t |(W(t)\phi)(a + h) - (W(t)\phi)(a)| \, da$$

$$= \int_{-h}^t |\ell(a + h, t) - \ell(a,t)| \, da + \int_0^{-h} |\ell(a,t)| \, da$$

$$\le \int_h^t |\ell(a + h, t) - \ell(a,t)| \, da + |h| r_1 r_0$$

Then, (3.74) follows from (3.78), since for $0 \le a \le t$, $0 \le a + h \le t$, we obtain from (2.50), (2.57), (2.58), (3.75) that

$$|(W(t)\phi)(a + h) - (W(t)\phi)(a)| = |\ell(a + h, t) - \ell(a,t)|$$

$$= |U(t, t - a - h; a + h - t, \ell)B(t - a - h)$$

$$- U(t, t - a; a - t, \ell)B(t - a)|$$

$$\le |[U(t, t - a - h; a + h - t, \ell) - U(t, t - a; a + h - t, \ell)]B(t - a - h)|$$

$$+ |[U(t, t - a; a + h - t, \ell) - U(t, t - a; a - t, \ell)]B(t - a - h)|$$

$$+ |U(t, t - a; a - t, \ell)[B(t - a - h) - B(t - a)]|$$

$$\le |h| r_2 r_1 r_0 + |h| c_4(r) r_1 t r_0$$

$$+ r_1 |B(t - a - h) - B(t - a)|$$

The conclusion now follows immediately from (3.74) and Lemma 2.3. □

REMARK 3.5 Observe that for $\phi, \hat{\phi} \in L^1$,

$$\left| \|\phi\|_{L^1} - \|\hat{\phi}\|_{L^1} \right| = \left| \sum_{i=1}^n \|\phi_i\|_{L^1} - \sum_{i=1}^n \|\hat{\phi}_i\|_{L^1} \right|$$

$$\le \sum_{i=1}^n \left| \|\phi_i\|_{L^1} - \|\hat{\phi}_i\|_{L^1} \right| \le \sum_{i=1}^n \|\phi_i - \hat{\phi}\|_{L^1}$$

$$= \|\phi - \hat{\phi}\|_{L^1}$$

Consequently, if $\beta$ satisfies (iii) in (3.73), then for $\phi, \hat{\phi} \in L^1$, $\|\phi\|_{L^1}, \|\hat{\phi}\|_{L^1} \le r$, $a \ge 0$, $\beta$ satisfies

$$|\beta(a,\phi) - \beta(a,\hat{\phi})| \le c_9(r)\|\phi - \hat{\phi}\|_{L^1}$$

but $\beta$ may not satisfy

$$|\beta(a,\phi) - \beta(a,\hat{\phi})| \le c_9(r)\,|\,\|\phi\|_{L^1} - \|\hat{\phi}\|_{L^1}|$$

Furthermore, for a bounded subset M of $L^1_+$, (3.77) implies that

(3.79) $$\lim_{h\to 0} |\,\|\ell(\cdot, s + h)\|_{L^1} - \|\ell(\cdot,s)\|_{L^1}| = 0 \text{ uniformly for } \phi \in M,\ s \in [0,t]$$

(but not conversely), and (3.77) is implied by

(3.80) $$\lim_{h\to 0} \|\ell(\cdot, s + h) - \ell(\cdot,s)\|_{L^1} = 0 \text{ uniformly for } \phi \in M,\ s \in [0,t]$$

(but not conversely). The property (3.80), however, will not be satisfied by the trajectories of the strongly continuous nonlinear semigroup $S(t)$, $t \ge 0$.

We now prove the main result of this section.

THEOREM 3.5 Let (2.1), (2.2), (2.22), (2.23), (2.49), (3.64), (3.73) hold, let $T_\phi = \infty$ for all $\phi \in L^1_+$, let $S(t)$, $t \ge 0$, be the strongly continuous nonlinear semigroup in $L^1_+$ as in Theorem 3.1, and let $S(t)$, $t \ge 0$, have the property that if $t > 0$ and M is a bounded subset of $L^1_+$, then there exists $r > 0$ such that $\|S(s)\phi\|_{L^1} \le r$ for all $\phi \in M$, $0 \le s \le t$. If $\phi \in L^1_+$ and $\{S(t)\phi : t \ge 0\}$ is bounded in $L^1$, then $\{S(t)\phi : t \ge 0\}$ has compact closure in $L^1$.

*Proof.* The proof follows immediately from Propositions 3.13, 3.14, and 3.16 by defining $U(t)$ as in (3.62) and $W(t)$ as in (3.63). □

REMARK 3.6 We note that the birth functions F discussed in Examples 3.1, 3.2, and 3.3 satisfy the condition (3.73) (since the function $\beta$ in (2.17), (2.19), (2.20) is continuously differentiable). Also, the aging function G discussed in these examples satisfies (2.49) (since the function $\mu$ in (2.18) is continuously differentiable). Suppose, in addition, that the function $\mu$ in (2.18) satisfies

(3.81) There exists a constant $\underline{\mu} > 0$ such that $\mu(P) \geq \underline{\mu}$ for all $P \geq 0$.

Then, the aging function G of these three examples satisfies (3.64), and so Theorem 3.5 can be applied to conclude that bounded trajectories are precompact.

In the case that the birth function F and the aging function G are bounded linear operators we can prove

PROPOSITION 3.17 Let F be a bounded linear operator from $L^1$ to $R^n$, let G be a bounded linear operator from $L^1$ into $L^1$, let $S(t)$, $t \geq 0$, be the strongly continuous semigroup of bounded linear operators in $L^1$ as in Proposition 3.2, let F satisfy (3.73), let G satisfy (2.49), and for each $t \geq 0$, $\phi \in L^1$, let $W(t)\phi$ be defined as in (3.63). Then, for each $t \geq 0$, $W(t)$ maps bounded sets of $L^1$ into sets with compact closure in $L^1$.

*Proof.* The proof is very similar to the proof of Proposition 3.16, where one uses the estimate (3.7). □

PROPOSITION 3.18 Let F be a bounded linear operator from $L^1$ into $R^n$, let G be a bounded linear operator from $L^1$ into $L^1$, let $S(t)$, $t \geq 0$, be the strongly continuous semigroup of bounded linear operators in $L^1$ as in Proposition 3.2, let F satisfy (3.73), and let G satisfy (2.49) and (3.67). If $\phi \in L^1$ and $\{S(t)\phi: \ t \geq 0\}$ is bounded in $L^1$, then $\{S(t)\phi: \ t \geq 0\}$ has compact closure in $L^1$.

*Proof.* For $t \geq 0$, $\phi \in L^1$, define $U(t)\phi$ as in (3.62) and $W(t)\phi$ as in (3.63). The proof follows immediately from Propositions 3.13, 3.15, and 3.17. □

## 3.5 APPROXIMATION OF THE SOLUTIONS

The numerical approximation of the solutions of age-dependent population models is of considerable importance in physical applications. As with many other evolution equations, the semigroup approach to this problem provides a convenient and efficient framework for approximation theory. The main idea of this approach may be summarized as follows: (1) choose a sequence $\{X^N\}$ of finite dimensional subspaces of the Banach space setting $L^1$ of the problem; (2) choose sequences $\{F^N\}$, $\{G^N\}$, with $F^N$: $X^N \to R^n$, $G^N$: $X^N \to X^N$, such that $F^N$ approximates the birth function F and $G^N$ approximates the aging function G; (3) define a sequence of nonlinear operators $\{B^N\}$, $B^N$: $X^N \to X^N$, where $B^N$ is defined in terms of $F^N$ and $G^N$, such that $B^N$ approximates the infinitesimal generator B of the nonlinear strongly continuous semigroup S(t), t ≥ 0, associated with (ADP); (4) show that $(I + \tau_N B^N)^{[t/\tau_N]}$ is uniformly Lipschitz continuous in N and bounded intervals of t, where $\{\tau_N\}$ is a sequence of time steps converging to 0; (5) show that $(I + \tau_N B^N)^{[t/\tau_N]}\phi$ converges to $S(t)\phi$ as N converges to ∞ for $\phi \in L^1$, t ≥ 0.

We will carry out these ideas for two different types of approximating schemes. The first scheme involves averaging approximations and the second scheme involves finite difference approximations. The crucial ingredients in proving the convergence of both the numerical schemes lie in establishing the consistency of the schemes (part (3) above) and the stability of the schemes (part (4) above). For the approximation theory of (ADP) there is the additional difficulty that the solutions are $L^1$-valued functions, and hence defined on the infinite interval [0,∞). Consequently, it is necessary to employ a truncation technique in order to deal with approximating solutions defined on only finite intervals.

For the averaging approximation scheme we make the following definitions:

(3.82) For M and N positive integers, let $\chi_k^{M,N}$: $[0,\infty) \to R$, k = 1, ..., MN, where $\chi_k^{M,N} \overset{\text{def}}{=} \chi_{[(k-1)/N,k/N)}$ is the characteristic function of [(k - 1)/N, k/N), let $X^{M,N}$ be the subspace of $L^1$ consisting of functions $\phi$ of the form $\phi = \Sigma_{k=1}^{MN} \chi_k^{M,N} h_k$, with $h_k \in R^n$, k = 1, ..., MN, and let $P^{M,N}$: $L^1 \to X^{M,N}$ be the projection defined by $P^{M,N}\phi \overset{\text{def}}{=} \Sigma_{k=1}^{MN} \chi_k^{M,N} h_k$, with $h_k \overset{\text{def}}{=} N \int_{(k-1)/N}^{k/N} \phi(a)\, da$, $\phi \in L^1$, k = 1, ..., MN.

We observe that for $\phi \in L^1$,

$$(3.83) \qquad \|P^{M,N}\phi\|_{L^1} = \sum_{k=1}^{MN} \|\chi_k^{MN}\|_{L^1} |h_k|$$
$$\leq \sum_{k=1}^{MN} \frac{1}{N} \left( N \int_{(k-1)/N}^{k/N} |\phi(a)| \, da \right) = \|\phi\|_{L^1}$$

Furthermore, if $\phi \in L^1$ such that $\phi$ is absolutely continuous on $[0,\infty)$ and $\phi \in L^1$, then

(3.84) $\lim_{M,N\to\infty} \|P^{M,N}\phi - \phi\|_{L^1} = 0$ (in the sense that if $\varepsilon > 0$, there exists $M_1$ and $N_1$ such that if $M > M_1$, $N > N_1$, then $\|P^{M,N}\phi - \phi\|_{L^1} < \varepsilon$).

The proof of (3.84) follows from the uniform continuity of $\phi$ on $[0,\infty)$, since for positive integers $M > M_1$, $N > N_1$,

$$\|P^{M,N}\phi - \phi\|_{L^1} \leq \int_0^{M_1} |(P^{M,N}\phi)(a) - \phi(a)| \, da$$
$$+ \int_{M_1}^{M} |(P^{M,N}\phi)(a)| \, da + \int_{M_1}^{\infty} |\phi(a)| \, da$$
$$\leq \int_0^{M_1} \left| \sum_{k=1}^{M_1N} \chi_k^{M,N}(a) \left( N \int_{(k-1)/N}^{k/N} \phi(b) \, db \right) - \phi(a) \right| da$$
$$+ \int_{M_1}^{M} \sum_{k=M_1N+1}^{MN} \chi_k^{M,N}(a) \left( N \int_{(k-1)/N}^{k/N} |\phi(b)| \, db \right) da + \int_{M_1}^{\infty} |\phi(a)| \, da$$
$$= \sum_{j=1}^{M_1N} \int_{(j-1)/N}^{j/N} \left| \sum_{k=1}^{M_1N} \chi_k^{M,N}(a) \left( N \int_{(k-1)/N}^{k/N} \phi(b) \, db \right) - \phi(a) \right| da$$
$$+ \sum_{k=M_1N+1}^{MN} \frac{1}{N} \left( N \int_{(k-1)/N}^{k/N} |\phi(b)| \, db \right) + \int_{M_1}^{\infty} |\phi(a)| \, da$$
$$= \sum_{j=1}^{M_1N} \int_{(j-1)/N}^{j/N} \left| \chi_j^{M,N}(a) \left( N \int_{(j-1)/N}^{j/N} \phi(b) \, db \right) - \phi(a) \right| da$$

$$+ \sum_{k=M_1N+1}^{MN} \int_{(k-1)/N}^{k/N} |\phi(b)|\, db + \int_{M_1}^{\infty} |\phi(a)|\, da$$

$$= \sum_{j=1}^{M_1N} \int_{(j-1)/N}^{j/N} \chi_j^{M,N}(a) \left| N \int_{(j-1)/N}^{j/N} [\phi(b) - \phi(a)]\, db \right| da$$

$$+ \int_{M_1}^{M} |\phi(b)|\, db + \int_{M_1}^{\infty} |\phi(a)|\, da$$

$$\leq \sum_{j=1}^{M_1N} \int_{(j-1)/N}^{j/N} \chi_j^{M,N}(a) \Big( \sup_{(j-1)/N \leq b, c \leq j/N} |\phi(b) - \phi(c)| \Big)\, da$$

$$+ 2 \int_{M_1}^{\infty} |\phi(a)|\, da$$

$$\leq \Big( \sup_{b, c \geq 0, |b-c| \leq 1/N_1} |\phi(b) - \phi(c)| \Big) M_1 + 2 \int_{M_1}^{\infty} |\phi(a)|\, da$$

For a given $\phi \in L^1$, $P^{M,N}\phi$ is the averaging approximation step function with step width $1/N$ for the truncation of $\phi$ to $[0,M]$. Now suppose that (2.1), (2.2), (2.22), (2.23) hold, $S(t)$, $t \geq 0$, is the strongly continuous nonlinear semigroup in $L^1_+$ associated with (ADP) as in Theorem 3.1, and $B$ is its infinitesimal generator as in Theorem 3.2. Corresponding to these averaging approximations, we require that

(3.85) For $M$ and $N$ positive integers, let $F^{M,N}\colon X^{M,N} \to R^n$ such that for some constant $c_1$ (independent of $M$ and $N$), $|F^{M,N}(\psi) - F^{M,N}(\hat{\psi})| \leq c_1 \|\psi - \hat{\psi}\|_{L^1}$ for $\psi, \hat{\psi} \in X^{M,N}$, and for each $\phi \in D(B)$, $\lim_{M,N\to\infty} F^{M,N}(P^{M,N}\phi) = F(\phi)$.

(3.86) For $M$ and $N$ positive integers, let $G^{M,N}\colon X^{M,N} \to X^{M,N}$ such that for some constant $c_2$ (independent of $M$ and $N$), $\|G^{M,N}(\psi) - G^{M,N}(\hat{\psi})\|_{L^1} \leq c_2 \|\psi - \hat{\psi}\|_{L^1}$ for $\psi, \hat{\psi} \in X^{M,N}$, and for each $\phi \in D(B)$, $\lim_{M,N\to\infty} G^{M,N}(P^{M,N}\phi) = G(\phi)$.

(3.87) For M and N positive integers, let $B^{M,N}: X^{M,N} \to X^{M,N}$ such that for $\phi \in X^{M,N}$, $B^{M,N}\phi \overset{def}{=} -\Sigma_{k=1}^{MN} \chi_k^{M,N} N(h_k - h_{k-1}) + G^{M,N}(\phi)$, where $\phi = \Sigma_{k=1}^{MN} \chi_k^{M,N} h_k$ and $h_0 \overset{def}{=} F^{M,N}(\phi)$.

THEOREM 3.6 Let (2.1), (2.2), (2.22), (2.23), (3.82), (3.85), (3.86), (3.87) hold, let $T_\phi = \infty$ for all $\phi \in L_+^1$, let $S(t)$, $t \geq 0$, be the strongly continuous nonlinear semigroup associated with (ADP) as in Theorem 3.1, let B be its infinitesimal generator as in Theorem 3.2, let $\phi \in D(B)$, let $T > 0$, let the function $t \to S(t)\phi$ be continuously differentiable in t from $[0,T]$ to $L^1$, and let $\{\tau_n\}$ be a sequence of positive numbers such that $\tau_n \leq 1/N$. Then, uniformly for $t \in [0,T]$,

$$\lim_{M,N\to\infty} \|(I + \tau_n B^{M,N})^{[t/\tau_n]} P^{M,N}\phi - S(t)\phi\|_{L^1} = 0 \tag{3.88}$$

Before giving the proof of Theorem 3.6 we first prove two lemmas, each under the hypothesis of the theorem.

LEMMA 3.1 (Stability) Let M and N be positive integers, let $\psi$, $\hat{\psi} \in X^{M,N}$, and let $\lambda > 0$ such that $\lambda N \leq 1$. Then,

$$\|(I + \lambda B^{M,N})\psi - (I + \lambda B^{M,N})\hat{\psi}\|_{L^1} \leq [1 + \lambda(c_1 + c_2)]\|\psi - \hat{\psi}\|_{L^1} \tag{3.89}$$

*Proof.* From (3.85), (3.86), (3.87), we have that for $\psi = \Sigma_{k=1}^{MN} \chi_k^{M,N} h_k$, $\hat{\psi} = \Sigma_{k=1}^{MN} \chi_k^{M,N} \hat{h}_k$,

$$\begin{aligned}
&\|(I + \lambda B^{M,N})\psi - (I + \lambda B^{M,N})\hat{\psi}\|_{L^1} \\
&= \Big\|\sum_{k=1}^{MN} \chi_k^{M,N}\{(1 - \lambda N)(h_k - \hat{h}_k) + \lambda N(h_{k-1} - \hat{h}_{k-1})\} \\
&\quad + \lambda[G^{M,N}(\psi) - G^{M,N}(\hat{\psi})]\Big\|_{L^1} \\
&\leq \sum_{k=1}^{MN} \|\chi_k^{M,N}\|_{L^1} |(1 - \lambda N)(h_k - \hat{h}_k) + \lambda N(h_{k-1} - \hat{h}_{k-1})| \\
&\quad + \lambda c_2\|\psi - \hat{\psi}\|_{L^1}
\end{aligned}$$

$$\leq \sum_{k=1}^{MN} \frac{1}{N} [(1 - \lambda N)|h_k - \hat{h}_k| + \lambda N|h_{k-1} - \hat{h}_{k-1}|] + \lambda c_2 \|\psi - \hat{\psi}\|_{L^1}$$

$$= \sum_{k=1}^{MN} \frac{1}{N} |h_k - \hat{h}_k| + \lambda|h_0 - \hat{h}_0| + \lambda c_2 \|\psi - \hat{\psi}\|_{L^1}$$

$$\leq \|\psi - \hat{\psi}\|_{L^1} + \lambda(c_1 + c_2)\|\psi - \hat{\psi}\|_{L^1} \quad \square$$

LEMMA 3.2 (Consistency) Let M and N be positive integers. Then,

$$\begin{aligned}(3.90)\qquad & \|P^{M,N}BS(t)\phi - B^{M,N}P^{M,N}S(t)\phi\|_{L^1} \\ & \leq \int_{1/N}^{M} \left| \left(\frac{d}{da} S(t)\phi\right)(a) - \left(\frac{d}{da} S(t)\phi\right)\left(a - \frac{1}{N}\right) \right| da \\ & + \int_0^{1/N} \left| \left(\frac{d}{da} S(t)\phi\right)(a) \right| da + |F(S(t)\phi) - F^{M,N}(P^{M,N}S(t)\phi)| \\ & + \|G(S(t)\phi) - G^{M,N}(P^{M,N}S(t)\phi)\|_{L^1}\end{aligned}$$

*Proof*. Let $\psi = S(t)\phi$, let $h_0 = F^{M,N}(P^{M,N}\psi)$, and for $k = 1, \ldots, MN$, let

$$h_k = N \int_{(k-1)/N}^{k/N} \psi(a)\, da$$

$$h_k' = N \int_{(k-1)/N}^{k/N} \psi'(a)\, da$$

$$= N\left\{\psi\left(\frac{k}{N}\right) - \psi\left(\frac{k - 1}{N}\right)\right\}$$

Recall from Remark 3.3 that $S(t)\phi \in D(B)$ and (3.23) holds. From (3.82) and (3.87) we have that

$$\|P^{M,N}B\psi - B^{M,N}P^{M,N}\psi\|_{L^1}$$

$$= \int_0^M \left| \sum_{k=1}^{MN} \chi_k^{M,N}(a)[-h_k' + N(h_k - h_{k-1})] + P^{M,N}G(\psi)(a) - G^{M,N}(P^{M,N}\psi)(a) \right| da$$

$$= \int_0^M \left| \sum_{k=1}^{MN} \chi_k^{M,N}(a) N\left[\psi\left(\frac{k}{N}\right) - \psi\left(\frac{k-1}{N}\right) - (h_k - h_{k-1})\right] \right.$$

$$\left. + P^{M,N}[G^{M,N}(P^{M,N}\psi)(a) - G(\psi)(a)] \right| da$$

$$\leq \left| \psi\left(\frac{1}{N}\right) - \psi(0) - N \int_0^{1/N} \psi(a)\ da - F^{M,N}(P^{M,N}\psi) \right|$$

$$+ \sum_{k=2}^{MN} \left| \psi\left(\frac{k}{N}\right) - \psi\left(\frac{k-1}{N}\right) - N\left(\int_{(k-1)/N}^{k/N} \psi(a)\ da - \int_{(k-2)/N}^{(k-1)/N} \psi(a)\ da\right) \right|$$

$$+ \|G^{M,N}(P^{M,N}\psi) - G(\psi)\|_{L^1}$$

Observe that

$$\left| \psi\left(\frac{1}{N}\right) - N \int_0^{1/N} \psi(a)\ da \right| \leq N \int_0^{1/N} \left| \psi\left(\frac{1}{N}\right) - \psi(a) \right| da$$

$$= N \int_0^{1/N} \left| \int_a^{1/N} \psi'(b)\ db \right| da$$

$$\leq N \int_0^{1/N} \int_0^b |\psi'(b)|\ da\ db$$

$$\leq \int_0^{1/N} |\psi'(b)|\ db$$

Also, since $\psi \in D(B)$,

$$|\psi(0) - F^{M,N}(P^{M,N}\psi)| = |F(\psi) - F^{M,N}(P^{M,N}\psi)|$$

Last, for $k = 2, \ldots, MN$, an integration by parts yields that

$$\left| \psi\left(\frac{k}{N}\right) - \psi\left(\frac{k-1}{N}\right) - N \int_{(k-1)/N}^{k/N} \left[\psi(a) - \psi\left(a - \frac{1}{N}\right)\right] da \right|$$

$$= \left| \psi\left(\frac{k}{N}\right) - \psi\left(\frac{k-1}{N}\right) + N\left[ \int_{(k-1)/N}^{k/N} \left(a - \frac{k-1}{N}\right) \right. \right.$$

$$\left. \left. \times \left(\psi'(a) - \psi'\left(a - \frac{1}{N}\right)\right) da - \frac{1}{N}\left(\psi\left(\frac{k}{N}\right) - \psi\left(\frac{k-1}{N}\right)\right)\right] \right|$$

$$= N \left| \int_{(k-1)/N}^{k/N} \left(a - \frac{k-1}{N}\right) \left(\psi'(a) - \psi'\left(a - \frac{1}{N}\right)\right) \right| da$$

$$\leq \int_{(k-1)/N}^{k/N} \left| \psi'(a) - \psi'\left(a - \frac{1}{N}\right) \right| da$$

The claim (3.90) now follows immediately. □

*Proof of Theorem 3.6.* Let $\varepsilon > 0$ and let $r > 0$ such that $\sup_{0 \leq t \leq T} \|S(t)\phi\|_{L^1} \leq r$. Since the function $t \to S(t)\phi$ is continuously differentiable from $[0,T]$ to $L^1$, there exists a finite set of points $\{t_1, \ldots, t_j\} \subset [0,T]$ such that if $t \in [0,T]$, then

$$\|S(t_k)\phi - S(t)\phi\|_{L^1} < \varepsilon$$

for some $k = 1, \ldots, j$. By (3.85) there exists $M_1$ and $N_1$ such that if $M > M_1$ and $N > N_1$, then

$$|F(S(t_k)\phi) - F^{M,N}(P^{M,N}S(t_k)\phi)| < \varepsilon$$

for each $k = 1, \ldots, j$. Thus, for $M > M_1$, $N > N_1$, and $t \in [0,T]$, there exists some $k = 1, \ldots, j$ such that

$$\begin{aligned}
&|F(S(t)\phi) - F^{M,N}(P^{M,N}S(t)\phi)| \\
&\leq |F(S(t)\phi) - F(S(t_k)\phi)| \\
&\quad + |F(S(t_k)\phi) - F^{M,N}(P^{M,N}S(t_k)\phi)| \\
&\quad + |F^{M,N}(P^{M,N}S(t_k)\phi) - F^{M,N}(P^{M,N}S(t)\phi)| \\
&< c_1(r)\varepsilon + \varepsilon + c_1\varepsilon
\end{aligned}$$

A similar argument using (3.86) shows that $M_1$ and $N_1$ can be chosen such that for $M > M_1$, $N > N_1$, $t \in [0,T]$,

$$\|G(S(t)\phi) - G^{M,N}(P^{M,N}S(t)\phi)\|_{L^1} < c_2(r)\varepsilon + \varepsilon + \varepsilon c_2$$

Since the function $t \to S(t)\phi$ is continuously differentiable from $[0,T]$ to $L^1$, the set $\{BS(t)\phi : 0 \leq t \leq T\}$ is compact in $L^1$ (see Remark 3.3). Consequently, the set $\{(d/da)(S(t)\phi) : 0 \leq t \leq T\}$

is compact in $L^1$. By Lemma 2.3 there exists $N_2$ such that for $M > M_1$, $N > N_2$, $t \in [0,T]$,

$$\int_{1/N}^{M} \left| \left(\frac{d}{da} S(t)\phi\right)(a) - \left(\frac{d}{da} S(t)\phi\right)\left(a - \frac{1}{N}\right) \right| da < \varepsilon$$

and

$$\int_{0}^{1/N} \left| \left(\frac{d}{da} S(t)\phi\right)(a) \right| da < \varepsilon$$

By Lemma 3.2 we have that for $M > M_1$, $N > \max\{N_1, N_2\}$, $t \in [0,T]$,

$$\|P^{M,N}BS(t)\phi - B^{M,N}P^{M,N}S(t)\phi\|_{L^1}$$
$$< [4 + c_1 + c_2 + c_1(r) + c_2(r)]\varepsilon$$

Let $\{\tau_N\}$ be a sequence of positive numbers such that $\tau_N \leq 1/N$. Since the function $t \to S(t)\phi$ is continuously differentiable on $[0,T]$ and satisfies (3.23), there exists $N_3$ such that if $N \geq N_3$, $t \in [0,T]$, and $k = 1, \ldots, [t/\tau_N]$, then

$$\|S(k\tau_N)\phi - S((k - 1)\tau_N)\phi - \tau_N BS((k - 1)\tau_N)\phi\|_{L^1}$$
$$= \left\| \int_{(k-1)\tau_N}^{k\tau_N} \left[\frac{d}{dt} S(t)\phi - \frac{d}{dt} S((k - 1)\tau_N)\phi\right] dt \right\|_{L^1} < \tau_N\varepsilon$$

Then, for $M > M_1$, $N > \max\{N_1, N_2, N_3\}$, $t \in [0,T]$, $m = [t/\tau_N]$, $\omega = c_1 + c_2$, Lemma 3.2 yields that

$$\|(I + \tau_N B^{M,N})^m P^{M,N}\phi - P^{M,N}S(m\tau_N)\phi\|_{L^1}$$
$$= \left\| \sum_{k=1}^{m} (I + \tau_N B)^{m-k+1} P^{M,N} S(\tau_N)^{k-1}\phi \right.$$
$$\left. - (I + \tau_N B^{M,N})^{m-k} P^{M,N} S(\tau_N)^k \phi \right\|_{L^1}$$
$$\leq \sum_{k=1}^{m} (1 + \tau_N\omega)^{m-k} \|(I + \tau_N B^{M,N})P^{M,N}S((k - 1)\tau_N)\phi - P^{M,N}S(k\tau_N)\phi\|_{L^1}$$

$$\le \sum_{k=1}^{m} (1 + \tau_N\omega)^{m-k}[\|P^{M,N}S((k-1)\tau_N)\phi - P^{M,N}S(k\tau_N)\phi$$

$$+ \tau_N P^{M,N}BS((k-1)\tau_N)\phi\|_{L^1}$$

$$+ \tau_N\|P^{M,N}BS((k-1)\tau_N)\phi - B^{M,N}P^{M,N}S((k-1)\tau_N)\phi\|_{L^1}]$$

$$< e^{\omega t}\sum_{k=1}^{m} \tau_N[5 + c_1 + c_2 + c_1(r) + c_2(r)]\varepsilon$$

$$\le e^{\omega t}t[5 + c_1 + c_2 + c_1(r) + c_2(r)]$$

The conclusion (3.88) now follows using (3.83), (3.84), and the uniform continuity of the function $t \to S(t)\phi$ from $[0,T]$ to $L^1$. □

REMARK 3.7 We remark that for a given $\phi \in D(B)$, a sufficient condition for the function $t \to S(t)\phi$ to be continuously differentiable from $[0,T]$ to $L^1$ is given by Theorem 2.10 (that is, F and G are continuously Fréchet differentiable). We also note that conditions (3.85) and (3.86) require uniformly global Lipschitz continuity conditions on $F^{M,N}$ and $G^{M,N}$, respectively. If F and G are only locally Lipschitz continuous, then a truncation device as in Proposition 3.10 and Theorem 3.3 may be required before applying Theorem 3.6.

For the finite difference approximation scheme we must assume more regularity of the solutions of (ADP). We define the following Banach spaces: for $\sigma \ge 0$, let $C_\sigma$ denote the Banach space of continuous functions $\phi$ from $[0,\infty)$ to $R^n$ such that $\sup_{a\ge 0} e^{\sigma a}|\phi(a)| < \infty$ with norm $\|\phi\|_\sigma \overset{\text{def}}{=} \sup_{a\ge 0} e^{\sigma a}|\phi(a)|$. We define

(3.91) For $\sigma \ge 0$ and M and N positive integers, let $X_\sigma^{M,N}$ denote the set of all piecewise linear continuous functions $\phi$ from $[1/N,M]$ to $R^n$ with knots at $k/N$, $k = 1, \ldots, MN$, where the norm of $X_\sigma^{M,N}$ is given by $\|\phi\|_\sigma^{M,N} \overset{\text{def}}{=} \sup_{1\le k\le MN} e^{\sigma k/N}|\phi(k/N)|$, and let $P^{M,N}\colon C_\sigma \to X_\sigma^{M,N}$ be the bounded linear operator defined by $(P^{M,N}\phi)(k/N) = \phi(k/N)$, $\phi \in C_\sigma$, $k = 1, \ldots, MN$.

We observe that

(3.92) $\|\phi\|_{L^1} \leq \|\phi\|_\sigma/\sigma$, $\phi \in C_\sigma$, $\sigma > 0$

(3.93) $\|P^{M,N}\phi\|_\sigma^{M,N} \leq \|\phi\|_\sigma$, $\phi \in C_\sigma$, $\sigma \geq 0$

(3.94) $\lim_{M\to\infty}[\lim_{N\to\infty}\{\sup_{1/N\leq a\leq M}|(P^{M,N}\phi)(a) - \phi(a)|\}] = 0$, $\phi \in C_\sigma$, $\sigma > 0$, in the sense that if $\varepsilon > 0$ there exists $M_1$ such that if $M > M_1$ then there exists $N_1 = N_1(M)$ (where $N_1$ depends on M) such that if $N > N_1(M)$, then $\sup_{1/N\leq a\leq M}$ $|(P^{M,N}\phi)(a) - \phi(a)| < \varepsilon$.

We now suppose that (2.1), (2.2), (2.22), (2.23) hold and $S(t)$, $t \geq 0$, is the strongly continuous nonlinear semigroup in $L^1_+$ associated with (ADP) as in Theorem 3.1 with infinitesimal generator B as in Theorem 3.2. We require that

(3.95) There exists $d_1 > 0$ such that for all $\phi$, $\hat{\phi} \in C_0 \cap L^1$, $|F(\phi) - F(\hat{\phi})| \leq d_1\|\phi - \hat{\phi}\|_{L^1}$.

(3.96) There exists $\sigma > d_1$ such that $G$: $C_\sigma \to C_\sigma$ and there exists $d_2 > 0$ such that for all $\phi$, $\hat{\phi} \in C_\sigma$, $\|G(\phi) - G(\hat{\phi})\|_\sigma \leq d_2\|\phi - \hat{\phi}\|_\sigma$.

(3.97) For M and N positive integers, let $F^{M,N}$: $X_\sigma^{M,N} \to R^n$ such that $|F^{M,N}(\psi) - F^{M,N}(\hat{\psi})| \leq (d_1/\sigma)\|\psi - \hat{\psi}\|_\sigma^{M,N}$ for all $\psi$, $\hat{\psi} \in X_\sigma^{M,N}$, and for $\phi \in D(B) \cap C_\sigma$, $T > 0$, $\lim_{M\to\infty}[\lim_{N\to\infty} N|F^{M,N}(P^{M,N}S(t)\phi) - F(S(t)\phi)|] = 0$ uniformly for $t \in [0,T]$.

(3.98) For M and N positive integers, let $G^{M,N}$: $X_\sigma^{M,N} \to X_\sigma^{M,N}$ such that $\|G^{M,N}(\psi) - G^{M,N}(\hat{\psi})\|_\sigma^{M,N} \leq d_2\|\psi - \hat{\psi}\|_\sigma^{M,N}$ for all $\psi$, $\hat{\psi} \in X_\sigma^{M,N}$, and for $\phi \in D(B) \cap C_\sigma$, $\lim_{M\to\infty}[\lim_{N\to\infty}\|P^{M,N}G(\phi) - G^{M,N}(P^{M,N}\phi)\|_\sigma^{M,N}] = 0$.

(3.99) For M and N positive integers, let $B^{M,N}$: $X_\sigma^{M,N} \to X_\sigma^{M,N}$ such that for $\phi \in X_\sigma^{M,N}$, $k = 2, \ldots, MN$, $(B^{M,N}\phi)(k/N) \overset{def}{=}$

$-N[\phi(k/N) - \phi((k-1)/N)] + G^{M,N}(\phi)(k/N)$ and $(B^{M,N}\phi)(1/N) \overset{\text{def}}{=} -N[\phi(1/N) - F^{M,N}(\phi)] + G^{M,N}(\phi)(1/N)$.

Notice that the definition of $B^{M,N}$ involves a left-sided difference approximation to the first derivative operator. For this approximation scheme we will prove

THEOREM 3.7 Let (2.1), (2.2), (2.22), (2.23), (3.91), (3.95), (3.96), (3.97), (3.98), (3.99) hold, let $T_\phi = \infty$ for all $\phi \in L^1_+$, let $S(t)$, $t \geq 0$, be the strongly continuous nonlinear semigroup associated with (ADP) as in Theorem 3.1, let B be its infinitesimal generator as in Theorem 3.2, let $\phi \in D(B) \cap C_\sigma$, let $T > 0$, let the function $t \to S(t)\phi$ be continuously differentiable from $[0,T]$ to $C_\sigma$ and satisfy $(d/dt) S(t)\phi = BS(t)\phi$ for $t \in [0,T]$, and let $\{\tau_N\}$ be a sequence of positive numbers such that $\tau_N \leq 1/N$. Then, uniformly for $t \in [0,T]$,

$$\text{(3.100)} \qquad \lim_{M\to\infty} [\lim_{N\to\infty} \|(I + \tau_N B^{M,N})^{[t/\tau_N]} P^{M,N}\phi - P^{M,N}S(t)\phi\|_\sigma^{M,N}] = 0$$

Before proving Theorem 3.7 we first prove two lemmas, each of which is under the hypothesis of the theorem.

LEMMA 3.3 (Stability) Let $\omega > \sigma + d_2$, let M and N be positive integers, let N be sufficiently large such that

$$\text{(3.101)} \qquad N(e^{\sigma/N} - 1) < \omega - d_2 \quad \text{and} \quad e^{\sigma/N} < \frac{\sigma}{d_1}$$

let $\psi, \hat{\psi} \in X_\sigma^{M,N}$, and let $\lambda > 0$ such that $\lambda N \leq 1$. Then,

$$\text{(3.102)} \qquad \|(I + \lambda B^{M,N})\psi - (I + \lambda B^{M,N})\hat{\psi}\|_\sigma^{M,N} \leq (1 + \lambda\omega)\|\psi - \hat{\psi}\|_\sigma^{M,N}$$

*Proof.* From (3.97), (3.98), (3.99), and (3.101) we have that for $k = 2, \ldots, MN$,

$$e^{\sigma k/N}\left|(I + \lambda B^{M,N})\psi\left(\frac{k}{N}\right) - (I + \lambda B^{M,N})\hat{\psi}\left(\frac{k}{N}\right)\right|$$
$$= e^{\sigma k/N}\left|(1 - \lambda N)\left[\psi\left(\frac{k}{N}\right) - \hat{\psi}\left(\frac{k}{N}\right)\right] + \lambda N\left[\psi\left(\frac{k-1}{N}\right) - \hat{\psi}\left(\frac{k-1}{N}\right)\right]\right.$$
$$\left. + \lambda\left[G^{M,N}(\psi)\left(\frac{k}{N}\right) - G^{M,N}(\hat{\psi})\left(\frac{k}{N}\right)\right]\right|$$

$$\le (1 - \lambda N)e^{\sigma k/N}\left|\psi\left(\frac{k}{N}\right) - \hat{\psi}\left(\frac{k}{N}\right)\right|$$

$$+ e^{\sigma k/N}\lambda N e^{-\sigma(k-1)/N} e^{\sigma(k-1)/N}\left|\psi\left(\frac{k-1}{N}\right) - \hat{\psi}\left(\frac{k-1}{N}\right)\right|$$

$$+ e^{\sigma k/N}\lambda\left|G^{M,N}(\psi)\left(\frac{k}{N}\right) - G^{M,N}(\hat{\psi})\left(\frac{k}{N}\right)\right|$$

$$\le (1 - \lambda N)\|\psi - \hat{\psi}\|_{\sigma}^{M,N} + e^{\sigma/N}\lambda N\|\psi - \hat{\psi}\|_{\sigma}^{M,N}$$

$$+ \lambda\|G^{M,N}(\psi) - G^{M,N}(\hat{\psi})\|_{\sigma}^{M,N}$$

$$\le \{[1 + \lambda N(e^{\sigma/N} - 1)] + \lambda d_2\}\|\psi - \hat{\psi}\|_{\sigma}^{M,N}$$

$$\le (1 + \lambda\omega)\|\psi - \hat{\psi}\|_{\sigma}^{M,N}$$

For $k = 1$ we have that

$$e^{\sigma/N}\left|(I + \lambda B^{M,N})\psi\left(\frac{1}{N}\right) - (I + \lambda B^{M,N})\hat{\psi}\left(\frac{1}{N}\right)\right|$$

$$= e^{\sigma/N}\left|(1 - \lambda N)\left[\psi\left(\frac{1}{N}\right) - \hat{\psi}\left(\frac{1}{N}\right)\right] + \lambda N[F^{M,N}(\psi) - F^{M,N}(\hat{\psi})]\right.$$

$$\left. + \lambda\left[G^{M,N}(\psi)\left(\frac{1}{N}\right) - G^{M,N}(\hat{\psi})\left(\frac{1}{N}\right)\right]\right|$$

$$\le (1 - \lambda N)\|\psi - \hat{\psi}\|_{\sigma}^{M,N} + \left(\lambda N e^{\sigma/N}\frac{d_1}{\sigma}\right)\|\psi - \hat{\psi}\|_{\sigma}^{M,N} + \lambda d_2\|\psi - \hat{\psi}\|_{\sigma}^{M,N}$$

$$\le (1 + \lambda\omega)\|\psi - \hat{\psi}\|_{\sigma}^{M,N} \qquad \square$$

LEMMA 3.4 (Consistency) Let M and N be positive integers. Then,

$$(3.103) \qquad \|P^{M,N}BS(t)\phi - B^{M,N}P^{M,N}S(t)\phi\|_{\sigma}^{M,N}$$

$$\le \sup_{k=1,\ldots,MN} e^{\sigma k/N}\left[\sup_{(k-1)/N\le a\le k/N}\left|\left(\frac{d}{da}S(t)\phi\right)(a)\right.\right.$$

$$\left.\left. - \left(\frac{d}{da}S(t)\phi\right)\left(\frac{k}{N}\right)\right|\right] + e^{\sigma/N}N|F(S(t)\phi) - F^{M,N}(P^{M,N}S(t)\phi)|$$

$$+ \|P^{M,N}G(S(t)\phi) - G^{M,N}(P^{M,N}S(t)\phi)\|_{\sigma}^{M,N}$$

*Proof.* Let $\psi = S(t)\phi$. Observe from the hypothesis that $\psi \in C_\sigma$ and $B\psi = -\psi' + G(\psi) \in C_\sigma$, which implies by (3.96) that $\psi' \in C_\sigma$. From (3.91) and (3.99) we have that for $k = 2, \ldots, MN$,

$$e^{\sigma k/N}\left|(P^{M,N}B\psi)\left(\frac{k}{N}\right) - (B^{M,N}P^{M,N}\psi)\left(\frac{k}{N}\right)\right|$$

$$= e^{\sigma k/N}\left|-\psi'\left(\frac{k}{N}\right) + G(\psi)\left(\frac{k}{N}\right) + N\left[\psi\left(\frac{k}{N}\right) - \psi\left(\frac{k-1}{N}\right)\right] - G^{M,N}(P^{M,N}\psi)\left(\frac{k}{N}\right)\right|$$

$$\leq e^{\sigma k/N}\left|N\int_{(k-1)/N}^{k/N}\left[\psi'(a) - \psi'\left(\frac{k}{N}\right)\right] da\right|$$

$$+ e^{\sigma k/N}\left|P^{M,N}G(\psi)\left(\frac{k}{N}\right) - G^{M,N}(P^{M,N}\psi)\left(\frac{k}{N}\right)\right|$$

$$\leq e^{\sigma k/N}\sup_{(k-1)/N\leq a\leq k/N}\left|\psi'(a) - \psi'\left(\frac{k}{N}\right)\right|$$

$$+ \|P^{M,N}G(\psi) - G^{M,N}(P^{M,N}\psi)\|_{\sigma}^{M,N}$$

Further, using the fact that $\psi(0) = F(\psi)$ [since $\psi \in D(B)$], we have that for $k = 1$,

$$e^{\sigma/N}\left|(P^{M,N}B\psi)\left(\frac{1}{N}\right) - (B^{M,N}P^{M,N}\psi)\left(\frac{1}{N}\right)\right|$$

$$= e^{\sigma/N}\left|-\psi'\left(\frac{1}{N}\right) + G(\psi)\left(\frac{1}{N}\right) + N\left[\psi\left(\frac{1}{N}\right) - F^{M,N}(P^{M,N}\psi)\right] - G^{M,N}(P^{M,N}\psi)\left(\frac{1}{N}\right)\right|$$

$$\leq e^{\sigma/N}\left|N\left[\psi\left(\frac{1}{N}\right) - \psi(0)\right] - \psi'\left(\frac{1}{N}\right)\right| + e^{\sigma/N}|N[F(\psi) - F^{M,N}(P^{M,N}\psi)]|$$

$$+ e^{\sigma/N}\left|P^{M,N}G(\psi)\left(\frac{1}{N}\right) - G^{M,N}(P^{M,N}\psi)\left(\frac{1}{N}\right)\right|$$

$$\leq e^{\sigma/N}\sup_{0\leq a\leq 1/N}\left|\psi'(a) - \psi'\left(\frac{1}{N}\right)\right| + e^{\sigma/N}N|F(\psi) - F^{M,N}(P^{M,N}\psi)|$$

$$+ \|P^{M,N}G(\psi) - G^{M,N}(P^{M,N}\psi)\|_{\sigma}^{M,N} \quad \square$$

*Proof of Theorem 3.7.* Let $\varepsilon > 0$. Since the function $t \to S(t)\phi$ is continuous from $[0,T]$ to $C_\sigma$, there is a finite set of points $\{t_1, \ldots, t_j\} \subset [0,T]$ such that if $t \in [0,T]$, then

$$\|S(t_k)\phi - S(t)\phi\|_\sigma < \varepsilon$$

for some $k = 1, \ldots, j$. By (3.98) there exists $M_1$ such that if $M > M_1$, then there exists $N_1 = N_1(M)$ (where $N_1$ depends on $M$) such that if $N > N_1(M)$, then

$$\|P^{M,N}G(S(t_k)\phi) - G^{M,N}(P^{M,N}S(t_k)\phi)\|_{\sigma}^{M,N} < \varepsilon$$

for all $k = 1, \ldots, j$. By (3.93), (3.96), (3.98) we have that for $M > M_1$, $N > N_1(M)$, and $t \in [0,T]$, there exists some $k = 1, \ldots, j$ such that

$$\begin{aligned}
&\|P^{M,N}G(S(t)\phi) - G^{M,N}(P^{M,N}S(t)\phi)\|_{\sigma}^{M,N} \\
&\leq \|P^{M,N}G(S(t)\phi) - P^{M,N}(S(t_k)\phi)\|_{\sigma}^{M,N} \\
&\quad + \|P^{M,N}G(S(t_k)\phi) - G^{M,N}(P^{M,N}S(t_k)\phi)\|_{\sigma}^{M,N} \\
&\quad + \|G^{M,N}(P^{M,N}S(t_k)\phi) - G^{M,N}(P^{M,N}S(t)\phi)\|_{\sigma}^{M,N} \\
&< d_2\varepsilon + \varepsilon + d_2\varepsilon
\end{aligned}$$

By (3.97) we can also choose $M_1$ such that for $M > M_1$, there exists $N_1 = N_1(M)$ (where $N_1$ depends on M) such that for $N > N_1(M)$, $t \in [0,T]$,

$$e^{\sigma/N}|F(S(t)\phi) - F^{M,N}(P^{M,N}S(t)\phi)| < \varepsilon$$

Let $M > M_1$. Since the function $t \to S(t)\phi$ is continuously differentiable from $[0,T]$ to $C_\sigma$ and $(d/dt)S(t)\phi = BS(t)\phi$ for $t \in [0,T]$ by hypothesis, the set $\{BS(t)\phi : 0 \leq t \leq T\}$ is compact in $C_\sigma$. Since G is continuous from $C_\sigma$ to $C_\sigma$ by (3.96) and $BS(t)\phi = (-d/da)S(t)\phi + G(S(t)\phi)$ by hypothesis, we obtain that the set $\{(d/da)S(t)\phi : 0 \leq t \leq T\}$ is compact in $C_\sigma$. By the Arzelà-Ascoli theorem (see [16], Theorem 26.7, p. 189) the set of functions $\{(d/da)\ S(t)\phi : 0 \leq t \leq T\}$ is uniformly equicontinuous on $[0,M]$. Thus, there exists $N_2 = N_2(M)$ (where $N_2$ depends on M) such that for $N > N_2(M)$,

$$\sup_{k=1,\ldots,MN} e^{\sigma k/N}\left[\sup_{(k-1)/N\leq a\leq k/N}\left|\left(\frac{d}{da}\,S(t)\phi\right)(a) - \left(\frac{d}{da}\,S(t)\phi\right)\left(\frac{k}{N}\right)\right|\right] < \varepsilon$$

By Lemma 3.4 we have that for $M > M_1$, $N > \max\{N_1(M), N_2(M)\}$, $t \in [0,T]$,

$$\begin{aligned}
&\|P^{M,N}BS(t)\phi - B^{M,N}P^{M,N}S(t)\phi\|_{\sigma}^{M,N} \\
&< 2d_2\varepsilon + 3\varepsilon
\end{aligned}$$

Let $\{\tau_N\}$ be a sequence of positive numbers such that $\tau_N \le 1/N$. Since the function $t \to S(t)\phi$ is continuously differentiable from $[0,T]$ to $C_\sigma$, there exists $N_3 = N_3(M)$ (where $N_3$ depends on M) such that if $N \ge N_3(M)$, $t \in [0,T]$, $k = 1, \ldots, [t/\tau_N]$, then

$$\|S(k\tau_N)\phi - S((k-1)\tau_N)\phi - \tau_N BS((k-1)\tau_N)\phi\|_\sigma$$

$$= \left\| \int_{(k-1)\tau_N}^{k\tau_N} \left[\frac{d}{dt} S(t)\phi - \frac{d}{dt} S((k-1)\tau_N)\phi\right] dt \right\|_\sigma$$

$$< \tau_N \varepsilon$$

Then, for $M > M_1$, $N > \max\{N_1(M), N_2(M), N_3(M)\}$ such that N satisfies (3.101), and $t \in [0,T]$, Lemma 3.3 and an argument similar to the one in the proof of Theorem 3.6 yield that

$$\left\| (I + \tau_N B^{M,N})^{[t/\tau_N]} P^{M,N}\phi - P^{M,N} S\left(\left[\frac{t}{\tau_N}\right]\tau_N\right)\phi \right\|_\sigma^{M,N}$$

$$< e^{\omega t} t(4 + 2d_2)\varepsilon$$

The conclusion (3.100) now follows immediately using (3.93) and the uniform continuity of the function $t \to S(t)\phi$ from $[0,T]$ to $C_\sigma$. □

REMARK 3.8 The reason for weighting the norm with $e^{\sigma k/N}$ in the spaces $X_\sigma^{M,N}$ in Theorem 3.7 arises in order to establish the stability of the approximation scheme as in Lemma 3.4. If this weighting of the norm in $X_\sigma^{M,N}$ is not present, then it may be necessary to establish a stability condition of the form

$$\|(I + \lambda B^{M,N})\psi - (I + \lambda B^{M,N})\hat{\psi}\|_0^{M,N} \le K(1 + \lambda\omega)\|\psi - \hat{\psi}\|_0^{M,N}$$

where $K > 1$, rather than the stability condition (3.102). If each $F^{M,N}$ satisfies a global Lipschitz continuity condition of nonexpansive form

$$|F^{M,N}(\psi) - F^{M,N}(\hat{\psi})| \le \|\psi - \hat{\psi}\|_0^{M,N}$$

then this weighting of the norm is not necessary (that is, we may take $\sigma = 0$), as may be seen from the proof of Lemma 3.3. The hypothesis in (3.97) that

$$\lim_{M\to\infty} [|F^{M,N}(P^{M,N}S(t)\phi) - F(S(t)\phi)| = o(1/N)]$$

uniformly for $t \in [0,T]$ requires a rapidly convergent sequence of approximations $F^{M,N}$ for F. The hypothesis of Theorem 3.7 requires global Lipschitz continuity conditions on the birth function F and the aging function G. If F and G are only locally Lipschitz continuous on $C_\sigma$, then a truncation device as in Proposition 3.10 and Theorem 3.3 may be used with Theorem 3.7.

We illustrate the numerical approximation results of Theorems 3.6 and 3.7 with the following example:

EXAMPLE 3.4 Let $R^n = R$. Let F and G be as in Examples 2.6 and 3.3 with

$$F\phi = 12 \int_0^\infty ae^{-2a}\phi(a)\, da \qquad \phi \in L^1_+$$

$$G(\phi)(a) = -(1 + P\phi)\phi(a) \qquad \phi \in L^1_+,\ \text{a.e. } a > 0$$

$$P\phi = \int_0^\infty \phi(a)\, da \qquad \phi \in L^1$$

For this birth function F and aging (mortality) function G the problem (ADP) has the form

$$D\ell(a,t) = -(1 + P\ell(\cdot,t))\ell(a,t)$$

$$\ell(0,t) = 12 \int_0^\infty ae^{-2a}\ell(a,t)\, da$$

$$\ell(a,0) = \phi(a)$$

Both the averaging approximation scheme of Theorem 3.6 and the finite difference approximating scheme of Theorem 3.7 were applied to this example. Three choices of the initial age distribution

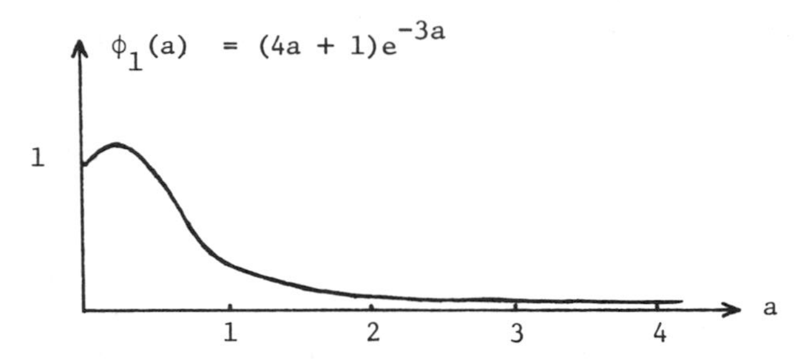


*Figure 3.1*

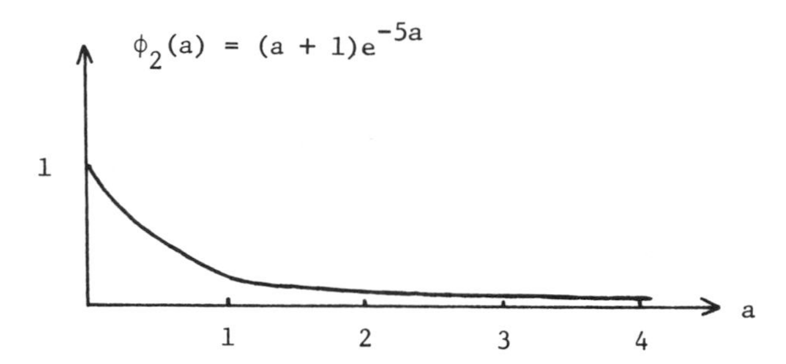


*Figure 3.2*

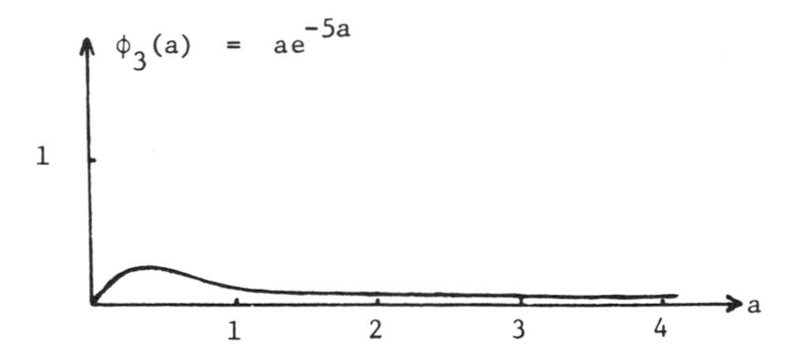


*Figure 3.3*

were selected: $\phi_1(a) = (4a + 1)e^{-3a}$, $a > 0$; $\phi_2(a) = (a + 1)e^{-5a}$, $a > 0$; and $\phi_3(a) = ae^{-5a}$, $a > 0$ (see Figures 3.1, 3.2, and 3.3).

Let $\ell_i(a,t)$ be the solution of (ADP) corresponding to the initial age distribution $\phi_i(a)$, $i = 1, 2, 3$, and let $P_i(t) \overset{\text{def}}{=} P\ell_i(\cdot,t)$ be the total population at time $t$, $i = 1, 2, 3$. In Section 5.4 it will be shown that for each i = 1, 2, 3,

$$\lim_{t\to\infty} P_i(t) = \sqrt{12} - 3 \overset{\text{def}}{=} \lambda_1 \sim .4641$$

$$\lim_{t\to\infty} \ell_i(a,t) = \lambda_1(1 + \lambda_1) \exp[-(1 + \lambda_1)a] \overset{\text{def}}{=} \hat{\phi}(a) \qquad a \geq 0$$

(in fact, the solutions of this (ADP) converge to the equilibrium solution $\hat{\phi}$ for any nonzero initial age distribution in $L_+^1$).

For the averaging approximation scheme of Theorem 3.6 the results for N = 300 and M = 4 are given in Table 3.1 below. For the

*TABLE 3.1* Averaging Approximation Scheme

| t | $P_1(t)$ | $P_2(t)$ | $P_3(t)$ |
|---|---|---|---|
| 0 | .7778 | .2400 | .0400 |
| 1 | .6526 | .3335 | .0703 |
| 2 | .5743 | .3893 | .1062 |
| 3 | .5299 | .4191 | .1055 |
| 4 | .5038 | .4384 | .2016 |
| 5 | .4900 | .4475 | .2565 |
| 6 | .4814 | .4549 | .3089 |
| 7 | .4761 | .4596 | .3540 |
| 8 | .4729 | .4626 | .3895 |
| 9 | .4709 | .4645 | .4156 |
| 10 | .4697 | .4657 | .4337 |
| 11 | .4689 | .4664 | .4459 |
| 12 | .4684 | .4669 | .4538 |
| 13 | .4681 | .4672 | .4589 |
| 14 | .4679 | .4673 | .4621 |
| 15 | .4678 | .4674 | .4642 |
| 16 | .4678 | .4675 | .4655 |
| 17 | .4677 | .4676 | .4663 |
| 18 | .4677 | .4676 | .4668 |
| 19 | .4677 | .4676 | .4671 |
| 20 | .4677 | .4676 | .4673 |

TABLE 3.2 Finite Difference Approximation Scheme

| a | $\ell_1(a,20)$ | $\hat{\phi}(a)$ |
|---|---|---|
| 0.25 | .4774 | .4712 |
| 0.50 | .3309 | .3268 |
| 0.75 | .2296 | .2266 |
| 1.00 | .1592 | .1572 |
| 1.25 | .1104 | .1090 |
| 1.50 | .0766 | .0756 |
| 1.75 | .0531 | .0524 |
| 2.00 | .0368 | .0363 |
| 2.25 | .0255 | .0252 |
| 2.50 | .0177 | .0175 |
| 2.75 | .0123 | .0121 |
| 3.00 | .0085 | .0084 |
| 3.25 | .0059 | .0058 |
| 3.50 | .0041 | .0040 |
| 3.75 | .0028 | .0028 |
| 4.00 | .0020 | .0019 |

finite difference approximation scheme of Theorem 3.7 the computations were made for the solution corresponding to the initial age distribution $\phi_1$. For N = 300, M = 4, and t = 20, the approximation for the total population was $P_1(20) = .4675$. The convergence in Theorem 3.7 is guaranteed to be pointwise in age, uniformly on bounded intervals. The results at time t = 20, N = 300, M = 4, for this scheme are given in Table 3.2.

For both numerical schemes an analysis of the approximations for varying values of N revealed that the convergence was of order $\mathcal{O}(1/N)$.

## 3.6 NOTES

The semigroup approach to age-dependent population dynamics first appeared in the papers [301], [303], [308], [309] of the author and in the papers [240], [241], [242] of J. Prüss. The infinitesimal generator of the semigroup associated with age-dependent population dynamics is closely related to the infinitesimal generator of the semigroup associated with functional differential equations, in that both involve the first derivative operator with a boundary condition.

The formulation of a functional differential equation in a space of continuous functions is as follows:

$$\dot{x}(t) = F(x_t) \qquad t \geq 0 \qquad x_0 = \phi \in C$$

where $C \overset{\text{def}}{=} C(-r,0;R^n)$, $F\colon\ C \to R^n$, $x\colon\ [-r,\infty) \to R^n$, and $x_t \in C$ is defined by $x_t(\theta) \overset{\text{def}}{=} x(t+\theta)$, $-r \leq \theta \leq 0$. The semigroup associated with this problem is given by $S(t)$, $t \geq 0$, where $S(t)\colon\ C \to C$ is defined by $S(t)\phi \overset{\text{def}}{=} x_t$, $t \geq 0$, $\phi \in C$, and its infinitesimal generator is $B\colon\ C \to C$, where

$$B\phi \overset{\text{def}}{=} \phi', \ D(B) \overset{\text{def}}{=} \{\phi \in C\colon\ \phi' \in C,\ \phi'^{-}(0) = F(\phi)\}$$

The formulation of a functional differential equation in a space of Lebesgue integrable functions is as follows:

$$\dot{x}(t) = F(x_t) \qquad t \geq 0,\ x_0 = \phi \in L^P,\ x(0) = h \in R^n$$

where $L^P \overset{\text{def}}{=} L^P(-r,0;R^n)$, $p \geq 1$, $F\colon\ L^P \to R^n$, $x\colon\ (-r,\infty) \to R^n$, and $x_t \in L^P$ is defined by $x_t(\theta) \overset{\text{def}}{=} x(t+\theta)$ for almost all $\theta \in (-r,0)$. The semigroup associated with this problem is given by $S(t)$, $t \geq 0$, where $S(t)\colon\ L^P \times R^n \to L^P \times R^n$ is defined by $S(t)\{\phi,h\} \overset{\text{def}}{=} \{x_t,x(t)\}$, $t \geq 0$, $\{\phi,h\} \in L^P \times R^n$, and its infinitesimal generator is $B\colon\ L^P \times R^n \to L^P \times R^n$, where

$$B\{\phi,h\} \overset{\text{def}}{=} \{\phi',F(\phi)\},\ D(B) \overset{\text{def}}{=} \{\{\phi,h\} \in L^P \times R^n\colon\ \phi' \in L^P \text{ and } \phi(0) = h\}$$

Semigroup methods have been applied to functional differential equations by many authors, and we refer the reader to our references for a listing of some of the relevant articles.

Proposition 3.13 was proved in [301]. The semigroup $S(t)$, $t \geq 0$, associated with nonlinear age-dependent population dynamics does not have the property that $S(t)$ is a compact operator for $t$ sufficiently large, except in certain cases in which the age range is a finite rather than an infinite interval. For functional differential equations on a finite interval $[-r,0]$ the semigroup $S(t)$, $t \geq 0$, defined by the solutions has the property that $S(t)$ is a compact operator whenever $t > r$ (which means that bounded trajectories are automatically precompact).

The numerical treatment of nonlinear age-dependent population dynamics by semigroup methods in Section 3.5 is similar to the numerical treatment of functional differential equations developed by a number of authors in recent years. The method of averaging approximations, as well as other methods employing a semigroup theoretic approach, has been used by such authors as H. Banks and J. Burns [10], [11], H. Banks and F. Kappel [12], J. Burns and E. Cliff [29], F. Kappel [158], F. Kappel and K. Kunisch [159], [160], F. Kappel and W. Schappacher [161], K. Kunish [175], and G. Reddien and G. Webb [244].

# 4

# Equilibrium Solutions and Their Stability

## 4.1 EXISTENCE OF NONTRIVIAL EQUILIBRIUM SOLUTIONS

The ultimate behavior of many biological populations reveals a convergence to a stable time-independent state as time evolves. The prediction and description of convergence to equilibrium states is one of the most valuable applications of mathematical population models. If the zero solution is an equilibrium solution of (ADP), then information about its stability properties can be used to predict whether or not the population becomes extinct. Information about the existence and stability of nonzero equilibrium solutions can be used to predict whether or not the population will stabilize to a nontrivial steady state. In the first section of this chapter we will consider the problem of existence of nontrivial equilibrium solutions to (ADP). In the next section we will consider the problem of the ultimate extinction of a population modelled by (ADP). In the remaining sections of this chapter we will consider the problem of the stability and instability of nontrivial equilibrium solutions of (ADP).

DEFINITION 4.1 Let (2.1), (2.2) hold, let $\phi \in L^1$, and let $\ell$ be the solution of (ADP) on $[0,T_\phi)$. Then, $\ell$ is an *equilibrium solution* of (ADP) if and only if $T_\phi = \infty$ and $\ell(\cdot,t) = \phi$ for all $t \geq 0$.

PROPOSITION 4.1 Let (2.1), (2.2), (2.22), (2.23) hold, let A be defined as in (3.16), let $\phi \in L^1_+$, and let $\ell$ be the solution of (ADP) on $[0,T_\phi)$. Then, $\ell$ is an equilibrium solution of (ADP) if and only if $A\phi = 0$.

*Proof.* Suppose that $\ell$ is an equilibrium solution of (ADP). Let $S(t)$, $t \geq 0$, be the strongly continuous nonlinear semigroup in $L^1_+$ associated with (ADP) as in Theorem 3.1. Then $\ell(\cdot,t) = S(t)\phi = \phi$ for all $t \geq 0$. Obviously, $\lim_{t\to 0^+} t^{-1}(S(t)\phi - \phi) = 0$. From Theorem 3.2 we have that $A\phi = 0$. Suppose that $A\phi = 0$. From (3.16) we see that $\phi'(a) = G(\phi)(a)$ for almost all $a > 0$ and $\phi(0) = F(\phi)$. Then, for all $t \geq 0$,

$$\phi(a) = \begin{cases} F(\phi) + \int_{t-a}^{t} \phi'(s + a - t)\, ds & \text{a.e. } a \in (0,t) \\ \phi(a - t) + \int_0^t \phi'(s + a - t)\, ds & \text{a.e. } a \in (t,\infty) \end{cases}$$

$$= \begin{cases} F(\phi) + \int_{t-a}^{t} G(\phi)(s + a - t)\, ds & \text{a.e. } a \in (0,t) \\ \phi(a - t) + \int_0^t G(\phi)(s + a - t)\, ds & \text{a.e. } a \in (t,\infty) \end{cases}$$

By the uniqueness of solutions to (ADP) (Theorem 2.1) and the equivalence of (ADP) and the integral equation (1.49) (Theorem 2.2), we must have that $\ell(\cdot,t) = \phi$ for all $t \geq 0$. Thus, $\ell$ is an equilibrium solution of (ADP). □

For a population divided into population subclasses it is sometimes natural to suppose that the birth and mortality rates of each subclass are influenced by the presence of the other subclasses, as well as its own subclass. For this case we require that the birth function F and the aging function G have the following forms:

(4.1) For $F$: $L^1 \to R^n$ and for each $i = 1, \ldots, n$, there exists a function $\beta_i$: $[0,\infty) \times L^1 \to [0,\infty)$ such that for all $\phi \in L^1$, $F(\phi)_i = \int_0^\infty \beta_i(a,\phi)\phi_i(a)\, da$, where $\beta_i$ satisfies: (i) there is an increasing function $c_{10}$: $[0,\infty) \to [0,\infty)$ (independent of i) such that if $\phi \in L^1$ and $a, \hat{a} \geq 0$, then $|\beta_i(a,\phi) - \beta_i(\hat{a},\phi)| \leq c_{10}(\|\phi\|_{L^1})|a - \hat{a}|$; (ii) there is an increasing function $c_{11}$: $[0,\infty) \to [0,\infty)$ (independent of i) such that $\beta_i(a,\phi) \leq c_{11}(\|\phi\|_{L^1})$ for all $\phi \in L^1$, $a \geq 0$; and (iii) there is an increasing function $c_{12}$: $[0,\infty) \to [0,\infty)$ (independent of i) such that $|\beta_i(a,\phi) - \beta_i(a,\hat{\phi})| \le c_{12}(r)\|\phi - \hat{\phi}\|_{L^1}$ for all $\phi$, $\hat{\phi} \in L^1$ with $\|\phi\|_{L^1}$, $\|\hat{\phi}\|_{L^1} \le r$ and all $a \ge 0$.

(4.2) For G: $L^1 \to L^1$ and for each i = 1, ..., n, there exists a function $\mu_i$: $[0,\infty) \times L^1 \to [0,\infty)$ such that for all $\phi \in L^1$ and almost all $a > 0$, $G(\phi)_i(a) = -\mu_i(a,\phi)\phi_i(a)$, where $\mu_i$ satisfies: (i) there is an increasing function $c_{13}$: $[0,\infty) \to [0,\infty)$ (independent of i) such that $|\mu_i(a,\phi) - \mu_i(\hat{a},\phi)| \le c_{13}(\|\phi\|_{L^1})|a - \hat{a}|$ for all $\phi \in L^1$, a, $\hat{a} \ge 0$; (ii) there is an increasing function $c_{14}$: $[0,\infty) \to [0,\infty)$ (independent of i) such that $\mu_i(a,\phi) \le c_{14}(\|\phi\|_{L^1})$ for all $a \ge 0$, $\phi \in L^1$; (iii) there exists an increasing function $c_{15}$: $[0,\infty) \to [0,\infty)$ (independent of i) such that $|\mu_i(a,\phi) - \mu_i(a,\hat{\phi})| \le c_{15}(r)\|\phi - \hat{\phi}\|_{L^1}$ for all $a \ge 0$, $\phi$, $\hat{\phi} \in L^1$ with $\|\phi\|_{L^1}$, $\|\hat{\phi}\|_{L^1} \le r$; and (iv) there exists a positive constant $\underline{\mu}$ (independent of i) such that $\mu_i(a,\phi) \ge \underline{\mu}$ for all $\phi \in L^1$, $a \ge 0$.

REMARK 4.1 We observe that if F satisfies (4.1), then F satisfies (2.1), (2.22), and (i), (ii) of (3.73). Also, if G satisfies (4.2), then G satisfies (2.2), (2.23), (2.49), and (3.64) (see Remark 3.4).

Suppose now that F and G have the forms (4.1) and (4.2), respectively, and suppose that $\hat{\phi}$ is an equilibrium solution of (ADP). Then, from Proposition 4.1 we have that

$$A\hat{\phi}(a) = \hat{\phi}'(a) - G(\hat{\phi})(a) = 0 \qquad \text{a.e. } a > 0$$
$$\hat{\phi}(0) = F(\hat{\phi})$$

which means

$$\hat{\phi}_i'(a) = -\mu_i(a,\hat{\phi})\hat{\phi}_i(a) \qquad \text{a.e. } a > 0,\ i = 1, \ldots, n$$

$$\hat{\phi}_i(0) = \int_0^\infty \beta_i(a,\hat{\phi})\hat{\phi}_i(a)\, da \qquad i = 1, \ldots, n$$

If we define $\Pi_i$: $[0,\infty) \times L^1 \to [0,\infty)$, $i = 1, \ldots, n$, by

$$(4.3) \qquad \Pi_i(a,\phi) \overset{\text{def}}{=} \exp\left[-\int_0^a \mu_i(b,\phi)\, db\right] \qquad a \geq 0,\ \phi \in L^1$$

then we see that

$$(4.4) \qquad \hat{\phi}_i(a) = \hat{\phi}_i(0)\Pi_i(a,\hat{\phi}) \qquad a \geq 0,\ i = 1, \ldots, n$$

$$(4.5) \qquad \hat{\phi}_i(0) = \int_0^\infty \beta_i(a,\hat{\phi})\hat{\phi}_i(0)\Pi_i(a,\hat{\phi})\, da \qquad i = 1, \ldots, n$$

In the theory of linear age-dependent population dynamics the net reproduction rate is a measure of the average number of offspring per individual over the entire lifespan of the individual. Analogously to the linear case, we define

DEFINITION 4.2 Let (4.1), (4.2) hold, let $\Pi_i$, $i = 1, \ldots, n$ be as in (4.3) and let $\phi \in L^1_+$. The *net reproduction rate* of the i-th subclass of the population having density $\phi$ is defined as

$$(4.6) \qquad N_i(\phi) \overset{\text{def}}{=} \int_0^\infty \beta_i(a,\phi)\Pi_i(a,\phi)\, da \qquad i = 1, \ldots, n$$

From (4.5) we see that a necessary condition for $\hat{\phi}$ to be an equilibrium solution of (ADP) [when F satisfies (4.1) and G satisfies (4.2)] is that the net reproduction rates satisfy

$$\hat{\phi}_i(0)[1 - N_i(\hat{\phi})] = 0 \qquad i = 1, \ldots, n$$

Thus, in this case we see that if $N_i(\phi) < 1$ for all $\phi \in L^1_+$ and all $i = 1, \ldots, n$, then the zero solution of (ADP) is the only equilibrium solution.

EXAMPLE 4.1 Let $R^n = R^1$. Let F and G be defined as in Example 2.4 and Example 3.1 (that is, F has the form (2.17) with $\beta$ continuously differentiable, G has the form (2.18) with $\mu$ continuously differentiable, and (2.30) holds). The infinitesimal generator B of the nonlinear strongly continuous semigroup in $L^1_+$ associated with this

(ADP) is given by (3.20), that is,

$$B\phi(a) = -\phi'(a) - \mu(P\phi)\phi(a) \qquad \text{a.e. } a > 0$$

$$D(B) = \{\phi \in L^1_+ : \ \phi \text{ is absolutely continuous on } [0,\infty),\ \phi' \in L^1,$$

$$\text{and } \phi(0) = \int_0^\infty \beta(P\phi)e^{-\alpha a}\phi(a)\ da\}$$

According to Theorem 3.2 and Proposition 4.1, $\hat{\phi} \in L^1_+$ is a nontrivial equilibrium solution of (ADP) if and only if $\hat{\phi} \neq 0$ and $B\hat{\phi} = 0$. If $B\hat{\phi} = 0$, then

$$\hat{\phi}(a) = \hat{\phi}(0)\ \exp[-a\mu(P\hat{\phi})] \qquad a \geq 0$$

$$\hat{\phi}(0) = \int_0^\infty \beta(P\hat{\phi})e^{-\alpha a}\hat{\phi}(0)\ \exp[-a\mu(P\hat{\phi})]\ da$$

Thus, a nontrivial equilibrium solution $\hat{\phi}$ must satisfy the equation $\beta(P\hat{\phi}) = \alpha + \mu(P\hat{\phi})$, or equivalently, $N(\hat{\phi}) = 1$. Conversely, if $\hat{P}$ is a positive solution of the equation

$$\alpha + \mu(\hat{P}) = \beta(\hat{P}) \tag{4.7}$$

and we define $\hat{\phi} \in L^1_+$ by

$$\hat{\phi}(a) = \mu(\hat{P})\hat{P}e^{-a\mu(\hat{P})} \qquad a \geq 0 \tag{4.8}$$

then $\hat{P} = P\hat{\phi} = \int_0^\infty \hat{\phi}(0)e^{-a\mu(\hat{P})}\ da$ and $B\hat{\phi} = 0$, so that $\hat{\phi}$ is a nontrivial equilibrium solution of (ADP) in $L^1_+$.

EXAMPLE 4.2 Let $R^n = R^1$. Let F and G be defined as in Example 2.5 and Example 3.2 (that is, F has the form (2.19) with $\beta$ continuously differentiable, G has the form (2.18) with $\mu$ continuously differentiable, and (2.30) holds). The infinitesimal generator B of the nonlinear strongly continuous semigroup in $L^1_+$ for this (ADP) is given by (3.21), that is,

$$B\phi(a) = -\phi'(a) - \mu(P\phi)\phi(a) \qquad \text{a.e. } a > 0$$

$$D(B) = \{\phi \in L^1_+ : \ \phi \text{ is absolutely continuous on } [0,\infty),\ \phi' \in L^1,$$

$$\text{and } \phi(0) = \int_0^\infty \beta(P\phi)(1 - e^{-\alpha a})\phi(a)\ da\}$$

As in the analysis of Example 4.1 we see that $\hat{\phi}$ is a nontrivial equilibrium solution of this (ADP) in $L^1_+$ if and only if there exists $\hat{P} > 0$ such that $\hat{P}$ satisfies the equation

$$(4.9) \qquad \mu(\hat{P})[\alpha + \mu(\hat{P})] = \alpha\beta(\hat{P})$$

and $\hat{\phi}$ is given by (4.8).

EXAMPLE 4.3 Let $R^n = R^1$. Let F and G be defined as in Example 2.6 and Example 3.3 (that is, F has the form (2.20) with $\beta$ continuously differentiable, G has the form (2.18) with $\mu$ continuously differentiable, and (2.31) holds). The infinitesimal generator B of the nonlinear strongly continuous semigroup in $L^1_+$ for this (ADP) is given by (3.22), that is,

$$B\phi(a) = \phi'(a) - \mu(P\phi)\phi(a) \qquad \text{a.e. } a > 0$$

$$D(B) = \{\phi \in L^1_+\colon \ \phi \text{ is absolutely continuous on } [0,\infty),\ \phi' \in L^1,$$
$$\text{and } \phi(0) = \int_0^\infty \beta(P\phi)ae^{-\alpha a}\phi(a)\ da\}$$

As in the analysis of Example 4.1 we see that $\hat{\phi}$ is a nontrivial equilibrium solution of this (ADP) in $L^1_+$ if and only if there exists $\hat{P} > 0$ such that $\hat{P}$ satisfies the equation

$$(4.10) \qquad [\alpha + \mu(\hat{P})]^2 = \beta(\hat{P})$$

and $\hat{\phi}$ is given by (4.8).

For many populations the birth and mortality moduli will satisfy the conditions $\beta_i(a,\phi) \le \beta_i(a,0)$ and $\mu_i(a,\phi) \ge \mu_i(a,0)$ for all $\phi \in L^1_+$, $a \ge 0$, $i = 1, \ldots, n$. Consequently, the net reproduction rates satisfy $N_i(\phi) \le N_i(0)$ for all $\phi \in L^1_+$, $i = 1, \ldots, n$. In this case a necessary condition that (ADP) have a nontrivial equilibrium solution is that $N_i(0) \ge 1$ for at least one $i = 1, \ldots, n$. A sufficient condition for the existence of a nontrivial equilibrium solution of (ADP) is given by the following result of J. Prüss:

THEOREM 4.1 (J. Prüss [240]) Let (4.1), (4.2) hold. Let $\{i\colon\ i = 1,\ldots,n\} = M \cup N$ such that (i) if $i \in M$, then $N_i(0) > 1$ [where $N_i$ is defined by (4.6)], and (ii) there exists $r > 0$ such that if $\phi \in L^1_+$, $i \in M$, and $\|\phi_i\|_{L^1((0,\infty);R)} \geq r$, then $N_i(\phi) \leq 1$. Then, there exists an equilibrium solution $\hat{\phi} \in L^1_+$ of (ADP) such that $\hat{\phi}_i \neq 0$ for some $i \in M$ and $\hat{\phi}_i = 0$ for all $i \in N$.

The proof of Theorem 4.1 uses the following fixed point theorem, a proof of which may be found in an article of H. Amann ([1], Theorem 12.3, p. 661):

PROPOSITION 4.2 Let X be a Banach space, let C be a closed convex cone in X (that is, C is a closed subset of X satisfying the property that if $x, y \in C$ and $\alpha, \beta \geq 0$, then $\alpha x + \beta y \in C$), let $\sigma > 0$, let $C_\sigma \overset{\text{def}}{=} \{x \in C: \|x\| \leq \sigma\}$, and let $f: C_\sigma \to C$ such that f is continuous and $f(C_\sigma)$ has compact closure in C. Suppose that

(4.11) $f(x) \neq \lambda x$ for all $x \in C_\sigma$ such that $\|x\| = \sigma$ and for all $\lambda > 1$.

(4.12) There exists $\tau \in (0,\sigma)$ and $x_1 \in C$, $x_1 \neq 0$, such that $x - f(x) \neq \lambda x_1$ for all $x \in C_\sigma$ such that $\|x\| = \tau$ and for all $\lambda > 0$.

Then, there exists $x_0 \in \{x \in C: \tau \leq \|x\| \leq \sigma\}$ such that $f(x_0) = x_0$.

*Proof of Theorem 4.1.* Without loss of generality let $M = \{1, \ldots, m\}$ and $N = \{m + 1, \ldots, n\}$. If $\phi \in L^{1,m}_+ \overset{\text{def}}{=} L^1_+(0,\infty;R^m)$, define $\hat{\phi} \in L^{1,n}_+ \overset{\text{def}}{=} L^1_+(0,\infty;R^n)$ by $\hat{\phi}_i = \phi_i$ for $1 \leq i \leq m$ and $\hat{\phi}_i = 0$ for $m < i \leq n$. Define the Banach space $X \overset{\text{def}}{=} L^1(0,\infty;R^m) \times R^m$ with norm

$$\|\{\phi,x\}\|_X \overset{\text{def}}{=} \sum_{i=1}^{m} (\|\phi_i\|_{L^1(0,\infty;R)} + |x_i|) \qquad \{\phi,x\} \in X$$

and define $X_+ \overset{\text{def}}{=} L^{1,m}_+ \times R^m_+$. Notice that $X_+$ is a closed convex cone in X. Define a mapping $f: X_+ \to X_+$ by

$$f(\{\phi,x\})_i \overset{\text{def}}{=} \{x_i \Pi_i(a,\hat{\phi}), x_i N_i(\hat{\phi})\} \qquad i = 1, \ldots, m$$

It is obvious that if $f(\{\phi,x\}) = \{\phi,x\}$ for some $\{\phi,x\} \in X_+$, then $\hat{\phi}$ satisfies (4.4) and (4.5), and thus, $\hat{\phi}$ is an equilibrium solution of (ADP).

Let $\sigma \overset{\text{def}}{=} mr[1 + c_{11}(mr)]$ and let $C_\sigma \overset{\text{def}}{=} \{\{\phi,x\} \in X_+ : \|\{\phi,x\}\|_X \le \sigma\}$. We claim that f is continuous from $C_\sigma$ to $X_+$ and $f(C_\sigma)$ is bounded. The continuity of f is proved as follows: Obviously, the mapping $\phi \to \hat{\phi}$ from $L_+^{1,m}$ to $L_+^{1,n}$ is continuous. For $\psi, \chi \in L_+^{1,n}$, $\|\psi\|_{L^1}, \|\chi\|_{L^1} \le \sigma$, $i = 1, \ldots, n$,

$$\begin{aligned} |N_i(\psi) - N_i(\chi)| &\le \int_0^\infty |\beta_i(a,\psi) - \beta_i(a,\chi)|\Pi_i(a,\psi)\, da \\ &\quad + \int_0^\infty \beta_i(a,\psi)|\Pi_i(a,\psi) - \Pi_i(a,\chi)|\, da \\ &\le c_{12}(\sigma)\|\psi - \chi\|_{L^1} \int_0^\infty e^{-\underline{\mu}a}\, da \\ &\quad + c_{11}(\sigma) \int_0^\infty \left[\int_0^a |\mu_i(b,\psi) - \mu_i(b,\chi)|\, db\right] e^{-\underline{\mu}a}\, da \\ &\le c_{12}(\sigma)\|\psi - \chi\|_{L^1}\underline{\mu}^{-1} \\ &\quad + c_{11}(\sigma) \int_0^\infty c_{15}(\sigma)\|\psi - \chi\|_{L^1} a e^{-\underline{\mu}a}\, da \\ &= [c_{12}(\sigma) + c_{11}(\sigma)c_{15}(\sigma)]\|\psi - \chi\|_{L^1}(\underline{\mu}^{-1} + \underline{\mu}^{-2}) \end{aligned}$$

The mapping $\psi \to N_i(\psi)$ is therefore continuous as a function from $L_+^{1,n}$ to $[0,\infty)$, and a similar argument proves that the mapping $\psi \to \exp[-M_i(a,\psi)]$ is continuous as a function from $L_+^{1,n}$ to $L_+^{1,1}$. The continuity of f now follows immediately. The boundedness of $f(C_\sigma)$ is proved as follows: For $\psi \in L_+^{1,n}$, $\|\psi\|_{L^1} \le \sigma$, $i = 1, \ldots, n$, we see that

$$\begin{aligned} N_i(\psi) &= \int_0^\infty \beta_i(a,\psi)\Pi_i(a,\psi)\, da \le c_{11}(\|\psi\|_{L^1}) \int_0^\infty e^{-\underline{\mu}a}\, da \\ &\le c_{11}(\sigma)\underline{\mu}^{-1} \end{aligned}$$

The boundedness of f on $C_\sigma$ now follows immediately.

We next claim that $f(C_\sigma)$ has compact closure in X. By Lemma 2.3 it suffices to show that (2.10) and (2.11) hold. For $\{\phi,x\} \in C_\sigma$, $\{\psi,y\} \overset{\text{def}}{=} f(\{\phi,x\})$, and i = 1, ..., m, we have that

$$|\psi_i(a)| = \left|x_i \exp\left[-\int_0^a \mu_i(b,\hat{\phi})\, db\right]\right|$$

$$\leq \sigma e^{-\underline{\mu} a} \qquad a \geq 0$$

$$|\psi_i'(a)| = \left|x_i \exp\left[-\int_0^a \mu_i(b,\hat{\phi})\, db\right]\mu_i(a,\hat{\phi})\right|$$

$$\leq \sigma c_{14}(\sigma) e^{-\underline{\mu} a} \qquad a \geq 0$$

Then, for h > 0,

$$\int_0^\infty |\psi_i(a) - \psi_i(a+h)|\, da \leq \int_0^\infty \left[\left|\int_a^{a+h} \psi_i'(b)\, db\right|\right] da$$

$$\leq \int_0^\infty h\sigma c_{14}(\sigma) e^{-\underline{\mu} a}\, da$$

$$\leq h\sigma c_{14}(\sigma)\underline{\mu}^{-1}$$

and for h < 0,

$$\int_0^\infty |\psi_i(a) - \psi_i(a+h)|\, da \leq \int_0^{-h} |\psi_i(a)|\, da + |h|\sigma c_{14}(\sigma)\underline{\mu}^{-1}$$

$$\leq |h|\sigma + |h|\sigma c_{14}(\sigma)\underline{\mu}^{-1}$$

(where $\psi_i(a) \overset{\text{def}}{=} 0$ for a < 0). Also, for h > 0,

$$\int_h^\infty |\psi_i(a)|\, da \leq \sigma e^{-\underline{\mu} h}$$

Thus, (2.10) and (2.11) must hold.

To establish (4.11) let $\{\phi,x\} \in C_\sigma$ such that $\|\{\phi,x\}\|_X = \sigma$ and assume there exists $\lambda > 1$ such that $f(\{\phi,x\}) = \lambda\{\phi,x\}$. Then, for i = 1, ..., m,

(4.13) $\lambda\phi_i(a) = x_i\Pi_i(a,\hat{\phi}) \qquad a \geq 0$

(4.14) $\lambda x_i = x_i N_i(\hat{\phi})$

Let i = 1, ..., m be such that $\|\phi_i\|_{L^1(0,\infty;R)} + |x_i| > 0$. If $x_i = 0$, then (4.13) implies that $\phi_i = 0$ (since $\lambda > 0$). Hence, $x_i \neq 0$ and then (4.14) implies that $N_i(\hat{\phi}) = \lambda > 1$. From the hypothesis of the theorem we must have $\|\hat{\phi}_i\|_{L^1(0,\infty;R)} < r$. From (4.13) and (4.14),

$$\sigma = \sum_{i=1}^{m} \|\phi_i\|_{L^1(0,\infty;R)} + |x_i| = \sum_{i=1}^{m} \|\hat{\phi}_i\|_{L^1(0,\infty;R)} + \sum_{i=1}^{m} |x_i|$$

$$< mr + \sum_{i=1}^{m} |x_i\lambda^{-1}N_i(\hat{\phi})| = mr + \sum_{i=1}^{m} \left|x_i\lambda^{-1}\int_0^\infty \beta_i(a,\hat{\phi})\Pi_i(a,\hat{\phi})\,da\right|$$

$$= mr + \sum_{i=1}^{m} \left|\int_0^\infty \beta_i(a,\hat{\phi})\phi_i(a)\,da\right| \leq mr + \sum_{i=1}^{m} c_{11}(mr)r$$

$$= mr[1 + c_{11}(mr)] = \sigma$$

We therefore obtain a contradiction, and so (4.11) is established.

To prove (4.12) choose $\tau \in (0,\sigma)$ such that for i = 1, ..., m, $N_i(\hat{\phi}) > 1$ for all $\phi \in C_\sigma$ such that $\|\phi\|_{L^1} \leq \tau$ (where we have used the fact that each $N_i(0) > 1$, i = 1, ..., m, and, as we have shown, each $N_i$, i = 1, ..., m, is continuous). Let $\bar{1} = (1,\ldots,1)^T \in R^m$ and assume that there exists $\lambda > 0$ and $\{\phi,x\} \in C_\sigma$ such that $\|\{\phi,x\}\|_X = \tau$ and $\{\phi,x\} - f(\{\phi,x\}) = \lambda\{0,\bar{1}\}$. Then, for i = 1, ..., m,

(4.15) $\phi_i(a) = x_i\Pi_i(a,\hat{\phi}) \qquad a \geq 0$

(4.16) $x_i[1 - N_i(\hat{\phi})] = \lambda$

Since $\|\hat{\phi}\|_{L^1} \leq \tau$, we must have that $N_i(\hat{\phi}) > 1$ for i = 1, ..., m. Since $\lambda > 0$, (4.16) implies that $x_i < 0$ for i = 1, ..., m. Then, (4.15) implies that $\phi_i(a) < 0$ for $a \geq 0$, i = 1, ..., m. But we then have a contradiction, since $\{\phi,x\} \in X_+$. Thus, (4.12) is established.

From Proposition 4.2 we have that there exists $\{\phi,x\} \in C_\sigma$, $\{\phi,x\} \neq \{0,0\}$, such that $f(\{\phi,x\}) = \{\phi,x\}$. Thus, $\{\phi,x\}$ is an equilibrium solution of (ADP). Further, if $\hat{\phi}_i = \phi_i = 0$ for $i = 1, \ldots, m$, then $x_i = 0$ for all $i = 1, \ldots, m$ (since $\phi_i(a) = x_i \Pi_i(a,\hat{\phi})$, $a \geq 0$). But then $\{\phi,x\} = \{0,0\}$, which is a contradiction. Hence, at least one integer $i = 1, \ldots, m$ has the property that $\hat{\phi}_i \neq 0$. □

## 4.2 LIAPUNOV FUNCTIONS AND THE INVARIANCE PRINCIPLE

A classical method for determining the asymptotic behavior of the solutions of differential equations involves the use of Liapunov functions. This method is also applicable to abstract differential equations in the setting of a Banach space. In this section we develop this approach for the problem (ADP). We first define

DEFINITION 4.3 Let $S(t)$, $t \geq 0$, be a strongly continuous nonlinear semigroup in the closed subset $C$ of the Banach space $X$ and let $\hat{x} \in C$. The *trajectory* of $\hat{x}$, denoted by $\gamma(\hat{x})$, is $\gamma(\hat{x}) \overset{\text{def}}{=} \{S(t)\hat{x} : t \geq 0\}$. If $\gamma(\hat{x}) = \{\hat{x}\}$, then $\hat{x}$ is an *equilibrium* for $S(t)$, $t \geq 0$. The trajectory $\gamma(\hat{x})$ is *stable* if and only if for each $\varepsilon > 0$, there exists $\delta > 0$ such that if $x \in C$ and $\|x - \hat{x}\| < \delta$, then $\|S(t)x - S(t)\hat{x}\| < \varepsilon$ for all $t \geq 0$. The trajectory $\gamma(\hat{x})$ is *unstable* if and only if it is not stable. The trajectory $\gamma(\hat{x})$ is *asymptotically stable* if and only if it is stable, and there exists $\delta > 0$ such that if $x \in C$ and $\|x - \hat{x}\| < \delta$, then $\lim_{t\to\infty} \|S(t)x - S(t)\hat{x}\| = 0$. The trajectory $\gamma(\hat{x})$ is *exponentially asymptotically stable* if and only if it is asymptotically stable, and there exists $\delta > 0$, $\omega > 0$, and $K > 0$ such that if $x \in C$ and $\|x - \hat{x}\| < \delta$, then $\|S(t)x - S(t)\hat{x}\| \leq Ke^{-\omega t}\|x - \hat{x}\|$ for all $t \geq 0$. If $\delta$ can be chosen arbitrarily large in each of these last two definitions, then the corresponding property is said to be *global*.

DEFINITION 4.4 Let $S(t)$, $t \geq 0$, be a strongly continuous nonlinear semigroup in the closed subset $C$ of the Banach space $X$ and let $C_1$

be a subset of C. If $S(t)(C_1) \subset C_1$ for all $t \geq 0$, then $C_1$ is said to be *positive invariant* under $S(t)$, $t \geq 0$. If there exists a mapping $U\colon R \times C_1 \to C_1$ such that $U(0,x) = x$ and $U(t + s, x) = S(t)U(s,x)$ for all $x \in C_1$, $s \in R$, and $t \geq 0$, then $C_1$ is said to be *invariant* under $S(t)$, $t \geq 0$.

We note that in [290], p. 166, it is shown that an invariant set under $S(t)$, $t \geq 0$, must be a positive invariant set under $S(t)$, $t \geq 0$, and that $C_1$ is invariant under $S(t)$, $t \geq 0$, if and only if $S(t)(C_1) = C_1$ (in $2^X$) for all $t \geq 0$.

DEFINITION 4.5 Let $S(t)$, $t \geq 0$, be a strongly continuous nonlinear semigroup in the closed set C of the Banach space X and let $C_1$ be a subset of C. A *Liapunov function* for $S(t)$, $t \geq 0$, on $C_1$ is a continuous function $V\colon C \to R$ such that for all $x \in C_1$

$$\dot{V}(x) \overset{\text{def}}{=} \limsup_{t \to 0^+} t^{-1}[V(S(t)x) - V(x)] \leq 0 \tag{4.17}$$

(where we allow the possibility that $\dot{V}(x) = -\infty$).

A proof of the following result, which is known as Liapunov's direct method, may be found in the monograph of J. Walker [290], Theorem 3.1, p. 157.

PROPOSITION 4.3 Let $S(t)$, $t \geq 0$, be a strongly continuous nonlinear semigroup in the closed subset C of the Banach space X, let $\hat{x}$ be an equilibrium for $S(t)$, $t \geq 0$, let $0 < r \leq \infty$, let $C_1 \overset{\text{def}}{=} \{x \in C\colon \|x - \hat{x}\| \leq r\}$, and let V be a Liapunov function for $S(t)$, $t \geq 0$, on $C_1$. Suppose there exists a strictly increasing continuous function $u\colon [0,\infty) \to [0,\infty)$ such that $u(0) = 0$ and u satisfies $V(x) \geq V(\hat{x}) + u(\|x - \hat{x}\|)$ for all x in $C_1$. Then, $\hat{x}$ is stable, and if $r = \infty$, then $\hat{x}$ is globally stable. Suppose, in addition, there exists a strictly increasing continuous function $v\colon [0,\infty) \to [0,\infty)$ such that $v(0) = 0$ and v satisfies $\dot{V}(x) \leq -v(\|x - \hat{x}\|)$ for all $x \in C_1$. Then, $\hat{x}$ is asymptotically stable, and if $r = \infty$, then $\hat{x}$ is globally asymptotically stable.

One possible choice for a Liapunov function for the semigroup associated with (ADP) is the $L^1$ norm. An illustration of this approach is given by the following result, which provides a sufficient condition for extinction in a multi-species population.

PROPOSITION 4.4 Let (2.1), (2.2), (2.22), (2.23) hold, let $T_\phi = \infty$ for all $\phi \in L^1_+$, let $S(t)$, $t \geq 0$, be the strongly continuous nonlinear semigroup associated with (ADP) as in Theorem 3.1, and let 0 be an equilibrium solution of (ADP). Let $0 < r \leq \infty$ and let there exist a strictly increasing continuous function $v$: $[0,\infty) \to [0,\infty)$ such that $v(0) = 0$ and

$$\text{(4.18)} \qquad \sum_{i=1}^{n} \left[F(\phi)_i + \int_0^\infty G(\phi)_i(a)\, da\right] \leq -v(\|\phi\|_{L^1}) \text{ for all } \phi \in L^1_+$$
$$\text{such that } \|\phi\|_{L^1} \leq r,\ i = 1, \ldots, n$$

Then, 0 is asymptotically stable, and if $r = \infty$, then 0 is globally asymptotically stable.

*Proof.* Define $u$: $[0,\infty) \to [0,\infty)$ by $u(x) = x$ for all $x \geq 0$ and define $V$: $L^1_+ \to [0,\infty)$ by $V(\phi) = \|\phi\|_{L^1}$ for all $\phi \in L^1_+$. Obviously, $V$ is continuous and $V(\phi) \geq u(\|\phi\|_{L^1})$ for all $\phi \in L^1_+$. For $\phi \in L^1_+$, $t > 0$, we have that

$$\begin{aligned}
&t^{-1}[V(S(t)\phi) - V(\phi)] \\
&= t^{-1} \sum_{i=1}^{n} \left(\int_{-t}^{\infty} (S(t)\phi)_i(t + c)\, dc - \int_0^\infty \phi_i(c)\, dc\right) \\
&= \sum_{i=1}^{n} \left(t^{-1} \int_{-t}^{0} (S(t)\phi)_i(t + c)\, dc\right. \\
&\quad \left. + \int_0^\infty t^{-1}[(S(t)\phi)_i(t + c)\, dc - \phi_i(c)]\, dc\right) \\
&\leq \sum_{i=1}^{n} \left(t^{-1} \int_0^t [|(S(t)\phi)_i(a) - F(\phi)_i| + F(\phi)_i]\, da\right. \\
&\quad \left. + \int_0^\infty \{|t^{-1}[(S(t)\phi)_i(a + t) - \phi_i(a)] - G(\phi)_i(a)| + G(\phi)_i(a)\}\, da\right)
\end{aligned}$$

Therefore, by (1.43), (1.44), and (4.18) we have that for $\phi \in C_1 \overset{\text{def}}{=} \{\phi \in L^1_+ : \|\phi\|_{L^1} \le r\}$,

$$\begin{aligned}\dot{V}(\phi) &= \limsup_{t\to 0^+} t^{-1}[V(S(t)\phi) - V(\phi)] \\ &\le \sum_{i=1}^{n} \left(F(\phi)_i + \int_0^\infty G(\phi)_i(a)\, da\right) \\ &\le -v(\|\phi\|_{L^1})\end{aligned}$$

Thus, V is a Liapunov function for $S(t)$, $t \ge 0$, on $C_1$ and u and v satisfy the hypothesis of Proposition 4.3. Consequently, 0 is asymptotically stable, and if $r = \infty$, then 0 is globally asymptotically stable. □

REMARK 4.2 Suppose that the birth function F satisfies (4.1) and the aging function G satisfies (4.2). Condition (4.18) is then implied by the condition

(4.19) There exists $r > 0$ and a nondecreasing continuous function $w: [0,\infty) \to [0,\infty)$ such that $w(0) > 0$ and $\beta_i(a,\phi) - \mu_i(a,\phi) \le -w(\|\phi\|_{L^1})$ for all $a \ge 0$, $\phi \in L^1_+$ such that $\|\phi\|_{L^1} \le r$, and $i = 1, \ldots, n$.

That (4.19) implies (4.18) is proved by taking $v(x) \overset{\text{def}}{=} w(x)x$ for $x \ge 0$, and observing that for all $\phi \in L^1_+$,

$$\begin{aligned}&\sum_{i=1}^{n} \left(F(\phi)_i + \int_0^\infty G(\phi)_i(a)\, da\right) \\ &= \sum_{i=1}^{n} \int_0^\infty [\beta_i(a,\phi) - \mu_i(a,\phi)]\phi_i(a)\, da \\ &\le -w(\|\phi\|_{L^1})\|\phi\|_{L^1} = -v(\|\phi\|_{L^1})\end{aligned}$$

The condition (4.19) is sufficient in order for the $L^1$ norm to be a Liapunov function for (ADP). In many cases, however, (4.19) is a much stronger condition than necessary for the asymptotic stability of the zero equilibrium solution. Notice that the inequality in (4.19) is required to hold uniformly for all $a \geq 0$. Notice also that if (4.19) holds, then the net reproduction rates $N_i(\phi)$ satisfy

$$\begin{aligned} N_i(\phi) &= \int_0^\infty \beta_i(a,\phi) \exp\left[-\int_0^a \mu_i(b,\phi)\, db\right] da \\ &< \int_0^\infty \mu_i(a,\phi) \exp\left[-\int_0^a \mu_i(b,\phi)\, db\right] da \\ &= \int_0^\infty -\frac{d}{da} \exp\left[-\int_0^a \mu_i(b,\phi)\, db\right] da \\ &= 1 . \end{aligned}$$

In the next result we provide some other sufficient conditions for extinction in a multi-species population. The hypothesis of this result requires that the net reproduction rate of the i-th species is less than the critical value 1. The proof of this result uses a weighted integral rather than the unweighted integral of the $L^1$ norm.

THEOREM 4.2 Let (4.1), (4.2) hold, let $T_\phi = \infty$ for all $\phi \in L^1_+$, let $S(t)$, $t \geq 0$, be the strongly continuous nonlinear semigroup in $L^1_+$ as in Theorem 3.1, let i be a fixed integer in $[1,n]$, and let there exist continuous functions $\beta_0$: $[0,\infty) \to [0,\infty)$ and $\mu_0$: $[0,\infty) \to [0,\infty)$ such that

$$\int_0^\infty e^{\delta a} \beta_0(a) \exp\left[-\int_0^a \mu_0(b)\, db\right] da < 1 \qquad \text{for some } \delta > 0 \tag{4.20}$$

$$\beta_i(a,\phi) \leq \beta_0(a) \text{ and } \mu_i(a,\phi) \geq \mu_0(a) \text{ for all } \phi \in L^1_+,\ a \geq 0 \tag{4.21}$$

Then, there exists $\omega > 0$ such that if $\phi \in L^1_+$ and $\phi$ has compact support in $[0,\infty)$, then for some constant $K = K(\phi)$ (depending on $\phi$) we have that:

$$\|(S(t)\phi)_i\|_{L^1(0,\infty;R)} \leq K(\phi) e^{-\omega t} \|\phi_i\|_{L^1(0,\infty;R)} \qquad \text{for } t \geq 0 \tag{4.22}$$

*Proof.* Let $\underline{\mu} > 0$ be as in (4.2) (iv). By (4.20) we may choose $\omega$ such that $0 < \omega < \underline{\mu}$ and

$$\int_0^\infty e^{\omega a}\beta_0(a)\ \exp\left[-\int_0^a \mu_0(b)\ db\right] da < 1 \tag{4.22}$$

Define the function $p\colon\ [0,\infty) \to R$ by

$$p(a) \overset{\text{def}}{=} \exp\left[-\omega a + \int_0^a \mu_0(b)\ db\right]$$
$$\times \left(1 - \int_0^a e^{\omega b}\beta_0(b)\ \exp\left[-\int_0^b \mu_0(\tau)\ d\tau\right] db\right) \qquad a \geq 0$$

Observe that $p(0) = 1$ and

$$p'(a) = p(a)\mu_0(a) - \beta_0(a) - \omega p(a) \qquad a \geq 0$$

By (4.2) (iv) and (4.22) there exists a constant $K_1 > 0$ such that

$$p(a) \geq \exp[(\underline{\mu} - \omega)a]\left(1 - \int_0^\infty e^{\omega b}\beta_0(b)\right. \tag{4.23}$$
$$\left.\times \exp\left[-\int_0^b \mu_0(\tau)\ d\tau\right] db\right) \geq K_1 \qquad \text{for all } a \geq 0$$

Let $\phi \in L^1_+$ such that $\phi$ has compact support in $[0,\infty)$ and let $a_0 > 0$ such that $\phi(a) = 0$ for $a \geq a_0$. Recall from Remark 4.1 that G satisfies (2.49), since G satisfies (4.1). By (2.70) we have that $(S(t)\phi)_i(a) = 0$ for $a > a_0 + t$. For $t \geq 0$, define

$$V(t) \overset{\text{def}}{=} \int_0^\infty (S(t)\phi)_i(a)p(a)\ da$$
$$= \int_{-t}^\infty (S(t)\phi)_i(t + c)p(t + c)\ dc$$

[this integral exists, since p is continuous and $(S(t)\phi)_i$ has compact support]. For $t \geq 0$ and $h > 0$, we have that

$$h^{-1}[V(t + h) - V(t)]$$

$$= h^{-1} \int_{-t-h}^{-t} (S(t + h)\phi)_i(t + h + c)p(t + h + c)\ dc$$

$$+ \int_{-t}^{\infty} h^{-1}[(S(t + h)\phi)_i(t + h + c)p(t + h + c)$$

$$- (S(t)\phi)_i(t + c)p(t + c)]\ dc$$

$$\overset{\text{def}}{=} L_1 + L_2$$

By (2.66), as $h \to 0^+$,

$$L_1 = h^{-1} \int_0^h (S(t + h)\phi)_i(a)p(a)\ da \to (S(t)\phi)_i(0)p(0) \tag{4.24}$$

$$= F(S(t)\phi)_i$$

$$= \int_0^{\omega} \beta_i(a,S(t)\phi)(S(t)\phi)_i(a)\ da$$

Next, consider $L_2$. By (2.67) and (2.68) there exists a set E of measure zero such that if $c \in R$, $c \notin E$, then for all $t \geq \max\{0,-c\}$,

$$\frac{d}{dt}(S(t)\phi)_i(t + c) = G(S(t)\phi)_i(t + c) \tag{4.25}$$

$$= -\mu_i(t + c,\ S(t)\phi)(S(t)\phi)_i(t + c)$$

Now let $t \geq 0$ be fixed, and observe that for $c > -t$, $c \notin E$, as $h \to 0^+$,

$$h^{-1}[(S(t + h)\phi)_i(t + h + c)p(t + h + c) \tag{4.26}$$

$$- (S(t)\phi)_i(t + c)p(t + c)]$$

$$\to -\mu_i(t + c,\ S(t)\phi)(S(t)\phi)_i(t + c)p(t + c)$$

$$+ (S(t)\phi)_i(t + c)\ p'(t + c)$$

Consider the difference quotient in (4.26). For $c \geq -t$, $c \notin E$, $h \in (0,1]$,

$$|h^{-1}[(S(t + h)\phi)_i(t + h + c)p(t + h + c) \tag{4.27}$$

$$- (S(t)\phi)_i(t + c)p(t + c)]|$$

$$\leq |p(t + h + c)h^{-1}[(S(t + h)\phi)_i(t + h + c) - (S(t)\phi)_i(t + c)]| + |(S(t)\phi)_i(t + c)h^{-1}[p(t + h + c) - p(t + c)]|$$

$$\leq |p(t + h + c)|\left|h^{-1}\int_t^{t+h} \frac{d}{ds}(S(s)\phi)_i(s + c)\, ds\right| + |(S(t)\phi)_i(t + c)|\left|h^{-1}\int_{t+c}^{t+h+c} p'(s)\, ds\right|$$

$$\leq |p(t + h + c)| \sup_{t\leq s\leq t+h} |\mu_i(s + c, S(s)\phi)(S(s)\phi)_i(s + c)| + |(S(t)\phi)_i(t + c)| \sup_{t+c\leq s\leq t+h+c} |p'(s)|$$

For $-t \leq c \leq 0$, we have by (2.66) that $(S(s)\phi)_i(s + c)$ is bounded by a constant independent of $c \in [-t,0]$, $h \in (0,1]$, and $s \in [t, t + h]$ (since $(S(\tau)\phi)_i(a)$ is continuous on the triangle $0 \leq a \leq \tau \leq t + 1$). For almost all $c > 0$, we have by (4.25) that $(S(s)\phi)_i(s + c) \leq (S(0)\phi)_i(c) = \phi_i(c)$ for all $s \in [t, t + h]$ (since $(d/ds)(S(s)\phi)_i(s + c) \leq 0$). Now use the facts that $p$, $p'$ are continuous and $(S(s)\phi)_i$ has support in $[0, a_0 + t + 1]$ for $0 \leq s \leq t + 1$, to argue that the last expression in (4.27) is bounded by $k\phi(c)$, where $k$ is a constant independent of $c > -t$, $c \notin E$, and $h \in (0,1]$. By the Lebesgue convergence theorem ([249], Theorem 15, p. 88) and by (4.21), as $h \to 0^+$,

$$L_2 \to \int_{-t}^{\infty} [-\mu_i(t + c, S(t)\phi)(S(t)\phi)_i(t + c)p(t + c) + (S(t)\phi)_i(t + c)p'(t + c)]\, dc \tag{4.28}$$

$$= \int_0^{\infty} [-\mu_i(a,S(t)\phi)(S(t)\phi)_i(a)p(a) + (S(t)\phi)_i(a)p'(a)]\, da$$

$$\leq \int_0^{\infty} [-p(a)\mu_0(a) + p'(a)](S(t)\phi)_i(a)\, da$$

$$= \int_0^{\infty} [-\beta_0(a) - \omega p(a)](S(t)\phi)_i(a)\, da$$

From (4.24), (4.28), and (4.21) we obtain

$$\limsup_{h \to 0^+} h^{-1}[V(t + h) - V(t)]$$

$$\leq \int_0^\infty [\beta_i(a,S(t)\phi) - \beta_0(a) - \omega p(a)](S(t)\phi)_i(a)\,da$$

$$\leq -\omega V(t)$$

Thus, $V(t) \leq e^{-\omega t}V(0)$ for $t \geq 0$ (see [183], Theorem 1.4.1, p. 15). By (4.23) we have that

$$\begin{aligned}\|(S(t)\phi)_i\|_{L^1(0,\infty;R)} &= \int_0^\infty (S(t)\phi)_i(a)\,da \\ &\leq K_1^{-1} \int_0^\infty (S(t)\phi)_i(a)p(a)\,da \\ &= K_1^{-1}V(t) \\ &\leq K_1^{-1}e^{-\omega t} \int_0^\infty \phi_i(a)p(a)\,da \\ &\leq e^{-\omega t}K_1^{-1}[\sup_{0\leq a\leq a_0} p(a)]\|\phi\|_{L^1(0,\infty;R)}\end{aligned}$$

Thus, (4.22) follows by taking $K(\phi) = K_1^{-1}[\sup_{0\leq a\leq a_0} p(a)]$. □

We illustrate Theorem 4.2 with the following examples:

EXAMPLE 4.4 Let $R^n = R^1$. Let F and G be defined as in Example 4.1 and, in addition, let

$$\mu(P) \geq \mu(0) \stackrel{\text{def}}{=} \underline{\mu} > 0 \text{ and } \beta(P) \leq \beta(0) \text{ for all } P \geq 0 \tag{4.29}$$

[so that (4.2) (iv) holds, and (4.21) holds with $\mu_0(a) \stackrel{\text{def}}{=} \mu(0)$, $\beta_0(a) \stackrel{\text{def}}{=} \beta(0)e^{-\alpha a}$, $a \geq 0$]. According to Theorem 4.2, a population with initial age distribution having compact support in $[0,\infty)$ will become extinct provided that there exists $\delta > 0$ such that

$$\int_0^\infty e^{\delta a}\beta(0)e^{-\alpha a}e^{-\mu(0)a}\,da = \frac{\beta(0)}{\alpha + \mu(0) - \delta} < 1$$

Thus, a sufficient condition for the extinction of a population whose initial age distribution has compact support is that

$$\beta(0) < \alpha + \mu(0)$$

EXAMPLE 4.5 Let $R^n = R^1$. Let F and G be defined as in Example 4.2 and, in addition, let (4.29) hold. Let $\beta_0(a) \overset{\text{def}}{=} \beta(0)(1 - e^{-\alpha a})$, $\mu_0(a) \overset{\text{def}}{=} \mu(0)$, $a \geq 0$. According to Theorem 4.2, a population with initial age distribution having compact support in $[0,\infty)$ will become extinct provided that there exists $\delta > 0$ such that

$$\int_0^\infty e^{\delta a}\beta(0)(1 - e^{-\alpha a})e^{-\mu(0)a}\, da$$

$$= \beta(0)[(\mu(0) - \delta)^{-1} - (\alpha + \mu(0) - \delta)^{-1}] < 1$$

that is, provided that

$$\beta(0)\alpha < \mu(0)[\alpha + \mu(0)]$$

EXAMPLE 4.6 Let $R^n = R^1$. Let F and G be defined as in Example 4.3 and, in addition, let (4.29) hold. Let $\beta_0(a) \overset{\text{def}}{=} \beta(0)ae^{-\alpha a}$, $\mu_0(a) \overset{\text{def}}{=} \mu(0)$, $a \geq 0$. According to Theorem 4.2, a population with initial age distribution having compact support in $[0,\infty)$ will become extinct provided that there exists $\delta > 0$ such that

$$\int_0^\infty e^{\delta a}\beta(0)ae^{-\alpha a}e^{-\mu(0)a}\, da = \frac{\beta(0)}{[\alpha + \mu(0) - \delta]^2} < 1$$

that is, provided that

$$\beta(0) < [\alpha + \mu(0)]^2$$

For many problems it is possible to find a Liapunov function, but not a Liapunov function satisfying the hypothesis of Proposition 4.3. For these problems there is another useful method for investigating asymptotic behavior. In this method one attempts to find the smallest closed set to which a trajectory will converge as time becomes infinite. The identification of this set is known as the invariance principle of J. La Salle. Before stating the invariance principle, we require

DEFINITION 4.6 Let $S(t)$, $t \geq 0$, be a strongly continuous nonlinear semigroup in the closed subset of $C$ of the Banach space $X$ and let $x \in C$. The *omega-limit set* of $x$, denoted by $\Omega(x)$, is $\{x_1 \in X$: there exists a sequence $\{t_k\}_{k=1}^{\infty}$ of positive numbers such that $t_k \to \infty$ and $S(t_k)x \to x_1\}$.

The following proposition is proved in [290], Proposition 4.1, p. 166, and Theorem 4.1, p. 167.

PROPOSITION 4.5 Let $S(t)$, $t \geq 0$, be a strongly continuous nonlinear semigroup in the closed subset $C$ of the Banach space $X$ and let $x \in C$. Then, $\Omega(x)$ is closed, positive invariant, and a subset of the closure of $\gamma(x)$. If $\gamma(x)$ has compact closure, then $\Omega(x)$ is nonempty, compact, connected, and invariant. Moreover, $S(t)x$ approaches $\Omega(x)$ as $t$ approaches infinity in the sense that

$$\lim_{t\to\infty} \inf_{x_i \in \Omega(x)} \|S(t)x - x_1\| = 0$$

Further, $\Omega(x)$ is the smallest closed set that $S(t)x$ approaches as $t$ approaches infinity, in the sense that if $S(t)x$ approaches a set $C_1 \subset C$ as $t$ approaches infinity, then $\Omega(x) \subset \bar{C}_1$.

Proposition 4.5 assures the existence of the smallest closed set to which the trajectory approaches as time approaches infinity, namely, the omega-limit set, provided that the trajectory has compact closure. The following proposition, proved in [290], Theorem 4.2, p. 168, provides a method for identifying the omega-limit set.

PROPOSITION 4.6 (Invariance Principle) Let $S(t)$, $t \geq 0$, be a strongly continuous nonlinear semigroup in the closed subset $C$ of the Banach space $X$, let $C_1$ be a subset of $C$, and let $V$ be a Liapunov function for $S(t)$, $t \geq 0$, on $\bar{C}_1$. Let $x \in C$ be such that $\gamma(x) \subset C_1$ and $\gamma(x)$ has compact closure. Then, $\Omega(x) \subset M^+$, where $M^+$ is the largest positive invariant subset of $M_1 \stackrel{\text{def}}{=} \{x_1 \in \bar{C}_1: \dot{V}(x_1) = 0\}$ (in fact, $\Omega(x) \subset M^+ \cap \{x_1 \in \bar{C}_1: V(x_1) = k\}$ for some constant k). Further, $S(t)x$ approaches $M$ as t approaches infinity, where $M$ is the largest invariant subset of $M_1$.

In order to apply Propositions 4.5 and 4.6 it is necessary to know that the trajectory has compact closure. Theorem 3.5 assures that a trajectory has compact closure, provided that it is bounded. For population models the boundedness of the trajectories is connected to the effects of crowding and resource limitation. For many populations it is reasonable to expect that the population will cease to grow when the total population becomes large. Such a phenomenon is exhibited by the following result:

THEOREM 4.3 Let (2.1), (2.2), (2.22), (2.23) hold and let

(4.30) There exists $K > 0$ such that $\Sigma_{i=1}^n F(\phi)_i + \int_0^\infty G(\phi)_i(a)\, da \le 0$ for all $\phi \in L_+^1$ such that $\|\phi\|_{L^1} \ge K$.

Then, $T_\phi = \infty$ for all $\phi \in L_+^1$ and the strongly continuous nonlinear semigroup $S(t)$, $t \ge 0$, in $L_+^1$ as in Theorem 3.1 satisfies

(4.31) $\|S(t)\phi\|_{L^1} \le \max\{K, \|\phi\|_{L^1}\}$ for all $\phi \in L_+^1$ and $t \ge 0$.

*Proof.* Let $\phi \in L_+^1$. By Theorem 2.3 $T_\phi = \infty$ if (4.31) holds for $0 \le t < T_\phi$. Suppose that $\|\phi\|_{L^1} \le K$ and assume that there exists $\varepsilon > 0$ and $t \in (0,T_\phi)$ such that $\|S(t)\phi\|_{L^1} > K + \varepsilon$. Define

$$t_0 \stackrel{\text{def}}{=} \inf\{t \in [0,T_\phi): \|S(t)\phi\|_{L^1} > K + \varepsilon\}$$

Since the function $t \to \|S(t)\phi\|_{L^1}$ is continuous from $[0,T_\phi)$ to $[0,\infty)$, we have that $\|S(t_0)\phi\|_{L^1} \ge K + \varepsilon$ and there exists $r > 0$ such that $\|S(t)\phi\|_{L^1} > K$ for $t \in [t_0, t_0 + r)$. Define $V: [t_0, t_0 + r) \to [0,\infty)$ by

$$V(t) \stackrel{\text{def}}{=} \|S(t)\phi\|_{L^1} \qquad t_0 \le t < t_0 + r$$

For $t_0 \le t < t + h < t_0 + r$ we have that

$$h^{-1}[V(t + h) - V(t)]$$

$$= h^{-1} \sum_{i=1}^{n} \left( \int_{-t}^{\infty} (S(t + h)\phi)_i(t + h + c) - (S(t)\phi)_i(t + c) \right) dc$$

$$+ \int_{-t-h}^{-t} (S(t + h)\phi)_i(t + h + c)\, dc$$

$$\le \sum_{i=1}^{n} \left( \int_0^{\infty} [|h^{-1}((S(t + h)\phi)_i(a + h) - (S(t)\phi)_i(a))$$

$$- G(S(t)\phi)_i(a)| + G(S(t)\phi)_i(a)]\, da$$

$$+ h^{-1} \int_0^h [|(S(t + h)\phi)_i(a) - F(S(t)\phi)_i|$$

$$\left. + F(S(t)\phi)_i]\, da \right)$$

Thus, by (1.43), (1.44), and (4.30) we have that

$$\limsup_{h \to 0^+} h^{-1}[V(t + h) - V(t)] \le 0$$

for all $t \in [t_0, t_0 + r)$. This differential inequality implies that $V(t) \le V(t_0)$ for all $t \in [t_0, t_0 + r)$ (see Theorem 1.41, p. 15, in [183]). Thus, $\|S(t)\phi\|_{L^1} \le \|S(t_0)\phi\|_{L^1}$ for all $t \in [t_0, t_0 + r)$. Assume that $t_0 > 0$. Then, $\|S(t_0)\phi\|_{L^1} = K + \varepsilon$ and so $\|S(t)\phi\|_{L^1} \le K + \varepsilon$ for all $t \in [t_0, t_0 + r)$. Hence, $t_0$ is not the infimum of $t \in [0,T_\phi)$ such that $\|S(t)\phi\|_{L^1} > K + \varepsilon$. Thus, $t_0 = 0$. But then $K + \varepsilon \le \|S(t_0)\phi\|_{L^1} = \|\phi\|_{L^1} \le K$, which is a contradiction. Hence, if $\|\phi\|_{L^1} \le K$, then $\|S(t)\phi\|_{L^1} \le K$ for all $t \in [0,T_\phi)$.

Now suppose that $\|\phi\|_{L^1} > K$ and assume that there exists $t \in (0,T_\phi)$ such that $\|S(t)\phi\|_{L^1} > \|\phi\|_{L^1}$. Define

$$t_0 \overset{\text{def}}{=} \inf\{t \in (0,T_\phi): \|S(t)\phi\|_{L^1} > \|\phi\|_{L^1}\}$$

Then, $\|S(t_0)\phi\|_{L^1} = \|\phi\|_{L^1}$ and an argument similar to the one above shows that there exists $r > 0$ such that $\|S(t)\phi\|_{L^1} \le \|S(t_0)\phi\|_{L^1}$ for $t \in [t_0, t_0 + r)$. But $t_0$ cannot then be the infimum of $t \in (0,T_\phi)$ such that $\|S(t)\phi\|_{L^1} > \|\phi\|_{L^1}$. Hence, if $\|\phi\|_{L^1} > K$, then $\|S(t)\phi\|_{L^1} \le \|\phi\|_{L^1}$ for all $t \in [0,T_\phi)$. □

THEOREM 4.4 Let (2.1), (2.2), (2.22), (2.23), (2.49), (3.64), (3.73), and (4.30) hold. Then, $T_\phi = \infty$ for all $\phi \in L^1_+$, and if $S(t)$, $t \ge 0$, is the strongly continuous nonlinear semigroup in $L^1_+$ as in Theorem 3.1, then $S(t)\phi$ approaches $\Omega(\phi)$ as $t$ approaches infinity for all $\phi \in L^1_+$. Further, if $V$ is a Liapunov function for $S(t)$, $t \ge 0$, on $L^1_+$, then $S(t)\phi$ approaches $M$ as $t$ approaches infinity, where $M$ is the largest invariant subset of $M_1 \overset{\text{def}}{=} \{\psi \in L^1_+ : \dot{V}(\psi) = 0\}$.

*Proof.* From Theorem 4.3 we see that $T_\phi = \infty$ for all $\phi \in L^1_+$ and $S(t)$, $t \ge 0$, has the property that if $t > 0$ and $M$ is a bounded subset of $L^1_+$, then there exists $r > 0$ such that $\|S(s)\phi\|_{L^1} \le r$ for all $\phi \in M$, $s \in [0,t]$. From Theorem 3.5 we see that if $\phi \in L^1_+$ and $\gamma(\phi)$ is bounded, then $\gamma(\phi)$ has compact closure in $L^1_+$. From Theorem 4.4 we see that $\gamma(\phi)$ is bounded for all $\phi \in L^1_+$. From Propositions 4.5 and 4.6 we then obtain the conclusion. □

An illustration of the invariance principle is given by

THEOREM 4.5 Let (4.1), (4.2), (3.73), (4.30) hold and let $S(t)$, $t \ge 0$ be the strongly continuous nonlinear semigroup in $L^1_+$ as in Theorem 3.1. Let $i$ be a fixed integer in $[1,n]$ and let there exist continuous functions $\beta_0$: $[0,\infty) \to [0,\infty)$ and $\mu_0$: $[0,\infty) \to [0,\infty)$ satisfying (4.21) and

(4.32) $\beta_0$ is positive on $(0,\infty)$, $\lim_{a\to\infty} \beta_0(a) \overset{\text{def}}{=} \beta_\infty < \infty$, $\mu_0$ is nondecreasing on $[0,\infty)$, and $\lim_{a\to\infty} \mu_0(a) \overset{\text{def}}{=} \mu_\infty < \infty$.

(4.33) $\int_0^\infty e^{\omega a}\beta_0(a) \exp[-\int_0^a \mu_0(b)\, db]\, da = 1$ for some $\omega > 0$.

Then, $\lim_{t \to \infty} (S(t)\phi)_i = 0$ in $L^1(0,\infty;R)$ for all $\phi \in L^1_+$.

*Proof.* Define the function $p: [0,\infty) \to R$ by

$$p(a) \overset{\text{def}}{=} \exp\left[-\omega a + \int_0^a \mu_0(b)\, db\right] \times \left(1 - \int_0^a e^{\omega b} \beta_0(b) \exp\left[-\int_0^b \mu_0(\tau)\, d\tau\right] db\right) \qquad a \geq 0$$

From (4.32) and (4.33) we see that $p(a) > 0$ for all $a \geq 0$. We claim that $p$ is bounded on $[0,\infty)$. If $\mu_\infty \leq \omega$, then this claim follows from the fact that

$$\exp\left[-\omega a + \int_0^a \mu_0(b)\, db\right] \leq e^{(\mu_\infty - \omega)a} \leq 1$$

If $\mu_\infty > \omega$, then this claim follows from the fact that

$$\lim_{a \to \infty} \exp\left[-\omega a + \int_0^a \mu_0(b)\, db\right]$$

$$= \lim_{a \to \infty} \exp\left[-\omega a_0 + \int_0^{a_0} \mu_0(b)\, db\right] \exp\left[\int_{a_0}^a (\mu_0(b) - \omega)\, db\right]$$

$$= \infty$$

(where $a_0$ is chosen so that $\mu_0(b) - \omega > (\mu_\infty - \omega)/2$ for $b > a_0$) and the fact that

$$\lim_{a \to \infty} p(a) = \lim_{a \to \infty} \left\{\left[\left(1 - \int_0^a e^{\omega b} \beta_0(b) \exp\left[-\int_0^b \mu_0(\tau)\, d\tau\right] db\right)'\right] \times \left[\left(\exp\left[\omega a - \int_0^a \mu_0(b)\, db\right]\right)'\right]^{-1}\right\}$$

$$= \lim_{a \to \infty} \left\{-\frac{\beta_0(a)}{[\omega - \mu_0(a)]}\right\} = \frac{\beta_\infty}{\mu_\beta - \omega} < \infty$$

Observe that $p(0) = 1$ and

$$p'(a) = p(a)\mu_0(a) - \beta_0(a) - \omega p(a) \qquad a \geq 0$$

Observe also that p' is bounded on $[0,\infty)$, since p, $\mu_0$, and $\beta_0$ are bounded on $[0,\infty)$.

Define V: $L^1_+ \to R$ by

$$V(\phi) \stackrel{\text{def}}{=} \int_0^\infty \phi_i(a)p(a)\,da \qquad \phi \in L^1_+$$

Now use the argument of Theorem 4.2 and the fact that p and p' are bounded on $[0,\infty)$ to show that

$$\dot{V}(\phi) = \limsup_{h\to 0^+} h^{-1}[V(S(t)\phi) - V(\phi)]$$
$$\le \int_0^\infty [\beta_i(a,\phi) - p(a)\mu_i(a,\phi) + p'(a)]\phi_i(a)\,da$$

If $(S(t)\phi)_i$ is not 0 almost everywhere on $[0,\infty)$, then (4.21) and the fact that $p(a) > 0$ for $a \ge 0$ imply that

$$\dot{V}(\phi) \le \int_0^\infty [\beta_0(a) - p(a)\mu_0(a) + p'(a)]\phi_i(a)\,da$$
$$= -\omega V(\phi) < 0$$

Observe that V is continuous on $L^1_+$, since for $\phi, \hat{\phi} \in L^1_+$,

$$|V(\phi) - V(\hat{\phi})| \le \sup_{a\ge 0} p(a) \int_0^\infty |\phi_i(a) - \hat{\phi}_i(a)|\,da$$
$$\le \sup_{a\ge 0} p(a)\|\phi - \hat{\phi}\|_{L^1}$$

Thus, V is a Liapunov function for S(t), $t \ge 0$, on $L^1_+$. Further, $\dot{V}(\phi) < 0$ for all $\phi \in L^1_+$ such that $\phi_i \ne 0$ and $\dot{V}(\phi) = 0$ for all $\phi \in L^1_+$ such that $\phi_i = 0$. The conclusion of the theorem now follows from Theorem 4.4 with $M_1 = \{\phi \in L^1_+: \ \phi_i = 0\}$. □

The following examples provide applications of Theorem 4.5:

EXAMPLE 4.7 Let $R^n = R^1$. Let F and G be defined as in Example 4.4 (that is, F satisfies (2.17) with $\beta$ continuously differentiable,

G satisfies (2.18) with $\mu$ continuously differentiable, and (4.29) holds). Instead of (2.30) require that

(4.34) $\quad \beta(P) - \mu(P) \le 0$ for all $P \ge 0$ sufficiently large.

Observe that F satisfies (3.73) and that (4.34) implies (4.30). Define $\mu_0(a) \stackrel{\text{def}}{=} \mu(0)$ and $\beta_0(a) \stackrel{\text{def}}{=} \beta(0)e^{-\alpha a}$ for all $a \ge 0$, and observe that (4.32) holds. The hypothesis of Theorem 4.5 will be fulfilled provided that we take $\omega \stackrel{\text{def}}{=} \alpha + \mu(0) - \beta(0)$ in (4.33) and $\omega > 0$. Thus, a sufficient condition for the extinction of the population corresponding to any initial age distribution in $L^1_+$ is that

$$\beta(0) < \alpha + \mu(0)$$

EXAMPLE 4.8 Let $R^n = R^1$. Let F and G be defined as in Example 4.5 (that is, F satisfies (2.19) with $\beta$ continuously differentiable, G satisfies (2.18) with $\mu$ continuously differentiable, and (4.29) holds). Instead of (2.30) require that (4.34) holds. As in Example 4.7 we may apply Theorem 4.5 to conclude that a sufficient condition for the extinction of the population corresponding to any initial age distribution in $L^1_+$ is that

$$\beta(0)\alpha < \mu(0)[\alpha + \mu(0)]$$

EXAMPLE 4.9 Let $R^n = R^1$. Let F and G be defined as in Example 4.6 (that is, F satisfies (2.20) with $\beta$ continuously differentiable, G satisfies (2.18) with $\mu$ continuously differentiable, and (4.29) holds). Instead of (2.31) require that

(4.35) $\quad (\beta(P)/\alpha e) - \mu(P) \le 0$ for all $P \ge 0$ sufficiently large.

As in Example 4.7, Theorem 4.5 yields that a sufficient condition for the extinction of the population corresponding to any initial age distribution in $L^1_+$ is that

$$\beta(0) < [\alpha + \mu(0)]^2$$

Liapunov functions and the invariance principle provide one means of analyzing the asymptotic behavior of a population problem. Another approach to obtain such information involves the linearization of the nonlinear problem. The key idea of the method of linearization is to approximate the nonlinear problem by a more tractable linear problem. In Sections 4.4 and 4.5 we will apply the method of linearization to study the asymptotic behavior of the nonlinear problem (ADP). We first, however, specialize our efforts to the linear problem (ADP).

## 4.3 STABILITY AND INSTABILITY IN THE LINEAR CASE

In this section we restrict our attention to the case that the birth function F and the aging function G are both linear operators. For this case the asymptotic behavior of the associated strongly continuous semigroup of bounded linear operators in $L^1$ can be studied by analyzing the spectral properties of its infinitesimal generator. Our development in this section and in Sections 4.4 and 4.5 is modelled on similar treatments by J. Hale in the theory of functional differential equations ([144], [145]), by D. Henry in the theory of parabolic partial differential equations [150], and by J. Prüss in the theory of age-dependent population dynamics [240], [241], [242]. We begin with some definitions concerning the resolvent and spectrum of a closed linear operator in a Banach space (see [312], p. 209, [167], p. 172).

DEFINITION 4.7 Let T be a closed linear operator in the complex Banach space X. The *resolvent set* of T, denoted by $\rho(T)$, is the set of complex numbers $\lambda$ for which $(\lambda I - T)^{-1}$ exists and is an everywhere defined bounded linear operator in X. The *spectrum* of T, denoted by $\sigma(T)$, is the complement of $\rho(T)$ in the complex plane $\mathbb{C}$. The *continuous spectrum* of T, denoted by $C\sigma(T)$, is the set of complex numbers $\lambda$ such that $(\lambda I - T)^{-1}$ exists, is densely defined in X, but not bounded. The *residual spectrum* of T, denoted by $R\sigma(T)$, is the set of complex numbers $\lambda$ such that $(\lambda I - T)^{-1}$ exists, but is not densely defined. The *point spectrum* of T, denoted by $P\sigma(T)$, is the set of complex numbers $\lambda$ such that $Tx = \lambda x$ for some nonzero $x \in X$. If $\lambda \in P\sigma(T)$, then $\lambda$ is called an *eigenvalue* of T and a nonzero vector $x \in X$ such that $Tx = \lambda x$ is called an *eigenvector* of T corresponding to the eigenvalue $\lambda$. If $\lambda$ is an eigenvalue of T, then the null space $N(\lambda I - T)$ (that is, the subspace consisting of all $x \in X$ such that $(\lambda I - T)x = 0$) is called the *geometric eigenspace* of T with respect to $\lambda$, and its dimension is called the *geometric multiplicity* of $\lambda$.

REMARK 4.3 It is well known that for a closed linear operator T in a complex Banach space X, the resolvent set $\rho(T)$ is an open subset of the complex plane C and the mapping $\lambda \to (\lambda I - T)^{-1}$ is a holomorphic function of $\lambda$ in each component of $\rho(T)$. Moreover, $\rho(T)$, $C\sigma(T)$, $R\sigma(T)$, $P\sigma(T)$, are disjoint (possibly empty) sets whose union is the complex plane (see [312], p. 211).

DEFINITION 4.8 Let T be a closed linear operator in the complex Banach space X. If $\lambda \in \sigma(T)$, then the *generalized eigenspace* of T with respect to $\lambda$, denoted by $N_\lambda(T)$, is the smallest closed subspace of X containing $\cup_{k=1}^{\infty} N((\lambda I - T)^k)$ (see [144], p. 168). If $\lambda \in \sigma(T)$ and there exists a positive integer k such that $N((\lambda I - T)^k) = N((\lambda I - T)^{k+j})$ for all $j = 1, 2, \ldots$, then $\lambda$ is said to have *finite index* and the smallest such k for which this is true is called the *index* of $\lambda$ (see [94], p. 556-573, and also [269], p. 271).

DEFINITION 4.9 Let X be a Banach space. An everywhere defined bounded linear operator P in X is called a *projection* provided that $P^2 = P$ (see [94], p. 37). Let $M_1$ and $M_2$ be linear subspaces of X. Then, X is the *direct sum* of $M_1$ and $M_2$, denoted by $X = M_1 \oplus M_2$, provided that $M_1 \cap M_2 = \{0\}$ and for each $x \in X$ there exists the (necessarily unique) representation $x = x_1 + x_2$ where $x_1 \in M_1$, $x_2 \in M_2$ (see [94], p. 37).

The following proposition is proved in [167], pp. 155-156.

PROPOSITION 4.7 Let X be a Banach space. If P is a projection in X, then I - P is a projection in X, and $X = M_1 \oplus M_2$, where $M_1 = P(X)$, $M_2 = (I - P)(X)$, and $M_1$, $M_2$ are closed subspaces of X. Conversely, if $X = M_1 \oplus M_2$ is the direct sum of two closed subspaces $M_1$, $M_2$, then $P_1$, $P_2$ are projections in X, where $P_i x = x_i$, $x = x_1 + x_2$, $x_i \in M_i$, i = 1, 2.

DEFINITION 4.10 Let the Banach space X have the direct sum representation $X = M_1 \oplus M_2$, where $M_1$ and $M_2$ are closed subspaces of X. Let $P_1$, $P_2$ be the projections induced by $M_1$, $M_2$, that is, $P_i x = x_i$, where $x = x_1 + x_2$, $x_i \in M_i$, i = 1, 2. A closed linear operator T in X is said to be *completely reduced* by $M_1$ and $M_2$ provided that $T(M_1 \cap D(T)) \subset M_1$, $T(M_2 \cap D(T)) \subset M_2$, $P_1(D(T)) \subset D(T)$, and $P_2(D(T)) \subset D(T)$.

The case in which $\lambda_0$ is an isolated point of the spectrum of T is of particular interest in our development. In this case the holomorphic mapping $\lambda \to (\lambda I - T)^{-1}$ can be expanded in a Laurent series about $\lambda_0$. If $\lambda_0$ is also a pole of $(\lambda I - T)^{-1}$, then a useful direct sum decomposition of X results. The following results are proved in [312], p. 228, [269], p. 306, and [167], pp. 178-181:

PROPOSITION 4.8 Let T be a closed linear operator in the complex Banach space X and let $\lambda_0$ be an isolated point of $\sigma(T)$. Then,

$$(4.36) \qquad (\lambda I - T)^{-1} = \sum_{k=-\infty}^{\infty} (\lambda - \lambda_0)^k A_k$$

where for each integer k

$$(4.37) \qquad A_k \overset{\text{def}}{=} (2\pi i)^{-1} \int_\Gamma (\lambda - \lambda_0)^{-k-1} (\lambda I - T)^{-1} \, d\lambda$$

and $\Gamma$ is a positively oriented circle of sufficiently small radius such that no point of $\sigma(T)$ lies on or inside $\Gamma$. Further, $A_{-1}$ is a projection on X. If $\lambda_0$ is a pole of $(\lambda I - T)^{-1}$ of order m (that is, $A_{-m} \neq 0$ and $A_k = 0$ for all $k < -m$), then $\lambda_0$ is an eigenvalue of T with index m, $R(A_{-1}) = N((\lambda_0 I - T)^m)$, $R(I - A_{-1}) = R((\lambda_0 I - T)^k)$ for all $k \geq m$, $X = N((\lambda_0 I - T)^m) \oplus R((\lambda_0 I - T)^m)$, and T is completely reduced by the two linear subspaces occurring in this direct sum. Also, $R(A_{-1})$ is closed, $R(I - A_{-1})$ is closed, and T restricted to $R(A_{-1})$ is bounded with spectrum $\{\lambda_0\}$. Finally, if $R(A_{-1})$ is finite dimensional, then $\lambda_0$ is a pole of $(\lambda I - T)^{-1}$.

DEFINITION 4.11 Let T be a closed linear operator in the complex Banach space X and let $\lambda_0$ be an isolated point of $\sigma(T)$. If $A_{-1}$ is defined as in (4.37), then the *algebraic multiplicity* of $\lambda_0$ is the dimension of $R(A_{-1})$.

REMARK 4.4 In view of Proposition 4.7, if $\lambda_0$ is a pole of $(\lambda I - T)^{-1}$ of order m then $\lambda_0$ is an eigenvalue of T and $N_{\lambda_0}(T) = N((\lambda_0 I - T)^m) = R(A_{-1})$. In this case the geometric multiplicity of $\lambda_0$ is less than or equal to its algebraic multiplicity.

DEFINITION 4.12 Let T be a closed linear operator in the complex Banach space X. The *essential spectrum* of T, denoted by $E\sigma(T)$, is the set of $\lambda \in \sigma(T)$ such that at least one of the following holds: (i) $R(\lambda I - T)$ is not closed; (ii) $\lambda$ is a limit point of $\sigma(T)$; or (iii) $N_\lambda(T)$ is infinite dimensional (see [28], p. 107).

DEFINITION 4.13 Let T be a bounded linear operator in the complex Banach space X. The *spectral radius* of T, denoted by $r_\sigma(T)$, is the supremum of $\{|\lambda|: \lambda \in \sigma(T)\}$ (see [269], p. 262). The *essential spectral radius* of T, denoted by $r_{E\sigma}(T)$, is the supremum of $\{|\lambda|: \lambda \in E\sigma(T)\}$ (see [225], p. 473).

DEFINITION 4.14 Let T be a bounded linear operator in the Banach space X. The *measure of noncompactness* of T, denoted by $\alpha[T]$, is the infimum of $\varepsilon > 0$ such that $\alpha[T(M)] \leq \varepsilon\alpha[M]$ for all bounded sets M in X, where $\alpha[M]$ is the measure of noncompactness of M as in Definition 3.6 (see [225], p. 474).

The proof of the following proposition follows directly from Proposition 3.12.

PROPOSITION 4.9 Let X be a Banach space and let $T_1$, $T_2$ be bounded linear operators in X. Then,

(4.38) $\alpha[T_1] \leq |T_1|$

(4.39) $\alpha[T_1T_2] \leq \alpha[T_1]\alpha[T_2]$

(4.40) $\alpha[T_1 + T_2] \leq \alpha[T_1] + \alpha[T_2]$

(4.41) $\alpha[T_1] = 0$ if and only if $T_1$ is compact (that is, $T_1$ maps bounded sets into sets with compact closure).

PROPOSITION 4.10 Let T be a bounded linear operator in the complex Banach space X. The following hold:

(4.42) $r_\sigma(T) = \lim_{n\to\infty} |T^n|^{1/n}$

(4.43) $r_{E\sigma}(T) = \lim_{n\to\infty} (\alpha[T^n])^{1/n}$

*Proof.* The proof of (4.42), which is a classical result in the theory of linear operators, is given in [269], p. 262. The proof of (4.43), which is due to R. Nussbaum, is given in [225], p. 477. □

The term essential spectrum arises from the following result of F. Browder [28], Lemma 17, p. 110 (see also [255], Theorem 4.2, p. 148, for the proof in the case of a bounded linear operator).

PROPOSITION 4.11 Let T be a closed linear operator in the complex Banach space X and let $\lambda_0 \in \sigma(T) - E\sigma(T)$. Then, $N_{\lambda_0}(T) = R(A_{-1})$ [where $A_{-1}$ is as in (4.37)], $\lambda_0$ is a pole of $(\lambda I - T)^{-1}$, and $\lambda_0 \in P\sigma(T)$.

*Proof.* By virtue of Proposition 4.8 it suffices to show that $N_{\lambda_0}(T) = R(A_{-1})$, since $N_{\lambda_0}(T)$ is finite dimensional and $\lambda_0$ is isolated in $\sigma(T)$. Since $N_{\lambda_0}(T)$ is finite dimensional, $N_{\lambda_0}(T) = N((\lambda_0 I - T)^m)$ for some positive integer m, that is, $\lambda_0$ has index m.

We first show that $N_{\lambda_0}(T) \subset R(A_{-1})$. Observe that $N((\lambda_0 I - T)^0) = N(I) = \{0\} \subset R(A_{-1})$. Suppose for induction that for $1 \leq k \leq m$, $N((\lambda_0 I - T)^{k-1}) \subset R(A_{-1})$. Let $x \in N((\lambda_0 I - T)^k)$ and let $x_0 \overset{\text{def}}{=} (\lambda_0 I - T)x$. Then, $(\lambda_0 I - T)^{k-1} x_0 = (\lambda_0 I - T)^k x = 0$, and so $x_0 \in R(A_{-1})$ by the induction hypothesis. Let $\lambda \in \Gamma$, where $\Gamma$ is as in (1.37). Then,

$$x_0 = \lambda_0 x - Tx = (\lambda_0 - \lambda)x + (\lambda I - T)x$$

which implies that

$$x = (\lambda_0 - \lambda)^{-1} x_0 - (\lambda_0 - \lambda)^{-1}(\lambda I - T)x$$

and

$$(\lambda I - T)^{-1} x = (\lambda_0 - \lambda)^{-1}(\lambda I - T)^{-1} x_0 - (\lambda_0 - \lambda)^{-1} x$$

Since $x \in R(A_{-1})$, $x_0 = A_{-1} y_0$ for some $y_0 \in X$, and so $(\lambda I - T)^{-1} x_0 = (\lambda I - T)^{-1} A_{-1} y_0 = A_{-1}(\lambda I - T)^{-1} y_0 \in R(A_{-1})$. Now integrate with respect to $\lambda$ around $\Gamma$ and divide by $2\pi i$ to obtain

$$A_{-1}x = (2\pi i)^{-1} \int_\Gamma (\lambda_0 - \lambda)^{-1}(\lambda I - T)^{-1} x_0 \, d\lambda - (2\pi i)^{-1} \int_\Gamma (\lambda_0 - \lambda)^{-1} x \, d\lambda$$

which implies that

$$x = (2\pi i)^{-1} \int_\Gamma (\lambda_0 - \lambda)^{-1}(\lambda I - T)^{-1} x_0 \, d\lambda - A_{-1}x$$

Recall by Proposition 4.7 that $R(A_{-1})$ is closed, since $A_{-1}$ is a projection. Since $(\lambda I - T)^{-1} x_0 \in R(A_{-1})$ for all $\lambda \in \Gamma$, we must have $x \in R(A_{-1})$. Thus, $N((\lambda_0 I - T)^k) \subset R(A_{-1})$. By induction $N((\lambda_0 I - T)^m) = N_{\lambda_0}(T) \subset R(A_{-1})$.

Now assume for contradiction that $N_{\lambda_0}(T)$ is a proper subspace of $R(A_{-1})$. Define $T_0 \overset{\text{def}}{=} \lambda_0 I - T$ with $D(T_0) \overset{\text{def}}{=} R(A_{-1})$. By Proposition 4.8 $T_0$ is bounded in $R(A_{-1})$ with spectrum consisting only of $\{0\}$. By (4.42) $\lim_{k\to\infty}|T_0^k|^{1/k} = 0$. Further, $T_0(N_{\lambda_0}(T)) \subset N_{\lambda_0}(T)$, since $\lambda_0$ has finite index. Also, $N_{\lambda_0}(T)$ is closed, since it is finite dimensional by hypothesis. Let $X_1$ be the Banach quotient space $R(A_{-1})/N_{\lambda_0}(T)$ (see [255], p. 71) and let $T_1$ be the bounded linear operator in $X_1$ defined by $T_1(x + N_{\lambda_0}(T)) = T_0 x + N_{\lambda_0}(T)$, $x \in R(A_{-1})$. Observe that for all $k \geq 1$, $x \in R(A_{-1})$,

$$T_1^k(x + N_{\lambda_0}(T)) = T_0^k x + N_{\lambda_0}(T)$$

Also, for all $x \in R(A_{-1})$,

$$\begin{aligned} \|T_1^k(x + N_{\lambda_0}(T))\| &= \|T_0^k x + N_{\lambda_0}(T)\| \\ &\overset{\text{def}}{=} \inf\{\|T_0^k x - x_0\|: \ x_0 \in N_{\lambda_0}(T)\} \\ &\leq |T_0^k|\|x\| \end{aligned}$$

Thus, $\lim_{k\to\infty}|T_1^k|^{1/k} = 0$. We claim that $R(T_1)$ is a closed subspace of $X_1$. Observe that

$$R(T_1) = \{T_0 x + N_{\lambda_0}(T): \ x \in R(A_{-1})\}$$

Since $R(T_0)$ is closed and $N_{\lambda_0}(T)$ is finite dimensional by hypothesis, $R(T_0) + N_{\lambda_0}(T)$ is a closed subspace of X (see Lemma 4.3, p. 150, in [255]). By Theorem 5.2, p. 71, in [255], $[R(T_0) + N_{\lambda_0}(T)]/N_{\lambda_0}(T)$ is a Banach space and hence closed. Since $R(T_1) = [R(T_0) + N_{\lambda_0}(T)]/N_{\lambda_0}(T)$, $R(T_1)$ is closed. Further, $N(T_1) = \{0\}$, since $T_1(x + N_{\lambda_0}(T)) = N_{\lambda_0}(T)$

implies that $T_0 x \in N_{\lambda_0}(T)$, which implies that $0 = (\lambda_0 I - T)^m T_0 x = (\lambda_0 I - T)^{m+1} x = (\lambda_0 I - T)^m x$, which in turn implies that $x \in N_{\lambda_0}(T)$. Thus, $T_1$ is a one-to-one mapping of $X_1$ onto the closed subspace $R(T_1)$ of $X_1$. Since $X_1$ is nontrivial by assumption, there exists a constant $c > 0$ such that $\|T_1 z\| \geq c\|z\|$ for all $z \in X_1$ by the open mapping theorem (see [312], p. 75). Then, $\|T_1^k z\| \geq c^k \|z\|$ for all $z \in X_1$, $k \geq 1$, which means that $|T_1^k|^{1/k} \geq c$. Thus, a contradiction is obtained and so $N_{\lambda_0}(T) = R(A_{-1})$. □

REMARK 4.5 Notice that the proof of Proposition 4.11 shows that if $T$ is a closed linear operator in the complex Banach space $X$ and $\lambda_0$ is isolated in $\sigma(T)$, then $N_{\lambda_0}(t) \subset R(A_{-1})$. The converse of Proposition 4.11 is not true, as is seen by taking $X$ infinite dimensional, $T \overset{\text{def}}{=} 0$, and $\lambda_0 \overset{\text{def}}{=} 0$, and observing that $\lambda_0$ is a simple pole of $T$, but $\lambda_0 \in E\sigma(T)$, since $N_{\lambda_0}(T) = N(\lambda_0 I - T) = X$. In [28], Lemma 17, p. 110, F. Browder has shown that if $T$ is a closed linear operator in $X$, $\lambda_0$ is isolated in $\sigma(T)$, $N_{\lambda_0}(T)$ is finite dimensional, and $\lambda_0$ is a pole of $(\lambda I - T)^{-1}$, then $\lambda_0 \in \sigma(T) - E\sigma(T)$.

We are now prepared to connect the asymptotic behavior of a strongly continuous semigroup of bounded linear operators with the spectral properties of its infinitesimal generator. For such a semigroup $T(t)$, $t \geq 0$, there always exist constants $\omega \in R$ and $M \geq 1$ such that $|T(t)| \leq Me^{\omega t}$ for all $t \geq 0$. The determination of the values of $\omega$ for which such an estimate holds is very useful in studying the asymptotic behavior of the trajectories of $T(t)$, $t \geq 0$. For this purpose we require

PROPOSITION 4.12 Let $T(t)$, $t \geq 0$, be a strongly continuous semigroup of bounded linear operators in the Banach space $X$ and let $B$ be its infinitesimal generator. The following hold:

(4.44) $\quad \omega_0 \overset{\text{def}}{=} \lim_{t\to\infty} t^{-1} \log(|T(t)|)$ exists $(-\infty \leq \omega_0 < \infty)$.

(4.45) $\omega_1 \overset{\text{def}}{=} \lim_{t\to\infty} t^{-1} \log(\alpha[T(t)])$ exists $(-\infty \le \omega_1 < \infty)$.

(4.46) $\omega_1 \le \omega_0$

(4.47) If $\omega > \omega_0$, then there exists a constant $M(\omega) \ge 1$ such that $|T(t)| \le M(\omega)e^{\omega t}$ for all $t \ge 0$.

*Proof.* The existence of the limit in (4.44) is proved in [94], Corollary 5, p. 619. The existence of the limit in (4.45) follows immediately from Lemma 4, p. 618, in [94], if it can be established that the function $t \to \log(\alpha[T(t)])$ is subadditive. This subadditivity follows from (4.39), since $\alpha[T(t_1 + t_2)] = \alpha[T(t_1)T(t_2)] \le \alpha[T(t_1)]\alpha[T(t_2)]$ for $t_1, t_2 \ge 0$. The inequality (4.46) follows immediately from (4.38). The claim (4.47) is proved in [94], Corollary 5, p. 619. □

DEFINITION 4.15 Let $T(t)$, $t \ge 0$, be a strongly continuous semigroup of bounded linear operators in the Banach space $X$ and let $B$ be its infinitesimal generator. The limit in (4.44), denoted by $\omega_0(B)$, is called the *growth bound* of $T(t)$, $t \ge 0$, and the limit in (4.45), denoted by $\omega_1(B)$, is called the $\alpha$-*growth bound* of $T(t)$, $t \ge 0$.

REMARK 4.6 Henceforth, we will not distinguish between real and complex Banach spaces. If $X$ is a real Banach space and $T(t)$, $t \ge 0$, is a strongly continuous semigroup of bounded linear operators in $X$ with infinitesimal generator $B$, then $T(t)$, $t \ge 0$, and $B$ may be extended to the complexification of $X$ by the complexification of $T(t)$, $t \ge 0$, and $B$ (see [255], p. 153).

With the convention of Remark 4.6 the connection between the spectrum of a strongly continuous semigroup of bounded linear operators and the spectrum of the infinitesimal generator is given by

PROPOSITION 4.13 Let $T(t)$, $t \ge 0$, be a strongly continuous semigroup of bounded linear operators in the Banach space $X$ with infinitesimal generator $B$. The following hold:

(4.48) $\quad r_\sigma(T(t)) = \exp[\omega_0(B)t]$ for $t \ge 0$

(4.49) $\quad r_{E\sigma}(T(t)) = \exp[\omega_1(B)t]$ for $t > 0$

(4.50) $\quad \{e^{\lambda t}: \lambda \in \sigma(B)\} \subset \sigma(T(t))$ for $t \ge 0$

(4.51) $\quad \{e^{\lambda t}: \lambda \in P\sigma(B)\} = P\sigma(T(t)) - \{0\}$ for $t \ge 0$

(4.52) If $\mu \in P\sigma(T(t))$ for some fixed $t > 0$, $\mu \neq 0$, then there exists $\lambda \in P\sigma(B)$ such that $e^{\lambda t} = \mu$, and $N(e^{\lambda t}I - T(t))$ is the closed linear extension of the linearly independent subspaces $N((\lambda_k I - B))$, where $\lambda_k \in P\sigma(B)$ and $e^{\lambda_k t} = \mu$.

(4.53) If $\lambda \in \sigma(B)$, then $N_\lambda(B) \subset N_{e^{\lambda t}}(T(t))$ for $t \ge 0$.

(4.54) $\quad \{e^{\lambda t}: \lambda \in E\sigma(B)\} \subset E\sigma(T(t))$ for $t > 0$

(4.55) $\quad \sup_{\lambda \in \sigma(B)} \operatorname{Re} \lambda \le \omega_0(B)$

(4.56) $\quad \sup_{\lambda \in E\sigma(B)} \operatorname{Re} \lambda \le \omega_1(B)$

(4.57) $\quad \omega_0(B) = \max\{\omega_1(B), \sup_{\lambda \in \sigma(B) - E\sigma(B)} \operatorname{Re} \lambda\}$

*Proof.* To prove (4.48) let $t > 0$ and observe from (4.42) that

$$\omega_0(B) = \lim_{s\to\infty} s^{-1} \log(|T(s)|) = \lim_{n\to\infty} (nt)^{-1} \log(|T(nt)|)$$
$$= \lim_{n\to\infty} t^{-1} \log(|T(t)^n|^{1/n}) = t^{-1} \log(r_\sigma(T(t))$$

The proof of (4.49) follows in a similar fashion from (4.43). The proof of (4.50) is given in [151], Theorem 16.7.1, p. 467 and in [76], Theorem 2.16, p. 42. The proof of (4.51) and (4.52) is given in [151], Theorem 16.7.2, p. 467.

To prove (4.53) let $x \in N(\lambda I - B)$. Since $u(t) \overset{\text{def}}{=} e^{\lambda t}x$ satisfies the initial value problem

$$\frac{d}{dt} u(t) = Bu(t) \qquad t \ge 0,\ u(0) = x$$

and $T(t)x$ is the unique solution of this problem (see [167], p. 481), we have that $T(t)x = e^{\lambda t}x$ for $t \geq 0$. Thus, $N(\lambda I - B) \subset N(e^{\lambda t}I - T(t))$ for $t \geq 0$. Let k be a positive integer and suppose for induction that $N((\lambda I - B)^k) \subset N((e^{\lambda t}I - T(t))^k)$ for $t \geq 0$. Let $x \in N((\lambda I - B)^{k+1})$. Then, $(\lambda I - B)^k x \in N(\lambda I - B)$, which implies that $(\lambda I - B)^k x \in N(e^{\lambda t}I - T(t))$. Thus, $0 = (e^{\lambda t}I - T(t))(\lambda I - B)^k x = (\lambda I - B)^k(e^{\lambda t}I - T(t))x$, which by the induction hypothesis implies that $(e^{\lambda t}I - T(t))x \in N((e^{\lambda t}I - T(t))^k)$. Consequently, $N((\lambda I - B)^{k+1}) \subset N((e^{\lambda t}I - T(t))^{k+1})$ for all $t \geq 0$, $k = 1, 2, \ldots$ .

To prove (4.54) let $\lambda \in E\sigma(B)$ and let $r > 0$. By (4.50) $e^{\lambda r} \in \sigma(T(r))$. Assume $e^{\lambda r} \in \sigma(T(r)) - E\sigma(T(r))$. Then, $e^{\lambda r}$ is isolated in $\sigma(T(r))$, $\mathcal{N}_{e^{\lambda r}}(T(r))$ is finite dimensional, and $R(e^{\lambda r}I - T(r))$ is closed. We will prove that $\lambda$ is isolated in $\sigma(B)$, $\mathcal{N}_\lambda(B)$ is finite dimensional, and $R(\lambda I - B)$ is closed, which means that $\lambda \notin E\sigma(B)$, a contradiction.

By (4.53) $\mathcal{N}_\lambda(B) \subset \mathcal{N}_{e^{\lambda r}}(T(r))$, so $\mathcal{N}_\lambda(B)$ must be finite dimensional. To show that $\lambda$ is isolated in $\sigma(B)$, suppose there exists a sequence $\{z_k\} \subset \sigma(B)$, $z_k \neq \lambda$ for any k, such that $z_k \to \lambda$. By (4.50) $e^{z_k r} \in \sigma(T(r))$ for all k. Further, $e^{z_k r} \neq e^{\lambda r}$ for all k sufficiently large, since $e^{z_k r} = e^{\lambda r}$ if and only if $\operatorname{Re} z_k = \operatorname{Re} \lambda$ and $\operatorname{Im} z_k = \operatorname{Im} \lambda + 2j\pi$ for some integer j. Since $e^{z_k r} \to e^{\lambda r}$, $e^{\lambda r}$ is not isolated in $\sigma(T(r))$. Consequently, $\lambda$ must be isolated in $\sigma(B)$.

It remains to prove that $R(\lambda I - B)$ is closed. We will use the following fact proved in [255], Theorem 5.1, p. 70:

(4.58) Let T be a one-to-one closed linear operator in the Banach space X. A necessary and sufficient condition that $R(T)$ is closed in that there exists a constant c such that $\|x\| \leq c\|Tx\|$ for all $x \in X$.

Since $\mathcal{N}_{e^{\lambda r}}$ is finite dimensional, there exists a positive integer m such that $\mathcal{N}_{e^{\lambda r}}(T(r)) = N((e^{\lambda r}I - T(r))^m)$. By Proposition 4.8 and Proposition 4.11 $e^{\lambda r}$ is a pole of the mapping $\mu \to (\mu I - T(r))^{-1}$,

$e^{\lambda r} \in P\sigma(T(r))$, and $X = M_1 \oplus M_2$, where $M_1$, $M_2$ are the closed subspaces $M_1 \overset{\text{def}}{=} N_{e^{\lambda r}}(T(r)) = N((e^{\lambda r}I - T(r))^m)$, $M_2 \overset{\text{def}}{=} R((e^{\lambda r}I - T(r))^m)$. Let $P_1$, $P_2$ be the projections induced by this direct sum representation, that is, $P_i x = x_i$, $x = x_1 + x_2$, $x_i \in M_i$. For $t \geq 0$ $T(t)$ commutes with $(e^{\lambda r}I - T(r))^m$, and so $T(t)$ must commute with $P_1$, $P_2$ for each $t \geq 0$. In fact, for each $t \geq 0$, $T(t)$ is completely reduced by $M_1$ and $M_2$. It now follows by the definition of infinitesimal generator [see (3.8)] that $B$ is completely reduced by $M_1$ and $M_2$. Let $T_i(t)$, $t \geq 0$, and $B_i$, $i = 1, 2$, denote the restriction of $T(t)$, $t \geq 0$, and $B$ to $M_i$, $i = 1, 2$. Observe that $B_i$ is the infinitesimal generator of $T_i(t)$, $t \geq 0$, on $M_i$, $i = 1, 2$.

We claim that $e^{\lambda r}I - T_2(r)$ is one-to-one on $M_2$. Suppose that $N(e^{\lambda r}I - T_2(r)) \neq \{0\}$. By (4.52) there exists $\lambda_k \in P\sigma(B_2)$ such that $e^{\lambda r} = e^{\lambda_k r}$. Thus, there exists $x \in M_2$, $x \neq 0$, such that $\lambda_k x = B_2 x$. But by (4.52), $x \in N(\lambda_k I - B_2) \subset N(e^{\lambda r}I - T_2(r)) \subset N(e^{\lambda r}I - T(r)) \subset M_1$. Hence, $x \in M_1 \cap M_2 = \{0\}$, which yields a contradiction.

We next claim that $R(e^{\lambda r}I - T_2(r))$ is closed in $M_2$. Let $\{y_k\} \subset R(e^{\lambda r}I - T_2(r))$ such that $y_k \to y_0 \in M_2$. Since $R(e^{\lambda r}I - T(r))$ is closed in $X$, there exists $x_0 \in X$ such that $y_0 = (e^{\lambda r}I - T(r))x_0 = (e^{\lambda r}I - T(r))(P_1x_0 + P_2x_0) = (e^{\lambda r}I - T(r))P_1x_0 + (e^{\lambda r}I - T(r))P_2x_0$. By the uniqueness of the direct sum representation $(e^{\lambda r}I - T(r)) P_1x_0 = 0$, and hence $(e^{\lambda r}I - T_2(r))P_2x_0 = y_0$.

We next claim that there exists a constant $c_1$ such that $\|(e^{\lambda r}I - T(r))x\| \leq c_1\|(\lambda I - B)x\|$ for all $x \in D(B)$. Let $x \in D(B)$ and define

$$u(t) \overset{\text{def}}{=} (e^{\lambda t}I - T(t))x \qquad t \geq 0$$

$$v(t) \overset{\text{def}}{=} \int_0^t e^{\lambda(t-s)}T(s)(\lambda I - B)x\,ds \qquad t \geq 0$$

Observe that $u$ and $v$ are both solutions of the initial value problem

$$\frac{d}{dt}w(t) = \lambda w(t) + (\lambda I - B)T(t)x \qquad t \geq 0,\ w(0) = 0$$

By uniqueness of the solution to this problem (see [167], p. 486), $u(t) = v(t)$ for $t \geq 0$. By the uniform boundedness of $T(t)$ from $[0,r]$ to $B(X)$, there exists a constant $c_1$ such that for all $x \in D(B)$,

$$(4.59) \qquad \|(e^{\lambda r}I - T(r))x\| = \left\| \int_0^r e^{\lambda(r-s)}T(s)(\lambda I - B)x\, ds \right\|$$
$$\leq c_1\|(\lambda I - B)x\|$$

Now use (4.58) to obtain a constant $c_2$ such that for all $x \in M_2$, $\|x\| \leq c_2\|(e^{\lambda r}I - T(r))x\|$. Then, from (4.59) we have that for $x \in M_2$

$$\|x\| \leq c_1c_2\|(\lambda I - B_2)x\|$$

By (4.58) $R(\lambda I - B_2)$ is closed in $M_2$. We now argue that $R(\lambda I - B)$ is closed in $X$. Let $\{y_k\} \subset R(\lambda I - B)$ such that $y_k \to y_0$. For $k = 1, 2, \ldots$, there exists $x_k \in D(B)$ such that $y_k = (\lambda I - B)x_k$. Then, $P_iy_k = (\lambda I - B)P_ix_k = (\lambda I - B_i)P_ix_k \to P_iy_0$, $i = 1, 2$. Since $R(\lambda I - B_2)$ is closed in $M_2$, there exists $x_{0,2} \in M_2$ such that $(\lambda I - B_2)x_{0,2} = P_2y_0$. Since $M_1$ is finite dimensional, $R(\lambda I - B_1)$ is closed in $M_1$ and there exists $x_{0,1} \in M_1$ such that $(\lambda I - B_1)x_{0,1} = P_1y_0$. Thus, $(\lambda I - B)(x_{0,1} + x_{0,2}) = P_1y_0 + P_2y_0 = y_0$. Hence, $R(\lambda I - B)$ is closed, and our original assumption must be false. Thus, (4.54) is proved.

The proof of (4.55) follows immediately from (4.50) and (4.48). The proof of (4.56) follows immediately from (4.54) and (4.49). To prove (4.57) define

$$\omega_2(B) \stackrel{\text{def}}{=} \max\{\omega_1(B), \sup_{\lambda \in \sigma(B) - E\sigma(B)} \operatorname{Re} \lambda\}$$

From (4.46) and (4.55) we see immediately that $\omega_2(B) \leq \omega_0(B)$. To prove that $\omega_0(B) \leq \omega_2(B)$, it suffices by (4.48) to prove that for some $t > 0$, $r_\sigma(T(t)) \leq \exp[\omega_2(B)t]$. Let $t > 0$ and let $\mu \in \sigma(T(t))$ such that $\mu \neq 0$. If $\mu \in E\sigma(T(t))$, then by (4.49) $|\mu| \in \exp[\omega_1(B)t] \leq \exp[\omega_2(B)t]$. If $\mu \in \sigma(T(t)) - E\sigma(T(t))$, then by Proposition 4.11 $\mu \in P\sigma(T(t))$ and by (4.52) there exists $\lambda \in P\sigma(B)$ such that $e^{\lambda t} = \mu$.

By (4.54) $\lambda \in \sigma(B) - E\sigma(B)$. Thus, $|\mu| \leq e^{\mathrm{Re}\lambda t} \leq \exp[\omega_2(B)t]$. Hence, $r_\sigma(T(t)) \leq \exp[\omega_2(B)t]$. □

REMARK 4.7 In [151], Theorem 16.7.3, p. 469, it is proved that if $\mu \in R\sigma(T(t))$ for some fixed $t > 0$, $\mu \neq 0$, then at least one of the solutions $\lambda$ of the equation $e^{\lambda t} = \mu$ lies in $R\sigma(B)$ and none lies in $P\sigma(B)$. On the other hand, an example is given in [151], Theorem 16.7.4, p. 469, for which $\mu \in C\sigma(T(t))$, $\mu \neq 0$, and all the solutions $\lambda$ of the equation $e^{\lambda t} = \mu$ lie in $\rho(B)$. Thus, it is not necessarily possible to account for the continuous spectrum of $T(t)$ in terms of the spectrum of $B$.

REMARK 4.8 Notice that if $T(t)$ is compact for some $t > 0$, then $\alpha[T(t)] = 0$ and $\omega_1(B) = -\infty$. Thus, by (4.53), $\omega_0(B) = \sup_{\lambda \in \sigma(B) - E\sigma(B)} \mathrm{Re}\,\lambda$. The ideas of Proposition 4.13 for this case are due to J. Hale, who used them in the study of functional differential equations on finite history intervals. For age-dependent population dynamics the ideas of Proposition 4.13 were first used by J. Prüss. The problem of relating the growth bound $\omega_0(B)$ to the spectrum of $B$ is of much interest. If $X$ is finite dimensional, then $\omega_0(B) = \sup_{\lambda \in \sigma(B)} \mathrm{Re}\,\lambda$. If $X$ is infinite dimensional, then strict inequality may hold in (4.55), and examples are given in [151], p. 665, [313], and [76], p. 44. If $X$ is infinite dimensional and $B$ is bounded, or $T(t)$, $t > 0$, is a holomorphic semigroup, or $T(t)$, $t \geq 0$, is a differentiable semigroup for $t > t_0 \geq 0$, then equality does hold. For a discussion of cases in which $\omega_0(B) = \sup_{\lambda \in \sigma(B)} \mathrm{Re}\,\lambda$ we refer the reader to [94], p. 26, [278], [81], [125], and [119].

We illustrate the preceding results with the following simple example of linear age-dependent population dynamics:

EXAMPLE 4.10 Let $X = L^1(0,\infty;R) = L^1$, let $F\colon L^1 \to R$ be defined by $F\phi \overset{\mathrm{def}}{=} 0$ for all $\phi \in L^1$, and let $G\colon L^1 \to L^1$ be defined by $G\phi \overset{\mathrm{def}}{=} -\mu\phi$ for all $\phi \in L^1$, where $\mu$ is a positive constant. According to

Proposition 3.2, Proposition 3.7, (1.12), and (2.62),

$B\phi \overset{\text{def}}{=} -\phi' + G(\phi) = -\phi' - \mu\phi$, $D(B) = \{\phi \in L^1$: $\phi$ is absolutely continuous on $[0,\infty)$, $\phi' \in L^1$, and $\phi(0) = F\phi = 0\}$

is the infinitesimal generator of the strongly continuous semigroup of bounded linear operators $S(t)$, $t \geq 0$, in $L^1$ given by

$$(S(t)\phi)(a) = \begin{cases} 0 & a < t \\ \phi(a - t)e^{-\mu t} & a > t \end{cases}$$

for all $\phi \in L^1$, $t \geq 0$. Obviously, we have that for all $t \geq 0$, $|S(t)| = \alpha[S(t)] = e^{-\mu t}$. Consequently, $\omega_0(B) = \omega_1(B) = -\mu$, and by (4.55), $\lambda \in \rho(B)$ for all $\lambda \in \mathbb{C}$ such that $\operatorname{Re}\lambda > -\mu$.

We claim that if $\lambda \in \mathbb{C}$ such that $\operatorname{Re}\lambda \leq -\mu$, then $\lambda \in R\sigma(B)$. Suppose that $\operatorname{Re}\lambda \leq -\mu$, $\phi \in D(B)$, and $\psi \in L^1(0,\infty;\mathbb{C})$ such that $(\lambda I - B)\phi = \psi$. Then,

$$\phi(a) = \int_0^a e^{-(\lambda+\mu)(a-b)}\psi(b)\, db \qquad a \geq 0$$

If $\psi = 0$, then $\phi = 0$, and hence $\lambda I - B$ is one-to-one. Observe that $\lim_{a\to\infty} \phi(a) = 0$, since if $0 < a_1 < a_2$, then

$$|\phi(a_2) - \phi(a_1)| = \left|\int_{a_1}^{a_2} \phi'(b)\, db\right|$$

$$\leq \int_{a_1}^{a_2} |\phi'(b)|\, db \to 0 \text{ as } a_1, a_2 \to \infty$$

Thus,

$$\lim_{a\to\infty} |e^{-(\lambda+\mu)a}| \left|\int_0^a e^{(\lambda+\mu)b}\psi(b)\, db\right| = 0$$

and since $|e^{-(\lambda+\mu)a}| = e^{-(\operatorname{Re}\lambda+\mu)a} \geq 1$, $a \geq 0$,

$$\int_0^\infty e^{(\lambda+\mu)b}\psi(b)\, db = 0$$

Now suppose that $\psi$ belongs to the closure of $R(\lambda I - B)$ and let $\{\psi_n\} \subset R(\lambda I - B)$ such that $\psi_n \to \psi$. Since $|e^{(\lambda+\mu)b}| = e^{(\mathrm{Re}\lambda+\mu)b} \leq 1$ for $b \geq 0$, we have that

$$\lim_{n\to\infty} \int_0^\infty |e^{(\lambda+\mu)b}[\psi_n(b) - \psi(b)]| \, db$$

$$= \lim_{n\to\infty} \int_0^\infty |\psi_n(b) - \psi(b)| \, db = 0$$

Thus, for all $\psi$ in the closure of $R(\lambda I - B)$,

$$\int_0^\infty e^{(\lambda+\mu)b}\psi(b) \, db = 0$$

Take $\psi_0(b) \overset{\text{def}}{=} e^{-\mu b}$, $b \geq 0$, and observe that

$$\int_0^\infty e^{(\lambda+\mu)b}\psi_0(b) \, db = -\lambda^{-1}$$

Thus, this $\psi_0$ is not in the closure of $R(\lambda I - B)$. Hence, $\sigma(B) = R\sigma(B) = E\sigma(B) = \{\lambda \in \mathbb{C}: \ \mathrm{Re}\,\lambda \leq -\mu\}$.

We next consider the spectrum of $S(t)$. Let $\lambda \in \mathbb{C}$, $\lambda \neq 0$, and observe that for $\phi \in L^1(0,\infty;\mathbb{C})$, $t > 0$,

$$(\lambda I - S(t))\phi(a) = \begin{cases} \lambda\phi(a) & a < t \\ \lambda\phi(a) - e^{-\mu t}\phi(a - t) & a > t \end{cases}$$

Suppose that $(\lambda I - S(t))\phi = 0$. Then, $\phi(a) = 0$ for $0 < a < t$, and $\phi(a) = e^{-\mu t}\lambda^{-1}\phi(a - t)$ for $a > t$. Thus, $\phi = 0$, and so $(\lambda I - S(t))$ is one-to-one.

Suppose now that $(\lambda I - S(t))\phi = \psi$ for $\phi, \psi \in L^1(0,\infty;\mathbb{C})$, $t > 0$. Then

$$\phi(a) = \begin{cases} \lambda^{-1}\psi(a) & a < t \\ \lambda^{-1}[e^{-\mu t}\phi(a - t) + \psi(a)] & a > t \end{cases}$$

Thus, for $m = 0, 1, 2, \ldots$, $mt < a < (m + 1)t$,

$$\phi(a) = \lambda^{-1} \sum_{k=0}^{m} (e^{-\mu t}\lambda^{-1})^k \psi(a - kt)$$

Then,

$$\int_0^\infty \phi(a)\, da = \lambda^{-1} \sum_{m=0}^{\infty} \int_{mt}^{(m+1)t} \sum_{k=0}^{m} (e^{-\mu t}\lambda^{-1})^k \psi(a - kt)\, da$$

$$= \lambda^{-1} \sum_{m=0}^{\infty} \sum_{k=0}^{m} \int_{(m-k)t}^{(m+1-k)t} (e^{-\mu t}\lambda^{-1})^k \psi(\tau)\, d\tau$$

$$= \lambda^{-1} \sum_{k=0}^{\infty} \sum_{m=k}^{\infty} \int_{(m-k)t}^{(m+1-k)t} (e^{-\mu t}\lambda^{-1})^k \psi(\tau)\, d\tau$$

$$= \lambda^{-1} \left[ \sum_{k=0}^{\infty} (e^{-\mu t}\lambda^{-1})^k \right] \int_0^\infty \psi(a)\, da$$

Thus, if $|\lambda| \leq e^{-\mu t}$, $\lambda \neq 0$, $\int_0^\infty \psi(a)\, da = 0$ for all $\psi \in R(\lambda I - S(t))$, and, in fact, for all $\psi$ in the closure of $R(\lambda I - S(t))$. Consequently, the closure of $R(\lambda I - S(t))$ is a proper subset of $L^1(0,\infty;\mathbb{C})$ and $\lambda \in R\sigma(S(t))$ for $|\lambda| \leq e^{-\mu t}$, $\lambda \neq 0$. A similar argument shows that $0 \in R\sigma(S(t))$. If $|\lambda| > e^{-\mu t} = |S(t)|$, then $\lambda \in \rho(S(t))$ (see [312], Theorem 3, p. 211). Thus, $\sigma(S(t)) = R\sigma(S(t)) = E\sigma(S(t)) = \{\lambda \in \mathbb{C}: |\lambda| \leq e^{-\mu t}\}$ (see Remark 4.7).

REMARK 4.9 In [28], Lemma 19, p. 113, F. Browder proves that if T is a closed densely defined operator in the Banach space X, C is a compact linear operator defined on all of X, and $P\sigma(T)$ and $P\sigma(T + C)$ are nowhere dense, then $E\sigma(T) = E\sigma(T + C)$.

We next prove a result we will need in the stability analysis of nonlinear age-dependent population dynamics, namely, that the $\alpha$-growth bound of an infinitesimal generator is invariant under compact perturbations.

PROPOSITION 4.14 Let $T(t)$, $t \geq 0$, be a strongly continuous semigroup of bounded linear operators in the Banach space X with infinitesimal generator B and let C be a bounded linear operator in X.

Then, $B + C$ is the infinitesimal generator of a strongly continuous semigroup of bounded linear operators $S(t)$, $t \geq 0$, in $X$ and

$$(4.60) \qquad S(t)x = T(t)x + \int_0^t T(t - s)CS(s)x\, ds \qquad t \geq 0,\ x \in X$$

Moreover, if $C$ is also compact (that is, $C$ maps bounded sets into sets with compact closure), then $\omega_1(B) = \omega_1(B + C)$.

*Proof.* That $B + C$ is the infinitesimal generator of a strongly continuous semigroup of bounded linear operators in $X$ with the representation (4.60) is proved in [167], Theorem 2.1, p. 495. To prove the last statement of the proposition it suffices to prove that $\alpha[T(t)] = \alpha[S(t)]$ for each $t > 0$. For $t > 0$ define $K(t)\colon\ X \to X$ by

$$K(t)x = \int_0^t T(t - s)CS(s)x\, ds \qquad x \in X$$

By (4.40) and (4.60) it suffices to prove that $\alpha[K(t)] = 0$ for $t > 0$.

Let $M$ be a bounded set in $X$, let $t > 0$, and define

$$M_{t,0} \overset{\text{def}}{=} \{K(t)x\colon\ x \in M\}$$
$$M_{t,1} \overset{\text{def}}{=} \{CS(s)x\colon\ x \in M,\ 0 \leq s \leq t\}$$
$$M_{t,2} \overset{\text{def}}{=} \{T(t - s)x\colon\ x \in M_{t,1},\ 0 \leq s \leq t\}$$

Since $S(t)$, $t \geq 0$, is uniformly bounded on finite intervals, $\{S(s)x\colon x \in M,\ 0 \leq s \leq t\}$ is bounded. Since $C$ is compact, $\alpha[M_{t,1}] = 0$. Thus, $M_{t,1}$ is totally bounded; that is, for every $\varepsilon > 0$, there exists a finite set of points of $M_{t,1}$ such that every point of $M_{t,1}$ has distance less than $\varepsilon$ from at least one of these points (see [312], p. 13). Let $\varepsilon > 0$ and let $x_1, \ldots, x_m$ belong to $M_{t,1}$ such that if $x \in M_{t,1}$, then $\|x - x_i\| < \varepsilon$ for some $i \in [1,m]$. By the strong continuity of $T(t)$, $t \geq 0$, and its uniform boundedness on finite intervals, there exists a positive integer $k$ such that if $x \in M_{t,1}$ and $0 \leq s \leq t$, then

$$\left\|T(t - s)x - T\left(\frac{j(t - s)}{k}\right)x_i\right\| < \varepsilon$$

for some $j \in [1,k]$, $i \in [1,m]$. Thus, $M_{t,2}$ is totally bounded, and hence its closure is compact (see [312], p. 13). Consequently, $\alpha[M_{t,2}] = 0$ by (3.56). Since $M_{t,0} \subset \overline{co}\ tM_{t,2}$, $\alpha[M_{t,0}] = 0$ by (3.55) and (3.59). □

If the growth bound $\omega_0(B)$ of a linear semigroup $T(t)$, $t \geq 0$, is negative then (4.47) implies that 0 is an exponentially stable equilibrium. On the other hand, if the point spectrum of B contains some $\lambda$ with Re $\lambda > 0$, then 0 will not be a stable equilibrium, since for $x \in N(\lambda I - B)$, $\|T(t)x\| = \|e^{\lambda t}x\| = e^{t\mathrm{Re}\lambda}\|x\|$. In the next proposition we use the spectral properties of the infinitesimal generator to provide a decomposition of the Banach space X as $X = M \oplus M_0$, where the semigroup $T(t)$, $t \geq 0$, exhibits different stability properties on the subspaces M and $M_0$. If the 0 equilibrium is stable on $M_0$ and unstable on M, then the 0 equilibrium is called a *saddle point*.

PROPOSITION 4.15 Let $T(t)$, $t \geq 0$, be a strongly continuous semigroup of bounded linear operators in the Banach space X and let B be the infinitesimal generator of $T(t)$, $t \geq 0$. Let $\Lambda = \{\lambda_1,\ldots,\lambda_k\}$ be a finite set of points in $\sigma(B)$ such that Re $\lambda_j > \omega_1(B)$ for $j = 1, \ldots, k$. Let

$$\omega_{2,\Lambda}(B) \overset{\mathrm{def}}{=} \max\{\omega_1(B), \sup_{\lambda \in \sigma(B) - E\sigma(B) - \Lambda} \mathrm{Re}\ \lambda\}$$

and let $\omega \in R$ such that

$$\omega_{2,\Lambda}(B) < \omega < \min\{\mathrm{Re}\ \lambda_j\colon\ j = 1, \ldots, k\}$$

The following hold:

(4.16) Each $\lambda_j \in \sigma(B) - E\sigma(B)$ and is therefore isolated in $\sigma(B)$, and if $P_j \overset{\mathrm{def}}{=} (2\pi i)^{-1} \int_{\Gamma_j} (\lambda I - B)^{-1}\, d\lambda$, $1 \leq j \leq k$, where $\Gamma_j$ is a positively oriented closed curve in C enclosing $\lambda_j$, but no other point of $\sigma(B)$, and $M_j \overset{\mathrm{def}}{=} R(P_j)$, then $P_j$ is a projection in X, $P_jP_h = 0$ for $j \neq h$, and B restricted to $M_j$, denoted by $B_{M_j}$ is bounded with spectrum consisting of the single point $\lambda_j$.

(4.62) If $P \overset{\text{def}}{=} \Sigma_{j=1}^{k} P_j$, $P_0 \overset{\text{def}}{=} I - P$, and $M_0 \overset{\text{def}}{=} R(P_0)$, then $B$ restricted to $M_0$, denoted by $B_{M_0}$, has spectrum $\sigma(B) - \Lambda$, $P_j Bx = BP_j x$ for all $x \in D(B)$, $0 \le j \le k$, $X = M \oplus M_0$, where $M = M_1 \oplus \cdots \oplus M_k$, and $B$ is completely reduced by $M$ and $M_0$.

(4.63) If $t \ge 0$, then $T(t)P_j x = P_j T(t)x$ for all $x \in X$, $0 \le j \le k$, and $T(t)$ is completely reduced by $M$ and $M_0$.

(4.64) If for some $j$, $\lambda_j$ is a pole of $(\lambda I - B)^{-1}$ of order $m$, then $M_j = N((\lambda_j I - B)^m)$, $R(I - P_j) = R((\lambda_j I - B)^m)$, and $\lambda_j$ is an eigenvalue of $B$ with index $m$.

(4.65) There exists a constant $K \ge 1$ such that $\|T(t)P_0 x\| \le Ke^{\omega t}\|P_0 x\|$ for all $x \in X$, $t \ge 0$.

(4.66) The restriction of $B$ to $M$, denoted by $B_M$, is bounded with spectrum consisting of $\Lambda$, $PT(t)x = \exp[tB_M]Px$ for $x \in X$, $t \ge 0$, where $\exp[tB_M]$, $-\infty < t < \infty$, is the exponential of $tB_M$ in $M$ (see [73], p. 25), and there exists $K \ge 1$ such that $\|\exp[tB_M]Px\| \le Ke^{\omega t}\|Px\|$ for $x \in X$ and $t \le 0$.

*Proof.* That each $\lambda_j \in \sigma(B) - E\sigma(B)$ follows directly from (4.56). Since $\Lambda$ then consists of a finite set of isolated points in $\sigma(B)$, (4.61) and (4.62) now follow directly from the results in [167], pp. 178-181. Since $T(t)$ commutes with $(\lambda I - B)^{-1}$ for $t \ge 0$ and $\lambda \in \rho(B)$, the definition of $P_j$ implies that $T(t)$ commutes with $P_j$ for $j = 0, 1, \ldots, k$, $t \ge 0$. Thus, (4.63) is proved and (4.64) is proved by Proposition 4.8.

To prove (4.65) we will use Proposition 4.12, Proposition 4.13, (4.62), and (4.63). Let $T_0(t) \overset{\text{def}}{=} T(t)P_0$ be the restriction of $T(t)$ to $M_0$ for each $t \ge 0$. By (3.8) the infinitesimal generator $B_{M_0}$ of $T_0(t)$, $t \ge 0$, is $B$ restricted to $M_0$. Set

$$\omega_2(B_{M_0}) \overset{\text{def}}{=} \max\{\omega_1(B_{M_0}), \sup_{\lambda \in \sigma(B_{M_0}) - E\sigma(B_{M_0})} \operatorname{Re} \lambda\}$$

and observe by (4.57) that $\omega_0(B_{M_0}) = \omega_2(B_{M_0})$. We claim that $\omega_{2,\Lambda}(B) \geq \omega_2(B_{M_0})$.

To prove this claim observe that

$$\omega_1(B_{M_0}) = \lim_{t\to\infty} t^{-1} \log(\alpha[T_0(t)]) = \lim_{t\to\infty} t^{-1} \log(\alpha[T(t)])$$
$$= \omega_1(B) \leq \omega_{2,\Lambda}(B)$$

[see (3.55) and Definition 4.14]. Also, suppose that $\lambda \in \sigma(B_{M_0}) - E\sigma(B_{M_0})$. By Proposition 4.11, $\lambda \in P\sigma(B_{M_0})$, which implies that $\lambda \in P\sigma(B)$. If $\lambda \in E\sigma(B)$, then Re $\lambda \leq \omega_1(B) \leq \omega_{2,\Lambda}(B)$ by (4.56). If $\lambda \notin E\sigma(B)$, then $\lambda \in \sigma(B) - E\sigma(B) - \Lambda$ by (4.62). Thus, Re $\lambda \leq \omega_{2,\Lambda}(B)$. Therefore, $\omega_0(B_{M_0}) = \omega_2(B_{M_0}) \leq \omega_{2,\Lambda}(B) < \omega$ and (4.65) now follows from (4.47).

To prove (4.66) use Theorem 6.17, p. 178 in [167] to obtain that $B_M$ is bounded in M and $\sigma(B_M) = \Lambda$. The agreement of PT(t) and $\exp[tB_M]P$ on M follows from the fact that PT(t)x and $\exp[tB_M]Px$ both solve the initial value problem

$$\frac{d}{dt} u(t) = B_M u(t) \qquad t \geq 0 \qquad u(0) = x \in M$$

whose solution is unique (see [167], p. 481). The last claim in (4.66) follows from the facts that $-B_M$ is bounded in M, $\sigma(-B_M) = -\sigma(B_M)$, and $\sup\{\text{Re } \lambda: \ \lambda \in \sigma(-B_M)\} < -\omega$, which imply that there exists a constant $K \geq 1$ such that $\|\exp[t(-B_M)]x\| \leq Ke^{-\omega t}\|x\|$ for all $x \in M$, $t \geq 0$ (see [73], p. 27). □

We now specialize the preceding results to the case that T(t), $t \geq 0$, is the strongly continuous semigroup of bounded linear operators in $L^1$ associated with (ADP) with the birth function F and the aging function G both bounded linear operators. For this linear case we require the following conditions on F and G:

(4.67) F: $L^1 \to R^n$, F is bounded, linear, and has the form $F\phi = \int_0^\infty \beta(a)\phi(a)\, da$, $\phi \in L^1$, where $\beta$: $[0,\infty) \to B(R^n,R^n)$, $\beta$ is

Lipschitz continuous from $[0,\infty)$ to $B(R^n,R^n)$, and there exists a constant $\bar{\beta}$ such that $0 \le |\beta(a)| \le \bar{\beta}$ for all $a \ge 0$.

(4.68) $G$: $L^1 \to L^1$, $G$ is bounded, linear, and has the form $G(\phi)(a) = -\mu(a)\phi(a)$, $\phi \in L^1$, almost everywhere $a > 0$, where $\mu$: $[0,\infty) \to B(R^n,R^n)$, $\mu$ is Lipschitz continuous from $[0,\infty)$ to $B(R^n,R^n)$, there exists $\underline{\mu} > 0$ such that $\Sigma_{i=1}^n(\operatorname{sgn} x_i)(-\mu(a)x)_i \le -\underline{\mu}|x|$ for all $x \in R^n$, $a \ge 0$, and there exists $\bar{\mu}$ such that $|\mu(a)| \le \bar{\mu}$ for all $a \ge 0$.

REMARK 4.10 Notice that if $F$ satisfies (4.67), then $F$ satisfies (3.73), and if $G$ satisfies (4.68), then $G$ satisfies (2.49) and (3.67).

DEFINITION 4.16 Let $F$ satisfy (4.67), let $G$ satisfy (4.68), and let $\Pi(a,b)$, $0 \le b \le a$ be the fundamental solution (or evolution operator) associated with the linear $R^n$-vector differential equation

$$\frac{d}{dt}\phi(a) = -\mu(a)\phi(a) \qquad 0 \le b \le a \tag{4.69}$$

where $\phi$: $[b,a] \to R^n$. That is, $\Pi(a,b)$, $0 \le b \le a$, is the family of linear operators in $R^n$ (or $n \times n$ matrices) such that the solution of (4.69) corresponding to the initial value $\phi(b)$ is $\phi(a) = \Pi(a,b)\phi(b)$, $a \ge b$ (see [217], Proposition 5.1, p. 241, and also [43], p. 69). Define $\Delta$: $C \to B(R^n,R^n)$ by

$$\Delta(\lambda)x \overset{\text{def}}{=} x - F[e^{-\lambda a}\Pi(a,0)x] = x - \int_0^\infty e^{-\lambda a}\beta(a)\Pi(a,0)x\,da \tag{4.70}$$

where $x \in R^n$ and $\operatorname{Re}\lambda > -\underline{\mu}$.

The equation $\det \Delta(\lambda) = 0$ (where the notation det means determinant) is defined as the *characteristic equation* of (ADP). A complex number $\lambda$ such that $\operatorname{Re}\lambda > -\underline{\mu}$ and $\det \Delta(\lambda) = 0$ is defined as a *characteristic value* of (ADP).

REMARK 4.11 The formula in (4.70) is well defined, since $|\Pi(a,b)| \le \exp[-\underline{\mu}(a - b)]$ for $0 \le b \le a$ and thus, the integral in (4.70) exists for $\operatorname{Re}\lambda > -\underline{\mu}$. The claim that $|\Pi(a,b)| \le \exp[-\underline{\mu}(a - b)]$ for

$0 \le b \le a$ follows from Theorem 5.1, p. 238 in [217], provided we can establish that

$$\text{(4.71)} \quad \lim_{h \to 0^-} h^{-1}(|x - h\mu(a)x| - |x|) \le -\underline{\mu}|x| \qquad x \in R^n,\ a \ge 0$$

But (4.71) follows from (4.68), since for $x \in R^n$, $a \ge 0$, and $h < 0$ and sufficiently close to 0,

$$\begin{aligned} h^{-1}(|x - h\mu(a)x| - |x|) &= \sum_{i=1}^{n} h^{-1}(|x_i - h(\mu(a)x)_i| - |x_i|) \\ &\le \sum_{i=1}^{n} (\operatorname{sgn} x_i)(-\mu(a)x)_i \\ &\le -\underline{\mu}|x| \end{aligned}$$

Further, the mapping $\lambda \to \Delta(\lambda)$ is holomorphic from $\{\lambda: \ \operatorname{Re}\lambda > -\underline{\mu}\}$ to $B(C^n, C^n)$, since

$$\begin{aligned} \left|\frac{d}{d\lambda}\Delta(\lambda)\right| &= \left|\int_0^\infty \lambda e^{-\lambda a}\beta(a)\Pi(a,0)\, da\right| \\ &\le \bar{\beta}|\lambda| \int_0^\infty e^{-(\operatorname{Re}\lambda + \underline{\mu})a}\, da \\ &= \frac{\bar{\beta}|\lambda|}{\operatorname{Re}\lambda + \underline{\mu}} \end{aligned}$$

Thus, $\Delta(\lambda)$ has only isolated zeros of finite order in $\{\lambda: \ \operatorname{Re}\lambda > -\underline{\mu}\}$. Finally, $\det \Delta(\lambda)$ is holomorphic in $\{\lambda: \ \operatorname{Re}\lambda > -\underline{\mu}\}$ and $\Delta(\lambda)^{-1}$ and $1/\det \Delta(\lambda)$ are meromorphic in this domain.

THEOREM 4.6 Let (4.67), (4.68) hold and let $S(t)$, $t \ge 0$, be the strongly continuous semigroup of bounded linear operators in $L^1$ as in Proposition 3.2 with infinitesimal generator $B$ as in Proposition 3.7. The following hold:

(4.72) $\quad \omega_1(B) \le -\underline{\mu}$

(4.73) $\quad$ If $\operatorname{Re}\lambda > -\underline{\mu}$ and $\det \Delta(\lambda) = 0$, then $\lambda \in P\sigma(B)$.

(4.74) $\quad$ If $\operatorname{Re}\lambda > -\underline{\mu}$ and $\lambda \in \rho(B)$, then for $\psi \in L^1$, $a \ge 0$,

$$((\lambda I - B)^{-1}\psi)(a) = \int_0^a e^{\lambda(a-b)}\Pi(a,b)\psi(b)\,db$$

$$+ e^{-\lambda a}\Pi(a,0)\Delta(\lambda)^{-1}\int_0^\infty \beta(b)e^{-\lambda b}$$

$$\times \left[\int_0^b e^{\lambda\tau}\Pi(b,\tau)\psi(\tau)\,d\tau\right] db$$

*Proof.* To prove (4.72) let $S(t) = U(t) + W(t)$, where $U(t)$ is defined as in (3.62) and $W(t)$ as in (3.63). From Proposition 3.15, Proposition 3.17, and Proposition 4.9 we obtain that

$$\alpha[S(t)] \le \alpha[U(t)] + \alpha[V(t)] \le e^{-\underline{\mu}t} + 0$$

which implies (4.72).

To prove (4.73) let $\text{Re}\,\lambda > -\underline{\mu}$ such that $\det \Delta(\lambda) = 0$. Then, there must exist $x \in R^n$, $x \neq 0$, such that $\Delta(\lambda)x = 0$. Define $\phi(0) \overset{\text{def}}{=} x$ and $\phi(a) \overset{\text{def}}{=} e^{-\lambda a}\Pi(a,0)\phi(0)$, $a \ge 0$. Then, $\phi \neq 0$,

$$\phi(0) = F\phi = \int_0^\infty e^{-\lambda a}\beta(a)\Pi(a,0)\phi(0)\,da$$

$$\phi'(a) = -\lambda\phi(a) - \mu(a)\phi(a) \qquad a \ge 0$$

and hence $\lambda\phi = B\phi$.

To prove (4.74) let $\text{Re}\,\lambda > -\underline{\mu}$ and let $\lambda \in \rho(B)$. Since $\lambda \notin \sigma(B)$, (4.73) implies that $\Delta(\lambda)^{-1}$ exists. Let $\psi \in L^1$. There exists $\phi \in L^1$ satisfying $(\lambda I - B)\phi = \psi$ if and only if

$$\lambda\phi(a) + \phi'(a) + \mu(a)\phi(a) = \psi(a) \qquad a \ge 0 \tag{4.75}$$

$$\phi(0) = F\phi = \int_0^\infty \beta(a)\phi(a)\,da \tag{4.76}$$

The solution of (4.75) is given by

$$\phi(a) = e^{-\lambda a}\Pi(a,0)\phi(0) + \int_0^a e^{-\lambda(a-b)}\Pi(a,b)\psi(b)\,db \qquad a \ge 0$$

(see Proposition 5.2, p. 242, in [217]). Substitute this formula for $\phi$ into (4.76) to obtain

$$\phi(0) = \int_0^\infty \beta(a) e^{-\lambda a} \Pi(a,0)\phi(0)\, da$$

$$+ \int_0^\infty \beta(a) \left[ \int_0^a e^{-\lambda(a-b)} \Pi(a,b)\psi(b)\, db \right] da$$

Since $\Delta(\lambda)^{-1}$ exists,

$$\phi(0) = \Delta(\lambda)^{-1} \int_0^\infty \beta(a) \left[ \int_0^a e^{-\lambda(a-b)} \Pi(a,b)\psi(b)\, db \right] da$$

and therefore (4.74) is established. □

THEOREM 4.7 Let (4.67), (4.68) hold, let $S(t)$, $t \geq 0$, be the strongly continuous semigroup of bounded linear operators in $L^1$ as in Proposition 3.2 with infinitesimal generator $B$ as in Proposition 3.7, and let $\text{Re}\ \lambda_0 > -\underline{\mu}$. The following are equivalent:

(4.77) $\lambda_0 \in \sigma(B)$

(4.78) $\lambda_0 \in \sigma(B) - E\sigma(B)$

(4.79) $\lambda_0$ is a pole of $(\lambda I - B)^{-1}$ of order $m$.

(4.80) $\lambda_0$ is a pole of $1/\det \Delta(\lambda)$ of order $m$.

(4.81) $\lambda_0$ is a zero of $\det \Delta(\lambda)$ of order $m$.

*Proof.* That (4.77) implies (4.78) follows immediately from (4.61) and (4.72). That (4.78) implies (4.79) follows immediately from Proposition 4.11. To prove that (4.79) and (4.80) are equivalent use the formula for $(\lambda I - B)^{-1}$ given in (4.74) and the fact that $\Delta(\lambda)^{-1}$ can be represented as a holomorphic function from $\{\lambda: \text{Re}\ \lambda > -\underline{\mu}\}$ to $B(C^n, C^n)$ divided by $\det \Delta(\lambda)$. The equivalence of (4.80) and (4.81) is obvious. That (4.81) implies (4.77) follows immediately from (4.73). □

THEOREM 4.8 Let (4.67), (4.68) hold and let $\sup_{\det\Delta(\lambda)=0} \text{Re}\ \lambda < 0$. Then, the zero equilibrium of (ADP) is globally exponentially asymptotically stable.

*Proof.* The proof follows immediately from (4.47), (4.57), (4.72), and the equivalence of (4.78) and (4.81). □

THEOREM 4.9 Let (4.67), (4.68) hold, let $S(t)$, $t \geq 0$ be the strongly continuous semigroup of bounded linear operators in $L^1$ as in Proposition 3.2 with infinitesimal generator $B$ as in Proposition 3.7, and let there exist a characteristic value $\lambda_1$ of (ADP) such that $\lambda_1$ is real, $\lambda_1 > -\underline{\mu}$, $\sup_{\det\Delta(\lambda)=0,\lambda\neq\lambda_1} \mathrm{Re}\,\lambda < \lambda_1$, and $\lambda_1$ is a simple zero of $\det \Delta(\lambda)$. Let $P: L^1 \to L^1$ be defined by

$$(4.82) \qquad P\phi \overset{\text{def}}{=} (2\pi i)^{-1} \int_\Gamma (\lambda I - B)^{-1}\, d\lambda \qquad \phi \in L^1$$

where $\Gamma$ is a positively oriented closed curve in $\mathbb{C}$ enclosing $\lambda_1$, but no other point of $\sigma(B)$. Then

$$(4.83) \qquad \lim_{t\to\infty} e^{\lambda_1 t} S(t)\phi = P\phi \qquad \text{for all } \phi \in L^1$$

*Proof.* Let $\Lambda \overset{\text{def}}{=} \{\lambda_1\}$, let $\omega \in R$ such that

$$\max\{\omega_1(B), \sup_{\det\Delta(\lambda)=0,\lambda\neq\lambda_1} \mathrm{Re}\,\lambda\} < \omega < \lambda_1$$

Let $P_0 \overset{\text{def}}{=} I - P$, let $M_0 = R(P_0)$, and let $M = R(P)$. By Theorem 4.7 and Proposition 4.15, $L^1 = M \oplus M_0$ and there exists a constant $K \geq 1$ such that

$$(4.84) \qquad \|S(t)P_0\phi\|_{L^1} \leq Ke^{\omega t}\|P_0\phi\|_{L^1} \qquad \text{for all } \phi \in L^1,\ t \geq 0$$

Further, (4.66) implies that $B$ restricted to $M$, denoted by $B_M$, is bounded and $S(t)P = \exp[tB_M]P$ for $t \geq 0$. Also, since $\lambda_1$ is a simple root of $\det \Delta(\lambda)$, Theorem 4.7 implies that $\lambda_1$ is a simple pole of $(\lambda I - B)^{-1}$. Then, (4.64) implies that $M = N(\lambda_1 I - B)$. Thus $\phi \in M$ if and only if $B\phi = \lambda_1\phi$, and so $S(t)\phi = \exp[tB_M]\phi = e^{\lambda_1 t}\phi$ for all $\phi \in M$, $t \geq 0$. From (4.84) we then obtain that for $\phi \in L^1$, $t \geq 0$,

$$\|e^{-\lambda_1 t}S(t)\phi - P\phi\|_{L^1} = \|e^{-\lambda_1 t}S(t)P\phi + e^{-\lambda_1 t}S(t)P_0\phi - P\phi\|_{L^1}$$

$$\le Ke^{(\omega-\lambda_1)t} \|P_0\phi\|_{L^1}$$

which implies (4.83). □

THEOREM 4.10 Let $R^n = R$, let (4.67), (4.68) hold, let $\beta$ be nonnegative on $[0,\infty)$, and let $S(t)$, $t \ge 0$, be the strongly continuous semigroup of bounded linear operators in $L^1$ as in Proposition 3.2 with infinitesimal generator $B$ as in Proposition 3.7. If

$$\int_0^\infty \beta(a)\Pi(a,0)\, da < 1$$

then the zero equilibrium is globally exponentially asymptotically stable. If

$$\int_0^\infty \beta(a)\Pi(a,0)\, da > 1$$

then there exists a unique positive real number $\lambda_1$ such that

$$\int_0^\infty e^{-\lambda_1 a}\beta(a)\Pi(a,0)\, da = 1 \tag{4.85}$$

and for all $\phi \in L^1$

$$\lim_{t\to\infty} e^{-\lambda_1 t}(S(t)\phi)(a) = \frac{e^{-\lambda_1 a}\Pi(a,0) \int_0^\infty \beta(b)e^{-\lambda_1 b}\left[\int_0^b e^{\lambda_1 \tau}\Pi(b,\tau)\phi(\tau)\, d\tau\right] db}{\int_0^\infty \beta(b)be^{-\lambda_1 b}\Pi(b,0)\, db} \tag{4.86}$$

where the limit is taken in the norm of $L^1$.

*Proof.* Define $q$: $\{\lambda \in C\colon \operatorname{Re}\lambda > -\underline{\mu}\} \to C$ by

$$q(\lambda) \overset{\text{def}}{=} \int_0^\infty e^{-\lambda a}\beta(a)\Pi(a,0)\, da$$

Notice that for $\text{Re}\,\lambda > -\underline{\mu}$, $q(\lambda) = 1$ if and only if $\Delta(\lambda) = 0$, that is, $\lambda$ is a characteristic value.

Suppose that $q(0) < 1$ and there exists $\lambda$ such that $\text{Re}\,\lambda > -\underline{\mu}$ and $q(\lambda) = 1$. Then,

$$(4.87) \qquad 1 = \text{Re}\, q(\lambda) = \int_0^\infty e^{-\text{Re}\lambda a}\cos(\text{Im}\,\lambda a)\beta(a)\Pi(a,0)\, da$$
$$\leq \int_0^\infty e^{-\text{Re}\lambda a}\beta(a)\Pi(a,0)\, da = q(\text{Re}\,\lambda)$$

Since q restricted to $(-\underline{\mu},\infty)$ is continuous and strictly decreasing, $\text{Re}\,\lambda < 0$ and there must exist a real number $\lambda_1 \in [\text{Re}\,\lambda, 0)$ such that $q(\lambda_1) = 1$. Further, (4.87) shows that for any $\lambda$ such that $\text{Re}\,\lambda > -\underline{\mu}$ and $q(\lambda) = 1$, we must have that $\text{Re}\,\lambda \leq \lambda_1$. Thus, $\sup_{\Delta(\lambda)=0} \text{Re}\,\lambda \leq \lambda_1 < 0$. The global exponential asymptotic stability of the zero solution now follows from Theorem 4.8.

Suppose that $q(0) > 1$. Since $\lim_{\lambda\ \text{real}, \lambda\to\infty} q(\lambda) = 0$, there exists a unique positive number $\lambda_1$ satisfying (4.85). We claim that there can be only finitely many $\lambda$ with $\text{Re}\,\lambda > 0$ and $q(\lambda) = 1$. Assume to the contrary that there exists an infinite sequence $\{z_k\}$ such that $\text{Re}\, z_k > 0$ and $q(z_k) = 0$ for $k = 1, 2, \ldots$ . Since $\Delta(\lambda)$ is holomorphic for $\text{Re}\,\lambda > -\underline{\mu}$, its zeros cannot accumulate in this region (see [251], p. 209). From (4.87) we see that $0 < \text{Re}\, z_k < \lambda_1$ for all $k = 1, 2, \ldots$ . Thus, $\lim_{k\to\infty} |\text{Im}\, z_k| = \infty$. Observe that for each $k = 1, 2, \ldots$,

$$1 = \int_0^\infty e^{-\text{Re} z_k a}\cos(\text{Im}\, z_k a)\beta(a)\Pi(a,0)\, da$$
$$< \int_0^\infty \cos(\text{Im}\, z_k a)\beta(a)\Pi(a,0)\, da$$

But by the Riemann-Lebesgue theorem (see [249], p. 90),

$$\lim_{k\to\infty} \int_0^\infty \cos(\text{Im}\, z_k a)\beta(a)\Pi(a,0)\, da = 0$$

which is a contradiction.

Thus, $\sup_{\Delta(\lambda)=0,\lambda\neq\lambda_1} \operatorname{Re}\lambda < \lambda_1$. To finish the proof it suffices by Theorem 4.9 to show that $\lambda_1$ is a simple zero of $\Delta(\lambda)$ and $P\phi$ in (4.82) is given by the right-side of (4.86). The claim that $\lambda_1$ is a simple zero of $\Delta(\lambda)$ follows from the fact that

$$\Delta'(\lambda_1) = \int_0^\infty ae^{-\lambda_1 a}\beta(a)\Pi(a,0)\,da > 0$$

Further, the residue of $1/\Delta(\lambda)$ at $\lambda_1$ is

$$\frac{1}{\Delta'(\lambda_1)} = \frac{1}{\int_0^\infty ae^{-\lambda_1 a}\beta(a)\Pi(a,0)\,da}$$

(see [251], p. 215). The claim that $P\phi$ in (4.82) is given by the right-side of (4.86) now follows directly from (4.74) and Theorem 4.7. □

REMARK 4.12 Let $\lambda_1$ be as in Theorem 4.10, let $\phi \in L^1$, and define

$$M_{\lambda_1} \stackrel{\text{def}}{=} \int_0^\infty \beta(a)ae^{-\lambda_1 a}\Pi(a,0)\,da$$

$$V_{\lambda_1}(\phi) \stackrel{\text{def}}{=} \int_0^\infty \beta(a)e^{-\lambda_1 a}\left[\int_0^a e^{\lambda_1 b}\Pi(a,b)\phi(b)\,db\right]da$$

The quantity $\lambda_1$ is called the *intrinsic growth constant* of the population, the quantity $M_{\lambda_1}$ is called the *mean age of childbirth* for the stable (or persistent) age distribution [see (1.19)], and the quantity $V_{\lambda_1}(\phi)$ is called the *natural reproductive value* of the initial age distribution. Recall that the birth rate of the population is

$$B(t) \stackrel{\text{def}}{=} F(S(t)\phi) = \int_0^\infty \beta(a)(S(t)\phi)(a)\,da \qquad t > 0$$

By virtue of (4.85) and (4.86)

$$(4.88) \qquad \lim_{t\to\infty} e^{-\lambda_1 t} B(t) = \frac{V_{\lambda_1}(\phi)}{M_{\lambda_1}}$$

Further, from (1.12) and (2.62) we have that for $\phi \in L^1$,

$$(4.89) \qquad (S(t)\phi)(a) = \begin{cases} B(t - a)\Pi(a,0) & 0 < a < t \\ \phi(a - t)\Pi(a, a - t) & \text{a.e. } a > t \end{cases}$$

Thus, from (4.88) and (4.89) we obtain

$$(4.90) \qquad \lim_{t\to\infty} e^{-\lambda_1 t}(S(t)\phi)(a) = V_{\lambda_1}(\phi)e^{-\lambda_1 a}\Pi(a,0)/M_{\lambda_1}$$

uniformly in bounded intervals of a (see [127], p. 22, and [235], p. 33).

THEOREM 4.11 Let (4.67), (4.68) hold, let $S(t)$, $t \geq 0$, be the strongly continuous semigroup of bounded linear operators in $L^1$ as in Proposition 3.2 with infinitesimal generator B as in Proposition 3.7, for $j, k = 1, \ldots, n$ let

$$\sum_{h=1}^{n} \beta_{jh}(a)\Pi_{hk}(a,0) \geq 0$$

for $a \geq 0$, but not identically 0 for $a \geq 0$, and for some $j = 1, \ldots, n$ let

$$\int_0^\infty \sum_{h=1}^{n} \beta_{jh}(a)\Pi_{hj}(a,0)\, da > 1$$

There exists a characteristic value $\lambda_1$ of (ADP) such that $\lambda_1$ is real and positive, $\sup_{\det\Delta(\lambda)=0,\lambda\neq\lambda_1} \operatorname{Re}\lambda < \lambda_1$, and $\lambda_1$ is a simple zero of $\det \Delta(\lambda)$. Furthermore, if P is defined as in (4.82), then for all $\phi \in L^1$

$$\lim_{t\to\infty} e^{-\lambda_1 t} S(t)\phi = P\phi$$

*Proof.* We will use the following result, proved in [19], Theorem 8.3, p. 259:

(4.91) Let H be a continuous function from $[0,\infty)$ to $B(R^n,R^n)$ such that (i) for $j,k = 1, \ldots, n$, $H_{jk}(a) \geq 0$ for $a \geq 0$, but not identically 0 for $a \geq 0$; (ii) for some $j = 1, \ldots, n$, $\int_0^\infty H_{jj}(a)\,da > 1$; and (iii) for $j,k = 1, \ldots, n$ and some $\lambda > 0$, $\int_0^\infty e^{-\lambda a} H_{jk}(a)\,da < \infty$. Then, there exists a unique positive real number $\lambda_1$ such that $\lambda_1$ is the root of $\det(I - \int_0^\infty e^{-\lambda a} H(a)\,da) = 0$ with largest real part and $\lambda_1$ is a simple root.

Define $H(a) \overset{\text{def}}{=} \beta(a)\Pi(a,0)$ for $a \geq 0$ and observe that for $j,k = 1, \ldots, n$, $a \geq 0$,

$$H_{jk}(a) = \sum_{h=1}^{n} \beta_{jh}(a)\Pi_{hk}(a,0)$$

By the hypothesis of the theorem, (i), (ii), and (iii) of (4.91) are satisfied. Thus, there exists a unique positive real root $\lambda_1$ of the characteristic equation $\det \Delta(\lambda) = 0$ such that $\lambda_1$ is the root with largest real part, and $\lambda_1$ is simple. The conclusion of the theorem will follow immediately from Theorem 4.9, if we can establish that $\sup_{\det\Delta(\lambda)=0, \lambda\neq\lambda_1} \operatorname{Re} \lambda < \lambda_1$. This claim follows by an argument similar to the one in the proof of Theorem 4.10, together with the Riemann-Lebesgue theorem, to show that there can exist at most only finitely many roots of the characteristic equation $\det \Delta(\lambda) = 0$ with $0 < \operatorname{Re} \lambda < \lambda_1$. □

## 4.4 THE METHOD OF LINEARIZATION FOR THE SEMILINEAR CASE

In certain cases the local stability of an equilibrium solution of the nonlinear problem (ADP) can be analyzed by the method of linearization. For this approach the framework of strongly continuous semigroups is again useful. Formally, the method proceeds as follows:

Consider the infinitesimal generator B of the strongly continuous nonlinear semigroup $S(t)$, $t \geq 0$, associated with (ADP). From Theorem 3.2 we have that

$$B\phi = -\phi' + G(\phi) \qquad \phi(0) = F(\phi) \qquad \phi \in D(B)$$

If we linearize B about an equilibrium solution $\hat{\phi}$ of (ADP), we obtain a linear operator $\hat{B}$: $L^1 \to L^1$, where $\hat{B}$ is the infinitesimal generator of a strongly continuous semigroup of linear operators $\hat{S}(t)$, $t \geq 0$, in $L^1$. Formally, $\hat{B}$ is given by

$$\hat{B}\phi = -\phi' + G'(\hat{\phi})\phi \qquad \phi(0) = F'(\hat{\phi})\phi \qquad \phi \in D(B)$$

where $F'(\hat{\phi})$: $L^1 \to R^n$ and $G'(\hat{\phi})$: $L^1 \to L^1$ are the Fréchet derivatives of F and G, respectively, at $\hat{\phi}$. If we can show that the linear semigroup $\hat{S}(t)$, $t \geq 0$, is globally exponentially asymptotically stable at 0, that is, $\omega_0(\hat{B}) < 0$, then we can establish the local exponential asymptotic stability of the equilibrium solution $\hat{\phi}$. As we have seen in Section 4.3, the value of $\omega_0(\hat{B})$ can be determined from the spectral properties of $\hat{B}$.

The main difficulty in carrying out the approach we have outlined above is that the nonlinear operator B is not differentiable, in fact not even continuous nor everywhere defined. We must therefore overcome the technical problems involved in the linearization of a discontinuous, unbounded, and only densely defined nonlinear operator. If the birth function F is linear, then we can take advantage of the resultant semilinear form of (ADP). This case was treated by J. Prüss in [240] and [241]. We will treat the semilinear case in this section. Our approach will be modelled on a similar treatment by D. Henry in [150] for semilinear parabolic partial differential equations. In the next section we will treat the full nonlinear case, that is, the case that both the birth function F and the aging function G are nonlinear.

We begin with some results from the theory of abstract semilinear differential equations in Banach spaces. The following fundamental result in this theory is due to I. Segal (see [259] and also [217], Chapter 8, and [18], Sections 5.4 and 5.5).

PROPOSITION 4.16 Let $T(t)$, $t \geq 0$, be a strongly continuous semigroup of bounded linear operators in the Banach space $X$ with infinitesimal generator $B$. Let $K$ be a (nonlinear) operator from $X$ to $X$ such that $K$ is continuously Fréchet differentiable on $X$ (as in Definition 2.4). The following hold:

(4.92) For each $x \in X$ there exists a maximal interval of existence $[0,T_x)$ and unique continuous function $t \to u(t;x)$ from $[0,T_x)$ to $X$ such that $u(t;x) = T(t)x + \int_0^t T(t-s)K(u(s;x))\,ds$ for all $t \in [0,T_x)$, and either $T_x = \infty$ or $\limsup_{t\to T_x^-} \|u(t;x)\| = \infty$.

(4.93) $u(t;x)$ is a continuous function of $x$ in the sense that if $x \in X$ and $0 \leq t < T_x$, there exist positive constants $C$ and $\varepsilon$ such that if $\hat{x} \in X$ and $\|x - \hat{x}\| < \varepsilon$, then $t < T_{\hat{x}}$ and $\|u(s;x) - u(s;\hat{x})\| \leq C\|x - \hat{x}\|$ for all $0 \leq s \leq t$.

(4.94) If $x \in D(B)$, then $u(t;x) \in D(B)$ for $0 \leq t < T_x$ and the function $t \to u(t;x)$ is continuously differentiable and satisfies $d/dt\, u(t;x) = Bu(t;x) + K(u(t;x))$ on $[0,T_x)$.

*Proof.* Let $M \geq 1$ and let $\omega \in R$ such that $|T(t)| \leq Me^{\omega t}$ for $t \geq 0$. Since $K$ is continuously Fréchet differentiable in the sense of Definition 2.4, there exists a continuous increasing function $C_0$ from $[0,\infty)$ to $[0,\infty)$ such that if $x, \hat{x} \in X$ with $\|x\|, \|\hat{x}\| \leq r$, then

$$\|K(x) - K(\hat{x})\| \leq C_0(r)\|x - \hat{x}\|$$

(see(8.62), p. 162, in [93]). Consequently, for $x \in X$ with $\|x\| \leq r$,

$$\|K(x)\| \leq C_0(r)\|x\| + \|K(0)\|$$

To prove (4.92) let $x \in X$, let $r \overset{\text{def}}{=} \|x\| + 1$, let $C_1 \overset{\text{def}}{=} C_0(r)r + \|K(0)\|$, and let $t_1 > 0$ be chosen sufficiently small such that

$$\sup_{0 \le t \le t_1} \{\|T(t)x - x\| + M \exp[|\omega|t]C_1 t\} < 1$$

Let the complete metric space Z, with metric $\rho$, be defined by

$$Z \overset{\text{def}}{=} \{u: \text{ u is a continuous function from } [0,t_1] \text{ to X such that } \sup_{0 \le t \le t_1} \|u(s) - x\| \le 1\}$$

$$\rho(u,v) \overset{\text{def}}{=} \sup_{0 \le s \le t_1} \|u(s) - v(s)\| \qquad u, v \in Z$$

Define a mapping H on Z by

$$(Hu)(t) \overset{\text{def}}{=} T(t)x + \int_0^t T(t - s)K(u(s))\, ds \qquad u \in Z,\ 0 \le t \le t_1$$

Observe that for $u \in Z$, $0 \le t \le t_1$,

$$\|(Hu)(t) - x\| \le \|T(t)x - x\| + M \exp[|\omega|t]C_1 t \le 1$$

Further, for all $u \in Z$, $(Hu)(t)$ is continuous in t from $[0,t_1]$ to X, and consequently, H maps Z into Z. Also, H is a strict contraction on Z, since for $u, \hat{u} \in Z$, $0 \le t \le t_1$

$$\|(Hu)(t) - (H\hat{u})(t)\| \le M \exp[|\omega|t_1]C_0(r)t_1\rho(u,\hat{u})$$

By the contraction mapping theorem, H has a unique fixed point in Z which must also be the unique solution of the integral equation in (4.92).

To show the second claim in (4.92) let $x \in X$ and let $u(t;x)$ be the unique solution of this integral equation defined on its maximal interval of existence $[0,T_x)$. Assume that $T_x < \infty$ and $\sup_{0 \le t < T_x} \|u(t;x)\| \le r$ for some constant r. If $0 < t < t + h < T_x$, then

$$\begin{aligned} \|u(t + h; x) - u(t;x)\| \le{}& \|T(t + h)x - T(t)x\| \\ &+ \int_t^{t+h} \|T(s)u(t + h - s; x)\|\, ds \\ &+ \int_0^t \|T(s)[u(t + h - s; x) - u(t - s; x)]\|\, ds \end{aligned}$$

$$\le M \exp[|\omega|T_x]\|T(h)x - x\|$$
$$+ hM \exp[|\omega|T_x]r$$
$$+ M \exp[|\omega|T_x] \int_0^t \|u(\tau + h; x) - u(\tau;x)\| \, d\tau$$

By Gronwall's lemma we obtain $\lim_{t\to T_x^-} u(t;x)$ exists. Consequently, $u(t;x)$ can be extended past $T_x$, which contradicts the maximality of $[0,T_x)$.

To prove (4.93) let $x \in X$ and let $0 \le t < T_x$. Let $r \overset{\text{def}}{=} 1 + \sup_{0\le s\le t}\|u(s;x)\|$, let $C \overset{\text{def}}{=} M \exp[|\omega|t] \exp[M \exp[|\omega|t]C_0(r)t]$, and choose $\varepsilon \in (0,1)$ such that $C\varepsilon \le 1/2$. Let $\hat{x} \in X$ such that $\|x - \hat{x}\| < \varepsilon$ and let $\hat{t} \le \infty$ be the largest extended real number such that $\|u(s;\hat{x})\| \le r$ for $0 \le s < \hat{t}$. Assume that $\hat{t} < t$. If $0 \le s < \hat{t}$, then

$$\|u(s;x) - u(s;\hat{x})\|$$
$$\le M \exp[|\omega|t]\|x - \hat{x}\|$$
$$+ \int_0^s M \exp[\omega(s - \tau)]C_0(r)\|u(\tau;x) - u(\tau;\hat{x})\| \, d\tau$$

By Gronwall's lemma, for $0 \le s < \hat{t}$,

$$\|u(s;x) - u(s;\hat{x})\| \le C\|x - \hat{x}\| \tag{4.95}$$

which implies that for $0 \le s < \hat{t}$, $\|u(s;\hat{x})\| \le r - 1/2$. Since $u(s;\hat{x})$ is a continuous function of $s$, $\hat{t}$ cannot be the largest number such that $\|u(s;\hat{x})\| \le r$ for $0 \le s < \hat{t}$. Thus, $\hat{t} \ge t$ and by (4.92) $t < T_{\hat{x}}$. Further, $\|u(s;\hat{x})\| \le r$ for $0 \le s \le t$ and (4.93) follows from (4.95).

To prove (4.94) let $x \in D(B)$. Using the techniques in the proof of (4.92), one can show that for $t_1 \in (0,T_x)$ there exists a unique solution to the equation

$$v(t) = T(t)(Bx + K(x))$$
$$+ \int_0^t T(t - s)K'(u(s;x))v(s) \, ds \qquad 0 \le t \le t_1$$

Define $w\colon [0,t_1] \to X$ by

$$w(t) \overset{\text{def}}{=} x + \int_0^t v(s)\, ds \qquad 0 \le t \le t_1$$

By the chain rule (see [93], p. 151) we have that for $0 \le t \le t_1$,

$$\frac{d}{dt}\int_0^t T(t-s)K(w(s))\, ds$$

$$= \frac{d}{dt}\int_0^t T(s)K(w(t-s))\, ds$$

$$= \lim_{h\to 0}\Big[h^{-1}\int_t^{t+h} T(s)K(w(t+h-s))\, ds$$

$$+ \int_0^t T(s)h^{-1}[K(w(t+h-s)) - K(w(t-s))]\, ds\Big]$$

$$= T(t)K(x) + \int_0^t T(t-s)K'(w(s))v(s)\, ds$$

Integrate this last equation to obtain

$$\int_0^t T(t-s)K(w(s))\, ds \tag{4.96}$$

$$= \int_0^t T(s)K(x)\, ds$$

$$+ \int_0^t\Big[\int_0^s T(s-\tau)K'(w(\tau))v(\tau)\, d\tau\Big]\, ds \qquad 0 \le t \le t_1$$

Since $x \in D(B)$, we have that

$$x = T(t)x - \int_0^t T(t-s)Bx\, ds \qquad t \ge 0$$

(see [167], p. 481). Next, use this fact and (4.96) to obtain that for $0 \le t \le t_1$,

$$w(t) = T(t)x + \int_0^t T(t-s)K(w(s))\, ds$$

$$- \int_0^t \left[ \int_0^s T(s - \tau)K'(w(\tau))v(\tau)\, d\tau \right] ds$$

$$+ \int_0^t \left[ \int_0^s T(s - \tau)K'(u(\tau;x))v(\tau)\, d\tau \right] ds$$

Now use this formula for $w(t)$ and the formula for $u(t)$ in (4.92) to obtain

$$\|w(t) - u(t)\| \le \text{const} \int_0^t \|w(s) - u(s)\|\, ds \qquad 0 \le t \le t_1$$

By Gronwall's lemma, $w(t) = u(t)$ on $[0,t_1]$ and, consequently, $u'(t)$ exists on $[0,t_1]$. The claim (4.94) now follows from the following fact:

(4.97) Let $x \in D(B)$, let $t_1 > 0$, let $h$: $[0,t_1] \to X$ be continuous, and let $f$: $[0,t_1] \to X$ be defined by $f(t) \overset{\text{def}}{=} T(t)x + \int_0^t T(t - s)h(s)\, ds$, $0 \le t \le t_1$. If $g$: $[0,t_1] \to X$ such that $g(0) = x$ and $g'(t) = Bg(t) + h(t)$, $0 \le t \le t_1$, then $g(t) = f(t)$, $0 \le t \le t_1$. If $h$ is also continuously differentiable on $[0,t_1]$, then $f'(t) = Bf(t) + h(t)$, $0 \le t \le t_1$.

[For the proof of (4.97) see [167], p. 486.] □

The next proposition provides a sufficient condition for an equilibrium solution of the abstract semilinear differential equation $d/dt\, u(t) = Bu(t) + K(u(t))$ to be locally asymptotically stable.

PROPOSITION 4.17 Let $T(t)$, $t \ge 0$, be a strongly continuous semigroup of bounded linear operators in the Banach space $X$ with infinitesimal generator $B$. Let $K$ be a (nonlinear) operator from $X$ to $X$ such that $K$ is continuously Fréchet differentiable on $X$ (as in Definition 2.4). For each $x \in X$ let $u(t;x)$ be the solution of the integral equation in (4.92) on the maximal interval of existence $[0,T_x)$. Let $\hat{x} \in X$ satisfy $B\hat{x} + K(\hat{x}) = 0$, let $\hat{T}(t)$, $t \ge 0$, be the strongly continuous semigroup of bounded linear operators in $X$ with infinitesimal generator

$\hat{B} \overset{\text{def}}{=} B + K'(\hat{x})$, and let $\omega_0(\hat{B}) < 0$. Then, $\hat{x}$ is a locally exponentially asymptotically stable equilibrium in the following sense:

(4.98) There exists $\varepsilon > 0$, $M \geq 1$, and $\gamma < 0$ such that if $x \in X$ and $\|x - \hat{x}\| \leq \varepsilon$, then $T_x = \infty$ and $\|u(t;x) - \hat{x}\| \leq Me^{\gamma t}\|x - \hat{x}\|$ for all $t \geq 0$.

*Proof.* Observe first that $u(t;\hat{x}) = \hat{x}$ for all $t \geq 0$ by (4.92) and (4.94). From Proposition 4.14 we have that $\hat{B}$ is the infinitesimal generator of a strongly continuous semigroup of bounded linear operators $\hat{T}(t)$, $t \geq 0$, in $X$ with the representation

$$\hat{T}(t)x = T(t)x + \int_0^t T(t - s)K'(\hat{x})\hat{T}(s)x \, ds \qquad x \in X,\ t \geq 0 \tag{4.99}$$

We claim the following fact:

(4.100) Let $x \in X$, let $t_1 > 0$, let $h\colon [0,t_1] \to X$ be continuous, and let $f\colon [0,t_1] \to X$ satisfy $f(t) = T(t)x + \int_0^t T(t - s)(K'(\hat{x})f(s) + h(s))\, ds$, $0 \leq t \leq t_1$. Then $f(t) = \hat{T}(t)\hat{x} + \int_0^t \hat{T}(t - s)h(s)\, ds$, $0 \leq t \leq t_1$.

To prove (4.100) define $g(t) \overset{\text{def}}{=} \hat{T}(t)x + \int_0^t \hat{T}(t - s)h(s)\, ds$, $0 \leq t \leq t_1$. Consider first the case that $x \in D(B)$ and $h$ is continuously differentiable. Then, $g(t) = h(t)$ on $[0,t_1]$ by (4.97). If $x \notin D(B)$ and $h$ is not continuously differentiable, let $\{x_k\} \subset D(B)$ such that $x_k \to x$ and let $\{h_k\}$ be a sequence of continuously differentiable functions from $[0,t_1]$ to $X$ which converges uniformly to $h$. Then, for $k = 1, 2, \ldots$, $0 \leq t \leq t_1$,

$$\begin{aligned} f_k(t) &\overset{\text{def}}{=} \hat{T}(t)x_k + \int_0^t \hat{T}(t - s)h_k(s)\, ds \\ &= T(t)x_k + \int_0^t T(t - s)[K'(\hat{x})f_k(s) + h_k(s)]\, ds \end{aligned}$$

If $M_1 \overset{\text{def}}{=} \sup_{0\leq t\leq t_1}(|T(t)| + |\hat{T}(t)|)$, then for $k = 1, 2, \ldots$, $0 \leq t \leq t_1$,

$$p_k(t) \overset{\text{def}}{=} \|g(t) - f_k(t)\| + \|f_k(t) - f(t)\|$$

$$\leq |\hat{T}(t)|\|x - x_k\| + \int_0^t |\hat{T}(t - s)|\|h(s) - h_n(s)\| \, ds$$

$$+ |T(t)|\|x - x_k\| + \int_0^t |T(t - s)|[|K'(\hat{x})|$$

$$\times \|f_k(s) - f(s)\| + \|h(s) - h_k(s)\|] \, ds$$

$$\leq M_1\left[\|x - x_k\| + \int_0^t \|h(s) - h_k(s)\| \, ds\right.$$

$$\left. + \int_0^t |K'(\hat{x})| p_k(s) \, ds\right]$$

By Gronwall's lemma we obtain for $k = 1, 2, \ldots, 0 \leq t \leq t_1$,

$$\|g(t) - f(t)\|$$

$$\leq p_k(t) + M_1\left[\|x - x_k\| + \int_0^{t_1} \|h(s) - h_k(s)\| \, ds\right] \exp[M_1|K'(\hat{x})|t_1]$$

Let $k \to \infty$ to conclude that $g(t) = f(t)$ on $[0,t_1]$. Hence, (4.100) is established.

To prove (4.98) let $K(x) = K(\hat{x}) + K'(\hat{x})(x - \hat{x}) + o(x - \hat{x})$ for all $x \in X$, where $o$: $X \to X$, $\|o(x)\| \leq b(r)\|x\|$ for $\|x\| \leq r$, and $b$: $[0,\infty) \to [0,\infty)$ is an increasing continuous function such that $b(0) = 0$ (see Definition 2.4). Let $\omega \in (\omega_0(\hat{B}),0)$ and by (4.47) we may choose $M \geq 1$ such that $|\hat{T}(t)| \leq Me^{\omega t}$ for $t \geq 0$. Let $r > 0$ such that $\|o(x)\| \leq (-\omega/2M)\|x\|$ for all $x \in X$ such that $\|x\| < r$. Let $\varepsilon = r/M$, let $x \in X$ such that $\|x - \hat{x}\| < \varepsilon$, and let $t_1 \leq \infty$ be the largest extended real number such that $\|u(t;x) - \hat{x}\| \leq r$ for $0 \leq t < t_1$. Let $h(s) \overset{\text{def}}{=} K(\hat{x}) - K'(\hat{x})\hat{x} + o(u(s;x) - \hat{x})$ for $0 \leq s < t_1$. By (4.92) and (4.97) we have that for $0 \leq t < t_1$,

$$u(t;x) = T(t)x + \int_0^t T(t - s)K(u(s;x)) \, ds$$

$$= T(t)x + \int_0^t T(t - s)[K'(\hat{x})u(s;x) + h(s)] \, ds$$

$$= \hat{T}(t)x + \int_0^t \hat{T}(t - s)h(s)\, ds$$

Further, for $t \geq 0$,

$$\hat{x} = \hat{T}(t)\hat{x} + \int_0^t \hat{T}(t - s)[K(\hat{x}) - K'(\hat{x})\hat{x}]\, ds$$

Consequently, for $0 \leq t < t_1$,

$$\|u(t;x) - \hat{x}\| \leq Me^{\omega t}\|x - \hat{x}\| + \int_0^t Me^{\omega(t-s)}\|o(u(s;x) - \hat{x})\|\, ds$$

$$\leq Me^{\omega t}\left[\|x - \hat{x}\| + \int_0^t e^{-\omega s}\left(\frac{-\omega}{2M}\right)\|u(s;x) - \hat{x}\|\, ds\right]$$

which implies that

$$e^{-\omega t}\|u(t;x) - \hat{x}\| \leq M\|x - \hat{x}\| + \left(\frac{-\omega}{2}\right)\int_0^t e^{-\omega s}\|u(s;x) - \hat{x}\|\, ds$$

By Gronwall's lemma we have that for $0 \leq t < t_1$,

$$e^{-\omega t}\|u(t;x) - \hat{x}\| \leq M\|x - \hat{x}\| \exp[(-\omega/2)t]$$

which implies that

$$\|u(t;x) - \hat{x}\| \leq M\|x - \hat{x}\| \exp[(\omega/2)t] \leq M\varepsilon \exp[(\omega/2)t] \leq r$$

Thus, by (4.92) $t_1 = \infty$, and so (4.98) holds with $\gamma = \omega/2$. □

The next proposition provides a sufficient condition for an equilibrium solution of the abstract semilinear differential equation $d/dt\, u(t) = Bu(t) + K(u(t))$ to be unstable.

PROPOSITION 4.18 Let $T(t)$, $t \geq 0$, be a strongly continuous semigroup of bounded linear operators in the Banach space $X$ with infinitesimal generator $B$. Let $K$ be a (nonlinear) operator from $X$ to $X$ such that $K$ is continuously Fréchet differentiable on $X$ (as in Definition 2.4). For each $x \in X$ let $u(t;x)$ be the solution of the integral equation in (4.92) on the maximal interval of existence $[0,T_x)$. Let $\hat{x} \in X$ such that $B\hat{x} + K(\hat{x}) = 0$ and let $\hat{T}(t)$, $t \geq 0$, be the strongly continuous semigroup of bounded linear operators in X with infinitesimal generator $\hat{B} \overset{\text{def}}{=} B + K'(\hat{x})$. Let $X = X_1 \oplus X_2$ with $X_2 \neq \{0\}$, let $P_1$, $P_2$ be the projections of X onto $X_1$, $X_2$, respectively, let $\hat{T}(t)(X_i) \subset X_i$ for $t \geq 0$, $i = 1, 2$, and let $B + K'(\hat{x})$ restricted to $X_2$, denoted by $\hat{B}_{X_2}$, be bounded, so that $\hat{T}(t)x = \exp[t\hat{B}_{X_2}]x$ for all $x \in X_2$, $t \in R$. Let $M \geq 1$ and let $0 < \delta < \omega$ such that

$$\|\hat{T}(t)x\| \leq Me^{(\omega-\delta)t}\|x\| \qquad x \in X_1 \qquad t \geq 0$$

$$\|\hat{T}(t)x\| \leq Me^{(\omega+\delta)t}\|x\| \qquad x \in X_2 \qquad t \leq 0$$

Then, $\hat{x}$ is an unstable equilibrium in the following sense:

(4.101) There exists $\varepsilon > 0$ and a sequence $\{x_m\}$ in X such that $x_m \to \hat{x}$ and $\|u(m;x_m) - \hat{x}\| \geq \varepsilon$ for $m = 1, 2, \ldots,$ .

*Proof.* Let $K(x) = K(\hat{x}) + K'(\hat{x})(x - \hat{x}) + o(x - \hat{x})$ for $x \in X$, where $o\colon X \to X$, $\|o(x)\| \leq b(r)\|x\|$ for $\|x\| \leq r$, and $b\colon [0,\infty) \to [0,\infty)$ is an increasing continuous function such that $b(0) = 0$ (as in Definition 2.4). From Definition 2.4 it is seen that there exists an increasing continuous function $b_1\colon [0,\infty) \to [0,\infty)$ such that $b_1(0) = 0$ and $\|o(x_1) - o(x_2)\| \leq b_1(r)\|x_1 - x_2\|$ for $\|x_1\|, \|x_2\| \leq r$. Choose $r > 0$ such that $M(|P_1| + |P_2|)b(r)/\delta < 1/4M$ and $M(|P_1| + |P_2|)b_1(r)/\delta < 1/2$. Observe that

$$M(|P_1| + |P_2|)\,\frac{b(r)}{\delta} < \frac{1}{2}$$

$$M|P_1|\,\frac{b(r)r}{\delta} < \frac{r}{4M}$$

Choose $x_0 \in X_2$ such that

$$M|P_1|\,\frac{b(r)r}{\delta} < \frac{\|x_0\|}{2} < \frac{r}{4M}$$

We claim that for each $m = 1, 2, \ldots,$ there exists a unique function $v_m\colon (-\infty,m] \to X$ such that

$$(4.102)\qquad v_m(t) = \hat{T}(t - m)x_0 - \int_t^m \hat{T}(t - s)P_2 o(v_m(s))\, ds + \int_{-\infty}^t \hat{T}(t - s)P_1 o(v_m(s))\, ds \qquad t \le m$$

satisfying

$$(4.103)\qquad \|v_m(t)\| \le re^{\omega(t-m)} \qquad t \le m$$

To establish this claim let m be fixed and define the complete metric space Z, with metric $\rho$, by

$Z \overset{\text{def}}{=} \{v$: v is a continuous function from $(-\infty,m]$ to X such that $\sup_{-\infty<t\le m}\|v(t)\| \le re^{\omega(t-m)}\}$

$$\rho(u,v) \overset{\text{def}}{=} \sup_{t\le m} e^{-\omega(t-m)}\|u(t) - v(t)\| \qquad u, v \in Z$$

Define a mapping H on Z by

$$(Hv)(t) \overset{\text{def}}{=} \hat{T}(t - m)x_0 - \int_t^m \hat{T}(t - s)P_2 o(v(s))\, ds + \int_{-\infty}^t \hat{T}(t - s)P_1 o(v(s))\, ds \qquad v \in Z,\ t \le m$$

Then, H maps Z into Z, since for $v \in Z$, $t \le m$, $(Hv)(t)$ is continuous in t and

$$\begin{aligned}\|(Hv)(t)\| &\le Me^{(\omega+\delta)(t-m)}\|x_0\| \\ &\quad + \int_t^m Me^{(\omega+\delta)(t-s)}|P_2|b(r)e^{\omega(s-m)}\, ds \\ &\quad + \int_{-\infty}^t Me^{(\omega-\delta)(t-s)}|P_1|b(r)re^{\omega(s-m)}\, ds \\ &\le M\left(\|x_0\| + (|P_1| + |P_2|)\frac{b(r)r}{\delta}\right) e^{\omega(t-m)} \\ &\le re^{\omega(t-m)}\end{aligned}$$

Also, H is a strict contraction in Z, since for $v, \hat{v} \in Z$, $t \leq m$,

$$\|(Hv)(t) - (H\hat{v})(t)\|$$
$$\leq \int_t^m Me^{(\omega+\delta)(t-s)}|P_2|b_1(r)\|v(s) - \hat{v}(s)\| \, ds$$
$$+ \int_{-\infty}^t Me^{(\omega-\delta)(t-s)}|P_1|b_1(r)\|v(s) - \hat{v}(s)\| \, ds$$
$$\leq Mb_1(r)(|P_1| + |P_2|)e^{\omega(t-m)}\rho(v,\hat{v})$$
$$\times \left[\int_t^m e^{\delta(t-s)} \, ds + \int_{-\infty}^t e^{-\delta(t-s)} \, ds\right]$$
$$\leq \left(2Mb_1(r)(|P_1| + |P_2|) \frac{e^{\omega(t-m)}}{\delta}\right) \rho(v,\hat{v})$$

By the contraction mapping theorem, H has a unique fixed point $v_m \in Z$, which is also the unique solution of (4.102) satisfying (4.103).

We next claim that for $m = 1, 2, \ldots, 0 \leq t \leq m$,

$$v_m(t) = T(t)v_m(0) + \int_0^t T(t - s)[K'(\hat{x})v_m(s) + o(v_m(s))] \, ds \tag{4.104}$$

Define

$$z(t) \stackrel{\text{def}}{=} \hat{T}(t - m)x_0 - \int_t^m \hat{T}(t - s)P_2 o(v_m(s)) \, ds \qquad 0 \leq t \leq m$$

Since $\hat{T}(t) = \exp[t\hat{B}_{X_2}]$ on $X_2$, we have that

$$z'(t) = [B + K'(\hat{x})]z(t) + P_2 o(v_m(t)) \qquad 0 \leq t \leq m$$

By (4.97) we obtain for $0 \leq t \leq m$,

$$P_2 v_m(t) = T(t)P_2 v_m(0) + \int_0^t T(t - s)[K'(\hat{x})P_2 v_m(s) + P_2 o(v_m(s))] \, ds \tag{4.105}$$

Also, from (4.102) we obtain that for $t \leq m$,

$$P_1 v_m(t) = \int_{-\infty}^{t} \hat{T}(t - s) P_1 o(v_m(s))\, ds$$

Let $\{h_k(s)\}$ be a sequence of continuously differentiable functions from $(-\infty, m]$ to $D(B) \cap X_1$ such that $h_k$ has support in $[-k, m]$, $k = 1, 2, \ldots$, and $h_k(s)$ converges uniformly to $P_1 o(v_m(s))$ on $(-\infty, m]$ (the existence of such a sequence follows from the denseness of $D(B)$ in $X_1$, the continuity of $P_1 o(v_m(s))$ on $(-\infty, m]$, and the estimate $\|P_1 o(v_m(s))\| \le |P_1| b(r) r e^{\omega(s-m)}$, $s \le m$). For $k = 1, 2, \ldots$, define

$$z_k(t) \overset{\text{def}}{=} \int_{-\infty}^{t} \hat{T}(t - s) h_k(s)\, ds \qquad t \le m$$

and observe that $z_k(t)$ converges uniformly to $P_1 v_m(t)$ on $(-\infty, m]$. For $t \le m$, $k = 1, 2, \ldots$,

$$\lim_{h \to 0} h^{-1}[z_k(t + h) - z_k(t)]$$

$$= \lim_{h \to 0} \left[ \int_{-\infty}^{t} h^{-1}[\hat{T}(t + h - s) - \hat{T}(t - s)] h_k(s)\, ds \right.$$

$$\left. + h^{-1} \int_{t}^{t+h} \hat{T}(t + h - s) h_k(s)\, ds \right]$$

$$= (B + K'(\hat{x})) \int_{-\infty}^{t} \hat{T}(t - s) h_k(s)\, ds + h_k(t)$$

which implies that

$$z_k'(t) = (B + K'(\hat{x})) z_k(t) + h_k(t)$$

By (4.97) we have that for $t \le m$, $k = 1, 2, \ldots$,

$$z_k(t) = T(t) z_k(0) + \int_{0}^{t} T(t - s)[K'(\hat{x}) z_k(s) + h_k(s)]\, ds$$

Now let $k \to \infty$ to obtain for $t \le m$,

$$P_1 v_m(t) = T(t) P_1 v_m(0) + \int_{0}^{t} T(t - s)[K'(\hat{x}) P_1 v_m(s) + P_1 o(v_m(s))]\, ds \tag{4.106}$$

The claim (4.104) then follows from (4.105) and (4.106).

From (4.104) we obtain for $m = 1, 2, \ldots, 0 \le t \le m$,

$$v_m(t) = T(t)v_m(0) + \int_0^t T(t - s)[K(v_m(s) + \hat{x}) - K(\hat{x})]\, ds$$

$$= T(t)(v_m(0) + \hat{x}) - \hat{x} + \int_0^t T(t - s)K(v_m(s) + \hat{x})\, ds$$

By (4.92) we must have that $v_m(t) + \hat{x} = u(t;v_m(0) + \hat{x})$ for $0 \le t \le m$, $m = 1, 2, \ldots,$ . Let $x_m \overset{\text{def}}{=} v_m(0) + \hat{x}$ for $m = 1, 2, \ldots$ . By (4.103) we see that $x_m \to \hat{x}$. Further, for $m = 1, 2, \ldots, \sup_{0\le t\le m}\|u(t;x_m) - \hat{x}\| \ge \|x_0\|/2$, since

$$\|u(m;x_m) - \hat{x}\| = \|v_m(m)\|$$

$$= \|x_0\| + \int_{-\infty}^m \hat{T}(m - s)P_1 o(v_m(s))\, ds$$

$$\ge \|x_0\| - \int_{-\infty}^m Me^{(\omega-\delta)(m-s)}|P_1|b(r)re^{\omega(s-m)}\, ds$$

$$= \|x_0\| - M|P_1|\,\frac{b(r)r}{\delta}$$

Hence, (4.101) is proved with $\varepsilon = M|P_1|b(r)r/\delta$. □

Combining Proposition 4.17 and Proposition 4.18, together with the results of Section 4.3, we obtain

PROPOSITION 4.19 Let $T(t)$, $t \ge 0$, be a strongly continuous semigroup of bounded linear operators in the Banach space $X$ with infinitesimal generator $B$. Let $K$ be a (nonlinear) operator from $X$ to $X$ such that $K$ is continuously Fréchet differentiable on $X$ (as in Definition 2.4). For each $x \in X$ let $u(t;x)$ be the solution of the integral equation in (4.92) on the maximal interval of existence $[0,T_x)$. Let $\hat{x} \in X$ such that $B\hat{x} + K(\hat{x}) = 0$ and let $\hat{T}(t)$, $t \ge 0$, be the strongly continuous semigroup of bounded linear operators in $X$ with infinitesimal generator $\hat{B} \overset{\text{def}}{=} B + K'(\hat{x})$. The following hold:

(4.107) If $\max\{\omega_1(\hat{B}), \sup_{\lambda\in\sigma(\hat{B})-E\sigma(\hat{B})} \text{Re}\,\lambda\} < 0$, then $\hat{x}$ is a locally exponentially asymptotically stable equilibrium in the sense of (4.98).

(4.108) If there exists $\lambda_1 \in \sigma(\hat{B})$ such that $\text{Re}\,\lambda_1 > 0$ and $\max\{\omega_1(\hat{B}), \sup_{\lambda\in\sigma(\hat{B})-E\sigma(\hat{B}),\lambda\neq\lambda_1} \text{Re}\,\lambda\} < \text{Re}\,\lambda_1$, then $\hat{x}$ is an unstable equilibrium in the sense of (4.101).

*Proof.* The proof of (4.107) follows immediately from (4.47), (4.57), and Proposition 4.17. The proof of (4.108) follows immediately from Proposition 4.15 with $\Lambda \overset{\text{def}}{=} \{\lambda_1\}$ and Proposition 4.18. □

Proposition 4.19 can be applied directly to the age-dependent population problem (ADP) if the birth function F is a bounded linear operator from $L^1$ to $R^n$

THEOREM 4.12 Let (2.1), (2.2), (2.22), (2.23) hold, let F be a bounded linear operator from $L^1$ to $R^n$, let G be continuously Fréchet differentiable on $L^1$ (as in Definition 2.4), let $T_\phi = +\infty$ for all $\phi \in L^1_+$, and let $S(t)$, $t \geq 0$, be the strongly continuous nonlinear semigroup in $L^1_+$ associated with (ADP) (as in Theorem 3.1) with infinitesimal generator B (as in Theorem 3.2). Let $\hat{\phi} \in L^1_+$ be an equilibrium solution of (ADP), let $\hat{S}(t)$, $t \geq 0$, be the strongly continuous semigroup of bounded linear operators in $L^1$ with infinitesimal generator $\hat{B}$: $L^1 \to L^1$ defined by

$$B\hat{\phi} \overset{\text{def}}{=} -\phi' + G'(\hat{\phi})\phi$$

$$D(B) \overset{\text{def}}{=} \{\phi \in L^1 : \phi \text{ is absolutely continuous on } [0,\infty),\ \phi' \in L^1, \text{ and } \phi(0) = F\phi\}$$

(as in Proposition 3.2 and Proposition 3.7), and let $\omega_1(\hat{B}) < 0$. The following hold:

(4.109) If $\sup_{\lambda\in\sigma(\hat{B})-E\sigma(\hat{B})} \text{Re}\,\lambda < 0$, then $\hat{\phi}$ is a locally exponentially asymptotically stable equilibrium of (ADP).

(4.110) If there exists $\lambda_1 \in \sigma(\hat{B})$ such that $\text{Re}\,\lambda_1 > 0$ and

$\sup_{\lambda\in\sigma(\hat{B})-E\sigma(\hat{B})\lambda\neq\lambda_1}$ Re $\lambda$ < Re $\lambda_1$, then $\hat{\phi}$ is an unstable equilibrium of (ADP).

*Proof.* Define $B_0$: $L^1 \to L^1$ by $B_0\phi \overset{\text{def}}{=} -\phi'$, $\phi \in D(B_0) \overset{\text{def}}{=} D(\hat{B})$. By Proposition 3.2 and Proposition 3.7 $B_0$ is the infinitesimal generator of a strongly continuous semigroup of bounded linear operators $S_0(t)$, $t \geq 0$, in $L^1$. We claim that for all $\phi \in L^1_+$, $t \geq 0$,

$$S(t)\phi = S_0(t)\phi + \int_0^t S_0(t - s)G(S(s)\phi)\, ds \tag{4.111}$$

By (4.92) the integral equation (4.111) has a unique solution $u(t;\phi)$ defined on a maximal interval of existence. If $\phi \in D(B_0)$, then by (4.94) $u(t;\phi)$ is continuously differentiable, $u(t;\phi) \in D(B_0) = D(B)$, and $d/dt\ u(t;\phi) = B_0u(t;\phi) + G(u(t;\phi))$ on this interval. By Theorem 3.4 $u(t;\phi) = S(t)\phi$ for all t in this interval, which must be $[0,\infty)$ by (4.92). Thus, (4.111) holds for $\phi \in D(B_0)$. If $\phi \in L^1_+ - D(B_0)$, then there exists a sequence $\{\phi_m\} \subset D(B_0)$ such that $\phi_m \to \phi$. Observe that for $t \geq 0$, $m = 1, 2, \ldots,$

$$\begin{aligned}
&\left\|S(t)\phi - S_0(t)\phi - \int_0^t S_0(t - s)G(S(s)\phi)\, ds\right\|_{L^1} \\
&\leq \|S(t)\phi - S(t)\phi_m\|_{L^1} + \|S_0(t)(\phi - \phi_m)\|_{L^1} \\
&\quad + \int_0^t \|S_0(t - s)[G(S(s)\phi) - G(S(s)\phi_m)]\, ds\|_{L^1}
\end{aligned}$$

Now use the continuity properties of $S(t)$, $t \geq 0$, $S_0(t)$, $t \geq 0$, and G to show that (4.111) holds for all $\phi \in L^1_+$. The claims (4.109) and (4.110) now follow directly from Proposition 4.19. □

Applications of Theorem 4.12 to the study of stability of equilibrium solutions of age-dependent population problems are given in the following single species models with linear birth functions.

EXAMPLE 4.11 Let $R^n = R$. As in Examples 2.1, 2.4, 3.1, and 4.1, let F have the form (2.17) (except that we require $\beta(P) \equiv \beta_0$ for

$P \geq 0$, where $\beta_0$ is a positive constant), let G have the form (2.18) (except that we require $\mu$ to be twice continuously differentiable), and let (3.81) hold:

$$F\phi \overset{\text{def}}{=} \int_0^\infty \beta_0 e^{-\alpha a}\phi(a)\, da \qquad \phi \in L^1$$

$$G(\phi)(a) \overset{\text{def}}{=} -\mu(P\phi)\phi(a) \qquad \phi \in L^1,\ \text{a.e. } a > 0 \text{ (where } \mu(\hat{P}) \geq \underline{\mu} > 0 \text{ for } P \geq 0)$$

$$P\phi \overset{\text{def}}{=} \int_0^\infty \phi(a)\, da \qquad \phi \in L^1$$

Notice that this birth function F corresponds to a reproductive process which is not influenced by crowding effects. Also, F is a bounded linear operator from $L^1$ to R. Recall from Example 3.1 that the infinitesimal generator B of the strongly continuous nonlinear semigroup $S(t)$, $t \geq 0$, in $L^1_+$ associated with this (ADP) is

> $B\phi \overset{\text{def}}{=} -\phi' - \mu(P\phi)\phi$, $\phi \in D(B)$, $D(B) \overset{\text{def}}{=} \{\phi \in L^1_+$: $\phi$ is absolutely continous on $[0,\infty)$, $\phi' \in L^1$, and $\phi(0) = \beta_0 \int_0^\infty e^{-\alpha a}\phi(a)\, da\}$.

Recall from Remark 2.4 that G is continuously Fréchet differentiable on $L^1$ and from (2.85)

$$(G'(\hat{\phi})\phi)(a) = -\mu'(P\hat{\phi})P\phi\hat{\phi}(a) - \mu(P\hat{\phi})\phi(a) \qquad \phi, \hat{\phi} \in L^1,\ \text{a.e. } a > 0$$

Let $\hat{\phi} \in L^1_+$ such that $\hat{\phi}$ is an equilibrium solution of (ADP) or, equivalently, $B\hat{\phi} = 0$. From Example 4.1, $\hat{\phi}(a) = \mu(\hat{P})\hat{P}\exp[-\mu(\hat{P})a]$, $a \geq 0$, where $P\hat{\phi} = \hat{P}$, and either $\hat{P} = 0$ or $\hat{P} > 0$ and satisfies $\alpha + \mu(\hat{P}) = \beta_0$. Define the linear operator $\hat{B}$ in $L^1$ by

$$\hat{B}\phi \overset{\text{def}}{=} -\phi' + G'(\hat{\phi})\phi \qquad \phi \in D(\hat{B}) \overset{\text{def}}{=} D(B)$$

By Proposition 3.2 and Proposition 3.7, $\hat{B}$ is the infinitesimal generator of a strongly continuous semigroup of bounded linear operators in $L^1$.

Define the bounded linear operator C in $L^1$ by

$$C\phi \overset{\text{def}}{=} -\mu'(\hat{P})P\phi\hat{\phi} \qquad \phi \in D(C) \overset{\text{def}}{=} L^1$$

and observe that C is compact from $L^1$ to $L^1$. Define the linear operator $\hat{B}_0$ in $L^1$ by

$$\hat{B}_0\phi \overset{\text{def}}{=} -\phi' - \mu(P\hat{\phi})\phi \qquad \phi \in D(\hat{B}_0) \overset{\text{def}}{=} D(\hat{B})$$

and observe from Theorem 4.6 that $\omega_1(\hat{B}_0) \leq -\underline{\mu}$. By Proposition 4.14 $\omega_1(\hat{B}) = \omega_1(\hat{B}_0 + C) \leq -\underline{\mu}$. Thus, by Theorem 4.12, the stability of $\hat{\phi}$ is determined by $\sigma(\hat{B}) - E\sigma(\hat{B})$. Accordingly, let $\lambda \in \mathbb{C}$ and let $\hat{B}\phi = \lambda\phi$ for $\phi \in L^1$, $\phi \neq 0$.

From the definition of $\hat{B}$ we obtain

$$\phi'(a) + \lambda\phi(a) + \mu'(\hat{P})P\phi\mu(\hat{P})\hat{P}\exp[-a\mu(\hat{P})] + \mu(\hat{P})\phi(a) = 0 \qquad a \geq 0 \tag{4.112}$$

$$\phi(0) = \int_0^\infty \beta_0 e^{-\alpha a}\phi(a)\,da \tag{4.113}$$

From (4.112) we have that

$$\phi(a) = \exp[-(\lambda + \mu(\hat{P}))a]\phi(0) - \int_0^a \exp[-(\lambda + \mu(\hat{P}))(a - b)]\mu'(\hat{P})P\phi\mu(\hat{P})\hat{P}\exp[-b\mu(\hat{P})]\,db \qquad a \geq 0 \tag{4.114}$$

If $\hat{\phi} = 0$, then (4.113) and (4.114) yield that

$$\phi(0) = \int_0^\infty \beta_0 e^{-\alpha a}\phi(0)\exp[-(\lambda + \mu(0))a]\,da$$

Consequently, $\lambda = \beta_0 - \mu(0) - \alpha$, and if $\beta_0 < \mu(0) + \alpha$, then 0 is locally exponentially asymptotically stable by (4.109), and if $\beta_0 > \mu(0) + \alpha$, then 0 is unstable by (4.110).

If $\hat{\phi} \neq 0$ and $\lambda = 0$, then (4.113) and (4.114) imply that

$$\phi(a) = \exp[-\mu(\hat{P})a]\phi(0) - \mu'(\hat{P})P\phi\mu(\hat{P})\hat{P}a\exp[-\mu(\hat{P})a] \qquad a \geq 0 \tag{4.115}$$

$$(4.116) \qquad \phi(0) = \beta_0(\alpha + \mu(\hat{P}))^{-1}\phi(0) - \beta_0\mu'(\hat{P})P\phi\mu(\hat{P})\hat{P}(\alpha + \mu(\hat{P}))^{-2}$$

Since $\beta_0 = \alpha + \mu(\hat{P})$, (4.116) implies that $\mu'(\hat{P})P\phi = 0$. Since $P\phi \neq 0$ by virtue of (4.115), we must have $\mu'(\hat{P}) = 0$. Further, if $\mu'(\hat{P}) = 0$, then 0 is an eigenvalue of $\hat{B}$ with eigenvector $\phi(a) \overset{\text{def}}{=} \exp[-\mu(\hat{P})a]$, $a \geq 0$.

If $\hat{\phi} \neq 0$ and $\lambda \neq 0$, then (4.114) implies that

$$(4.117) \qquad \phi(a) = \exp[-(\lambda + \mu(\hat{P}))a](\phi(0) + \lambda^{-1}\mu'(\hat{P})P\phi\mu(\hat{P})\hat{P}) - \exp[-\mu(\hat{P})a](\lambda^{-1}\mu'(\hat{P})P\phi\mu(\hat{P})\hat{P}) \qquad a \geq 0$$

If Re $\lambda \leq -\mu(\hat{P})$, then (4.117) implies that $\phi(0) = -\lambda^{-1}\mu'(\hat{P})P\phi\mu(\hat{P})\hat{P}$ (since $\phi \in L^1$). Then, (4.117) implies that $P\phi = -\lambda^{-1}\mu'(\hat{P})P\phi\hat{P}$. Since (4.117) also implies that $P\phi \neq 0$, we must have $\lambda = -\mu'(\hat{P})\hat{P}$. If Re $\lambda > -\mu(\hat{P})$, then (4.117) implies that

$$P\phi = \phi(0)(\lambda + \mu(\hat{P}))^{-1} - \mu'(\hat{P})\hat{P}P\phi(\lambda + \mu(\hat{P}))^{-1}$$

which implies that

$$(4.118) \qquad (\lambda + \mu(\hat{P}) + \mu'(\hat{P})\hat{P})P\phi = \phi(0)$$

From (4.113), (4.117), and (4.118) we obtain

$$(\lambda + \mu(\hat{P}) + \mu'(\hat{P})\hat{P})P\phi = \beta_0[\phi(0)(\alpha + \lambda + \mu(\hat{P}))^{-1} - \lambda^{-1}\mu'(\hat{P})P\phi\mu(\hat{P})\hat{P}[(\alpha + \mu(\hat{P}))^{-1} - (\lambda + \alpha + \mu(\hat{P}))^{-1}]]$$

Now use (4.118) again and the fact that $\beta_0 = \alpha + \mu(\hat{P})$ to obtain

$$(\beta_0 + \lambda)(\lambda + \mu(\hat{P}) + \mu'(\hat{P}))P\phi = \beta_0(\lambda + \mu(\hat{P}) + \mu'(\hat{P}))P\phi - \mu'(\hat{P})P\phi\mu(\hat{P})\hat{P}$$

But $P\phi \neq 0$ by (4.118) and (4.117), and we thus obtain

$$\lambda^2 + \lambda[\mu(\hat{P}) + \mu'(\hat{P})\hat{P}] + \mu'(\hat{P})\mu(\hat{P})\hat{P} = 0$$

Solve this quadratic equation for $\lambda$ to obtain $\lambda = -\mu'(\hat{P})\hat{P}$ or $\lambda = -\mu(\hat{P})$. Further, the calculations above show that if $\lambda = -\mu'(\hat{P})\hat{P}$

or $\lambda = -\mu(\hat{P})$, $\lambda \neq 0$, then $\lambda$ is an eigenvalue of B.

Thus, $P\sigma(\hat{B}) = \{-\mu(\hat{P}), -\mu'(\hat{P})\hat{P}\}$ for $\hat{P} > 0$. Since $\mu(\hat{P}) > 0$, we conclude from Theorem 4.12 that if $\mu'(\hat{P}) > 0$, then $\hat{\phi}$ is locally exponentially asymptotically stable, and if $\mu'(\hat{P}) < 0$, then $\hat{\phi}$ is unstable.

EXAMPLE 4.12. Let $R^n = R$. As in Examples 2.2, 2.5, 3.2, and 4.2, let F have the form (2.19) (except that $\beta(P) \equiv \beta_0$ for $P \geq 0$, where $\beta_0$ is a positive constant), let G have the form (2.18) (except that $\mu$ is twice continuously differentiable), and let (3.81) hold:

$$F\phi \overset{\text{def}}{=} \int_0^\infty \beta_0(1 - e^{-\alpha a})\phi(a)\, da \qquad \phi \in L^1$$

$$G(\phi)(a) \overset{\text{def}}{=} -\mu(P\phi)\phi(a), \ \phi \in L^1, \text{ a.e. } a > 0 \text{ (where } \mu(\hat{P}) \geq \underline{\mu} > 0 \text{ for } P \geq 0)$$

$$P\phi \overset{\text{def}}{=} \int_0^\infty \phi(a)\, da \qquad \phi \in L^1$$

As in Example 4.11, this birth process is not influenced by crowding effects, and the birth function F is a bounded linear operator from $L^1$ to R. From Example 3.2 the infinitesimal generator B of the strongly continuous nonlinear semigroup $S(t)$, $t \geq 0$, in $L^1_+$ associated with this (ADP) is

$$B\phi \overset{\text{def}}{=} -\phi' - \mu(P\phi)\phi, \ \phi \in D(B), \ D(B) \overset{\text{def}}{=} \{\phi \in L^1_+: \ \phi \text{ is absolutely continuous on } [0,\infty), \ \phi' \in L^1, \text{ and } \phi(0) = \beta_0 \int_0^\infty (1 - e^{-\alpha a})\phi(a)\, da\}$$

Let $\hat{\phi} \in L^1_+$ such that $\hat{\phi}$ is an equilibrium solution of (ADP), that is, $B\hat{\phi} = 0$. From Example 4.2, $\hat{\phi}(a) = \mu(\hat{P})\hat{P} \exp[-\mu(\hat{P})a]$, $a \geq 0$, where $P\hat{\phi} = \hat{P}$, and either $\hat{P} = 0$ or $\hat{P} > 0$ and satisfies $\mu(\hat{P})(\alpha + \mu(\hat{P})) = \alpha\beta_0$. Define the linear operator $\hat{B}$ in $L^1$ by

$$\hat{B}\phi \overset{\text{def}}{=} -\phi' + G'(\hat{\phi})\phi \qquad \phi \in D(\hat{B}) \overset{\text{def}}{=} D(B)$$

where $G'(\hat{\phi})\phi$ is given by (2.85). As in Example 4.11, $\hat{B}$ is the infinitesimal generator of a strongly continuous semigroup of bounded linear operators in $L^1$ and $\omega_1(\hat{B}) \le -\underline{\mu}$. By Theorem 4.12, the stability of $\hat{\phi}$ is determined from $\sigma(\hat{B}) - E\sigma(\hat{B})$. Therefore, let $\lambda \in C$ and let $\phi \in L^1$, $\phi \ne 0$, such that $\hat{B}\phi = \lambda\phi$.

As in Example 4.11 we obtain (4.114) and

$$\phi(0) = \beta_0 \int_0^\infty (1 - e^{-\alpha a})\phi(a)\, da \tag{4.119}$$

If $\hat{\phi} = 0$, then (4.114) and (4.119) yield that

$$\phi(0) = \beta_0 \int_0^\infty (1 - e^{-\alpha a}) \exp[-(\lambda + \mu(0))a]\phi(0)\, da$$

Thus, $\text{Re}\,\lambda + \mu(0) + \alpha > 0$ and

$$(\lambda + \mu(0))(\lambda + \alpha + \mu(0)) = \alpha\beta_0$$

which means that $\lambda$ is real and

$$2\lambda_\pm \overset{\text{def}}{=} -(\alpha + 2\mu(0)) \pm (\alpha^2 + 4\beta_0\alpha)^{1/2}$$

If $\alpha\beta_0 < \mu(0)(\alpha + \mu(0))$, then $\lambda_\pm < 0$, and so 0 is locally exponentially stable by (4.109). If $\alpha\beta_0 > \mu(0)(\alpha + \mu(0))$, then $\lambda_+ > 0$, and so 0 is unstable by (4.110).

If $\hat{\phi} \ne 0$ and $\lambda = 0$, then (4.114) and (4.119) imply (4.115) and

$$\begin{aligned}\phi(0) = \beta_0\{&\phi(0)[\mu(\hat{P})^{-1} - (\alpha + \mu(\hat{P}))^{-1}] \\ &- \mu'(\hat{P})P\phi\mu(\hat{P})\hat{P}[\mu(\hat{P})^{-2} - (\alpha + \mu(\hat{P}))^{-2}]\}\end{aligned} \tag{4.120}$$

Now substitute $\alpha\beta_0 = \mu(\hat{P})(\alpha + \mu(\hat{P}))$ into (4.120) to obtain

$$\mu'(\hat{P})P\phi\hat{P}(\alpha + 2\mu(\hat{P}))(\alpha + \mu(\hat{P}))^{-1} = 0$$

Since $P\phi \ne 0$ by (4.115), we must have that $\mu'(\hat{P}) = 0$. Further, if $\mu'(\hat{P}) = 0$, then 0 is an eigenvalue of $\hat{B}$ with eigenvector $\phi(a) \overset{\text{def}}{=} \exp[-\mu(\hat{P})a]$, $a \ge 0$.

If $\hat{\phi} \ne 0$ and $\lambda \ne 0$, we obtain (4.117). As in Example 4.11, $\text{Re}\,\lambda \le -\mu(\hat{P})$ implies that $\lambda = -\mu'(\hat{P})\hat{P}$. If $\text{Re}\,\lambda > -\mu(\hat{P})$, then (4.117) implies (4.118). From (4.117), (4.118), and (4.119) we obtain

$$(\lambda + \mu(\hat{P}) + \mu'(\hat{P})\hat{P})P\phi = \beta_0 P\phi\alpha(\lambda + \mu(\hat{P}))^{-1}(\alpha + \lambda + \mu(\hat{P}))^{-1} \times [(\lambda + \mu(\hat{P}) + \mu'(\hat{P})\hat{P}) - \mu'(\hat{P})\hat{P} \times (\alpha + \lambda + 2\mu(\hat{P}))(\alpha + \mu(\hat{P}))^{-1}]$$

Since $P\phi \neq 0$ by (4.118) and (4.117), we may substitute $\alpha\beta_0 = \mu(\hat{P})(\alpha + \mu(\hat{P}))$ into this last equation to obtain

$$\lambda^2 + \lambda(\alpha + 2\mu(\hat{P}) + \mu'(\hat{P})\hat{P}) + 2\mu'(\hat{P})\hat{P}\mu(\hat{P}) + \mu'(\hat{P})\hat{P}\alpha = 0$$

Now solve this quadratic equation to obtain $\lambda = -\mu'(\hat{P})\hat{P}$ or $\lambda = -\alpha - 2\mu(\hat{P})$. Further, these calculations show that if $\lambda = -\mu'(\hat{P})\hat{P}$ or $\lambda = -\alpha - 2\mu(\hat{P})$, $\lambda \neq 0$, then $\lambda$ is an eigenvalue of $\hat{B}$.

Thus, $P\sigma(\hat{B}) = \{-\alpha - 2\mu(\hat{P}), -\mu'(\hat{P})\hat{P}\}$ for $\hat{P} > 0$. Since $\alpha > 0$ and $\mu(\hat{P}) > 0$, we conclude from Theorem 4.12 that $\hat{\phi}$ is locally exponentially asymptotically stable if $\mu'(\hat{P}) > 0$ and unstable if $\mu'(\hat{P}) < 0$.

EXAMPLE 4.13. Let $R^n = R$. As in Examples 2.3, 2.6, 3.3, and 4.3, let F have the form (2.20) (except that $\beta(P) \equiv \beta_0$ for $P \geq 0$, where $\beta_0$ is a positive constant), let G have the form (2.18) (except that $\mu$ is twice continuously differentiable), and let (3.81) hold:

$$F\phi \overset{\text{def}}{=} \int_0^\infty \beta_0 a e^{-\alpha a}\phi(a)\, da \qquad \phi \in L^1$$

$$G(\phi)(a) \overset{\text{def}}{=} -\mu(P\phi)\phi(a), \quad \phi \in L^1, \text{ a.e. } a > 0 \text{ (where } \mu(P) \geq \underline{\mu} > 0 \text{ for } P \geq 0)$$

$$P\phi \overset{\text{def}}{=} \int_0^\infty \phi(a)\, da \qquad \phi \in L^1$$

From Example 3.3 the infinitesimal generator B of the strongly continuous nonlinear semigroup $S(t)$, $t \geq 0$, in $L^1_+$ associated with this (ADP) is

$$B\phi \overset{\text{def}}{=} -\phi' - \mu(P\phi)\phi, \quad \phi \in D(B), \quad D(B) \overset{\text{def}}{=} \{\phi \in L^1_+ : \phi \text{ is absolutely continuous on } [0,\infty),\ \phi' \in L^1, \text{ and } \phi(0) = \beta_0 \int_0^\infty a e^{-\alpha a}\phi(a)\, da\}$$

Let $\hat{\phi} \in L^1_+$ be an equilibrium solution of (ADP), that is, $B\hat{\phi} = 0$. From Example 4.3, $\hat{\phi}(a) = \mu(\hat{P})\hat{P}\exp[-\mu(\hat{P})a]$, $a \geq 0$, where $P\hat{\phi} = \hat{P}$, and either $\hat{P} = 0$ or $\hat{P} > 0$ and satisfies $\beta_0 = (\alpha + \mu(\hat{P}))^2$. Define the linear operator $\hat{B}$ in $L^1$ as in Examples 4.11 and 4.12, so that $\hat{B}$ is the infinitesimal generator of a strongly continuous semigroup of bounded linear operators in $L^1$ with $\omega_1(\hat{B}) \leq -\underline{\mu}$. By Theorem 4.12, the stability of $\hat{\phi}$ is determined from $\sigma(\hat{B}) - E\sigma(\hat{B})$. Let $\lambda \in C$ and let $\phi \in L^1$, $\phi \neq 0$, such that $\hat{B}\phi = \lambda\phi$.

As in Example 4.11 we obtain (4.114) and

$$(4.121)\qquad \phi(0) = \beta_0 \int_0^\infty ae^{-\alpha a}\phi(a)\, da$$

If $\hat{\phi} = 0$, then (4.114) and (4.121) yield that

$$\phi(0) = \beta_0 \int_0^\infty ae^{-\alpha a} \exp[-(\lambda + \mu(0))a]\phi(0)\, da$$

Thus, $\mathrm{Re}\,\lambda + \mu(0) + \alpha > 0$ and $(\lambda + \mu(0) + \alpha)^2 = \beta_0$, so that $\lambda = \sqrt{\beta_0} - \mu(0) - \alpha$. If $\beta_0 < (\alpha + \mu(0))^2$, then 0 is locally exponentially asymptotically stable by (4.109). If $\beta_0 > (\alpha + \mu(0))^2$, then 0 is unstable by (4.110).

If $\hat{\phi} \neq 0$ and $\lambda = 0$, then (4.114) and (4.121) imply (4.115) and

$$(4.122)\qquad \phi(0) = \beta_0[\phi(0)(\alpha + \mu(\hat{P}))^{-2} - 2\mu'(\hat{P})P\phi\mu(\hat{P})\hat{P}(\alpha + \mu(\hat{P}))^{-3}]$$

Now substitute $\beta_0 = (\alpha + \mu(\hat{P}))^2$ into (4.122) to obtain $\mu'(\hat{P})P\phi\mu(\hat{P})\hat{P}/\sqrt{\beta_0} = 0$. Since $P\phi \neq 0$ by (4.115), $\mu'(\hat{P}) = 0$. Further, if $\mu'(\hat{P}) = 0$, then 0 is an eigenvalue of $\hat{B}$ with eigenvector $\phi(a) \overset{\text{def}}{=} \exp[-\mu(\hat{P})a]$, $a \geq 0$.

If $\hat{\phi} \neq 0$ and $\lambda \neq 0$, we obtain (4.117). As in Example 4.11, $\mathrm{Re}\,\lambda \leq -\mu(\hat{P})$ implies that $\lambda = -\mu'(\hat{P})\hat{P}$. If $\mathrm{Re}\,\lambda > -\mu(\hat{P})$, then (4.117) implies (4.118). From (4.117), (4.118), and (4.121) we obtain

$$\begin{aligned}(\lambda + \mu(\hat{P}) + \mu'(\hat{P})\hat{P})P\phi = \beta_0 P\phi\{&(\lambda + \mu(\hat{P}) + \mu'(\hat{P})\hat{P})(\alpha + \lambda + \mu(\hat{P}))^{-2} \\ &- \lambda^{-1}\mu'(\hat{P})\mu(\hat{P})\hat{P}[(\alpha + \mu(\hat{P}))^{-2} \\ &- (\alpha + \lambda + \mu(\hat{P}))^{-2}]\}\end{aligned}$$

Since $P\phi \neq 0$ by (4.118) and (4.117), we may substitute $\beta_0 = (\alpha + \mu(\hat{P}))^2$ into this last equation to obtain

$$(\lambda + 2\sqrt{\beta_0})(\lambda^2 + \lambda[\mu(\hat{P}) + \mu'(\hat{P})\hat{P}] + \mu'(\hat{P})\mu(\hat{P})\hat{P}) = 0$$

The roots of this cubic equation are $\lambda = -2\sqrt{\beta_0}$, $\lambda = -\mu(\hat{P})$, or $\lambda = -\mu'(\hat{P})\hat{P}$. Further, these calculations show that if $\lambda = -2\sqrt{\beta_0}$, $\lambda = -\mu(\hat{P})$, or $\lambda = -\mu'(\hat{P})\hat{P}$, then $\lambda$ is an eigenvalue of $\hat{B}$.

Thus, $P\sigma(\hat{B}) = \{-2\sqrt{\beta_0}, -\mu(\hat{P}), -\mu'(\hat{P})\hat{P}\}$ for $\hat{P} > 0$. We therefore conclude from Theorem 4.12 that $\hat{\phi}$ is locally exponentially asymptotically stable if $\mu'(P) > 0$ and unstable if $\mu'(\hat{P}) < 0$.

## 4.5 THE METHOD OF LINEARIZATION FOR THE NONLINEAR CASE

We now employ the linear approximation to study the stability of equilibria of (ADP) for the full nonlinear case, that is, when both the birth function F and the aging function G are nonlinear functions of the density. The development of this section is based upon the treatment of this problem by J. Prüss in [242]. We will suppose that F and G have special forms which allow convenient formulas for their Fréchet derivatives. We first present three lemmas concerning the existence, uniqueness, and representation of solutions to nonhomogeneous equations of linear age-dependent population dynamics.

We require the following hypotheses:

(4.123) $\hat{\beta}$: $[0,\infty) \to B(R^n,R^n)$, $\hat{\beta}$ is continuous and there exists $\beta_0 > 0$ such that $|\hat{\beta}(a)| \leq \beta_0$ for all $a \geq 0$.

(4.124) $\hat{\mu}$: $[0,\infty) \to B(R^n,R^n)$, $\hat{\mu}$ is continuous, its values commute, and there exists $\mu_0 > 0$ such that $|\hat{\mu}(a)| \leq \mu_0$ for all $a \geq 0$.

(4.125) f: $[0,\infty) \to R^n$, f is continuous.

(4.126) g: $[0,\infty) \times [0,\infty) \to R^n$, g is continuous, $g(\cdot,t) \in L^1$ for $t \geq 0$, and the mapping $t \to g(\cdot,t)$ is continuous from $[0,\infty)$ to $L^1$.

(4.127) $\phi$: $[0,\infty) \to R^n$, $\phi$ is continuous and $\phi \in L^1$.

LEMMA 4.1 Let (4.123), (4.124), (4.125), (4.126), (4.127) hold. There exists a unique continuous function $k$: $[0,\infty) \times [0,\infty) \to R^n$ such that

(4.128) $Dk(a,t) = -\hat{\mu}(a)k(a,t) + g(a,t)$, $a, t \geq 0$, $a \neq t$ [where D is defined as in (1.5)].

(4.129) $$k(0,t) = \int_0^\infty \hat{\beta}(a)k(a,t)\,da + f(t) \qquad t > 0$$

(4.130) $$k(a,0) = \phi(a) \qquad a \geq 0$$

*Proof.* We proceed as in Section 1.1. Suppose that $k$ is a solution of (4.128), (4.129), (4.130) and $c \in R$. Define the cohort function $k_c(t) \overset{\text{def}}{=} k(t + c, t)$, $t \geq t_c \overset{\text{def}}{=} \max\{0,-c\}$. Then

$$\frac{d}{dt} k_c(t) = Dk(t + c, t) = -\hat{\mu}(t + c)k_c(t) + g(t + c, t) \qquad t \geq t_c$$

which implies that

$$k_c(t) = \exp\left[-\int_{t_c}^t \hat{\mu}(\tau + c)\,d\tau\right]k_c(t_c) + \int_{t_c}^t \exp\left[-\int_s^t \hat{\mu}(\tau + c)\,d\tau\right]g(s + c, s)\,ds \qquad t \geq t_c$$

Set $c = a - t$, where $a \geq t$, to obtain

(4.131) $$k_c(t) = k(a,t) = \exp\left[-\int_0^t \hat{\mu}(\tau + c)\,d\tau\right]k_c(0) + \int_0^t \exp\left[-\int_s^t \hat{\mu}(\tau + c)\,d\tau\right]g(s + c, s)\,ds \qquad t \geq 0$$

Set $c = a - t$, where $a < t$, to obtain

$$(4.132) \quad k_c(t) = k(a,t) = \exp\left[-\int_{-c}^{t} \hat{\mu}(\tau + c)\, d\tau\right] k_c(-c) + \int_{-c}^{t} \exp\left[-\int_{s}^{t} \hat{\mu}(\tau + c)\, d\tau\right] \times g(s + c, s)\, ds \qquad t \geq -c$$

Define

$$(4.133) \quad \Pi(b,a) \overset{\text{def}}{=} \exp\left[-\int_{a}^{b} \hat{\mu}(s)\, ds\right] \qquad 0 \leq a \leq b$$

and combine (4.130), (4.131), (4.132) to obtain

$$(4.134) \quad k(a,t) = \begin{cases} \Pi(a,0)k(0, t - a) + \int_{t-a}^{t} \Pi(a, s + a - t) \times g(s + a - t, s)\, ds, & a < t \\ \Pi(a, a - t)\phi(a - t) + \int_{0}^{t} \Pi(a, s + a - t) \times g(s + a - t, s)\, ds, & a \geq t \end{cases}$$

Next, substitute the formula (4.134) into (4.129) to obtain

$$(4.135) \quad k(0,t) = \int_{0}^{t} \hat{\beta}(a)\Pi(a,0)k(0, t - a)\, da + h(t) \qquad t \geq 0$$

where

$$h(t) \overset{\text{def}}{=} \int_{0}^{t} \hat{\beta}(a)\left[\int_{t-a}^{t} \Pi(a, s + a - t)g(s + a - t, s)\, ds\right] da + \int_{t}^{\infty} \hat{\beta}(a)\left[\Pi(a, a - t)\phi(a - t) + \int_{0}^{t} \Pi(a, s + a - t) \times g(s + a - t, s)\, ds\right] da + f(t) \qquad t \geq 0$$

Conversely, the linear Volterra integral equation (4.135) has a unique continuous solution $B(t) \overset{\text{def}}{=} k(0,t)$, $t \geq 0$ (see [220], Theorem 1.1, p. 87, and Theorem 2.2, p. 95). The formula (4.134) then provides the unique continuous solution of (4.128), (4.129), (4.130). □

Let (4.123), (4.124) hold, let $\gamma \in R$, and define

(4.136) $S_\gamma(t)$, $t \geq 0$, is the strongly continuous semigroup of bounded linear operators in $L^1$ with infinitesimal generator $B_\gamma$, where $B_\gamma\phi \overset{\text{def}}{=} -\phi' - \hat{\mu}\phi - \gamma\phi$, $D(B_\gamma) = \{\phi \in L^1$: $\phi$ is absolutely continuous on $[0,\infty)$, $\phi' \in L^1$, and $\phi(0) = \int_0^\infty \hat{\beta}(a)\phi(a)\, da\}$.

(see Propositions 3.2 and 3.7). In addition, define

(4.137) $R_\gamma$: $[0,\infty) \to B(R^n,R^n)$ is the solution of the linear Volterra integral equation $R_\gamma(t) = \hat{\beta}_\gamma(t) + (\hat{\beta}_\gamma * R_\gamma)(t)$, $t \geq 0$, where $\hat{\beta}_\gamma(t) \overset{\text{def}}{=} \hat{\beta}(t)\Pi_\gamma(t,0)$, $t \geq 0$, $\Pi_\gamma(b,a) \overset{\text{def}}{=} \exp[-\gamma(b - a)]\Pi(b,a)$, $0 \leq a \leq b$, and $*$ denotes convolution, that is, $(\hat{\beta}_\gamma * R_\gamma)(t) \overset{\text{def}}{=} \int_0^t \hat{\beta}_\gamma(t - a)R_\gamma(a)\, da$, $t \geq 0$ (see [220], Chapter IV, p. 189).

Also, define

(4.138) $L_\gamma(t)$, $t \in R$, is the family of bounded linear operators given as follows: if $t \in R$ and $u \in C(-\infty,t;R^n)$ then $L_\gamma(t)u \in L^1$ is given by

$$(L_\gamma(t)u)(a) = \begin{cases} \Pi_\gamma(a,0)\,[u(t - a) + (R_\gamma * u)(t - a)] & \text{if } a < t \\ 0 & \text{if } a \geq t \end{cases}$$

REMARK 4.13 Since $\hat{\beta}$ is continuous, $R_\gamma$ is continuous. Further, for $t \geq 0$,

$$\begin{aligned} |R_\gamma(t)| &\leq |\hat{\beta}_\gamma(t)| + \int_0^t |\beta_\gamma(t - a)||R_\gamma(a)|\, da \\ &\leq \beta_0 \exp[(\mu_0 - \gamma)t] \\ &\quad + \beta_0 \int_0^t \exp[(\mu_0 - \gamma)(t - a)]|R_\gamma(a)|\, da \end{aligned}$$

which, by Gronwall's lemma, implies that

$$|R_\gamma(t)| \leq \beta_0 \exp[(\beta_0 + \mu_0 - \gamma)t] \qquad t \geq 0$$

Further, for $t \geq 0$, $u \in C(-\infty,t;R^n)$,

$$\|L_\gamma(t)u\|_{L^1} \leq \int_0^t e^{(\mu_0-\gamma)(t-a)} |u(a) + (R_\gamma * u)(a)| \, da$$

and since

$$\int_0^t e^{(\mu_0-\gamma)(t-a)} \left[ \int_0^a |R_\gamma(a-b)u(b)| \, db \right] da$$

$$\leq \beta_0 \int_0^t e^{(\mu_0-\gamma)(t-a)} \left[ \int_0^a e^{(\beta_0+\mu_0-\gamma)(a-b)} |u(b)| \, db \right] da$$

$$= \beta_0 \int_0^t e^{(\mu_0-\gamma)t} e^{-(\beta_0+\mu_0-\gamma)b} |u(b)| \left[ \int_b^t e^{\beta_0 a} da \right] db$$

$$\leq \int_0^t e^{(\mu_0-\gamma)t} e^{-(\beta_0+\mu_0-\gamma)b} e^{\beta_0 t} |u(b)| \, db$$

we obtain

$$\|L_\gamma(t)u\|_{L^1} \leq 2 \int_0^t e^{(\beta_0+\mu_0-\gamma)(t-a)} |u(a)| \, da \tag{4.139}$$

Also, $L_\gamma(t) = 0$ if $t \leq 0$, and for $t_0 \in R$, $u \in C((-\infty,t_0);R^n)$, the mapping $t \to L_\gamma(t)u$ is continuous from $(-\infty,t_0]$ to $L^1$.

LEMMA 4.2 Let (4.123), (4.124), (4.125), (4.126), (4.127), (4.136), (4.137), (4.138) hold. The solution k of (4.128), (4.129), (4.130) has the representation

$$k(\cdot,t) = S_\gamma(t)\phi + L_\gamma(t)f + \int_0^t S_\gamma(t-s)[g(\cdot,s) + \gamma k(\cdot,s)] \, ds \qquad t \geq 0, \ \gamma \in R \tag{4.140}$$

*Proof.* Recall from (2.62) and (2.66) that for $\psi \in L^1$, $t \geq 0$, $\gamma \in R$,

$$(S_\gamma(t)\psi)(0) = \int_0^\infty \hat{\beta}(a)(S_\gamma(t)\psi)(a) \, da$$

and for $\psi \in L^1$ such that $\psi$ is continuous, $a, t \geq 0$, $a \neq t$, $\gamma \in R$,

$$D(S_\gamma(t)\psi)(a) = -[\hat{\mu}(a) + \gamma](S(t)\psi)(a)$$

Define for $a, t \geq 0, \gamma \in R$,

$$j(a,t) \overset{\text{def}}{=} (S_\gamma(t)\phi)(a) + (L_\gamma(t)f)(a) + \int_0^t S_\gamma(t - s)[g(a,s) + \gamma k(a,s)]\, ds$$

Obviously, j satisfies (4.130).

Further, for $t > 0$,

$$\begin{aligned} j(0,t) &= \int_0^\infty \hat{\beta}(a)(S_\gamma(t)\phi)(a)\, da \\ &\quad + \int_0^\infty \hat{\beta}(a)\left[\int_0^t S_\gamma(t - s)[g(a,s) + \gamma k(a,s)]\, ds\right] da \\ &\quad + f(t) + (R_\gamma * f)(t) \\ &= \int_0^\infty \hat{\beta}(a)[j(a,t) - (L_\gamma(t)f)(a)]\, da + f(t) + (R_\gamma * f)(t) \end{aligned}$$

From (4.138) we have that

$$\begin{aligned} \int_0^\infty \hat{\beta}(a)(L_\gamma(t)f)(a)\, da &= (\hat{\beta}_\gamma * f)(t) + (\hat{\beta}_\gamma * (R_\gamma * f))(t) \\ &= (R_\gamma * f)(t) \qquad t \geq 0 \end{aligned}$$

where we have used the associativity of convolution (see [220], p. 20). Thus, j satisfies (4.129).

Also, for $a, t \geq 0$, $a \neq t$, $\gamma \in R$,

$$\begin{aligned} Dj(a,t) &= -[\hat{\mu}(a) + \gamma](S_\gamma(t)\phi)(a) \\ &\quad + \lim_{h\to 0}\left\{\int_t^{t+h} S_\gamma(t + h - s)[g(a + h, s) + \gamma k(a + h, s)]\, ds\right. \\ &\quad + \int_0^t (S_\gamma(t + h - s)[g(a + h, s) + \gamma k(a + h, s)] \\ &\quad - S_\gamma(t - s)[g(a,s) + k(a,s)])\, ds \\ &\quad \left. + [\Pi_\gamma(a + h, 0) - \Pi_\gamma(a,0)][f(t - a) + (R_\gamma * f)(t - a)]\right\} \end{aligned}$$

$$= -[\hat{\mu}(a) + \gamma](S_\gamma(t)\phi)(a) + g(a,t) + \gamma k(a,t)$$
$$- \int_0^t [\hat{\mu}(a) + \gamma]S_\gamma(t - s)[g(a,s) + \gamma k(a,s)]\, ds$$
$$- [\hat{\mu}(a) + \gamma](L_\gamma(t)f)(a)$$
$$= -\hat{\mu}(a)j(a,t) - \gamma[j(a,t) - k(a,t)] + g(a,t)$$

Thus,

$$D[j(a,t) - k(a,t)]$$
$$= -[\hat{\mu}(a) + \gamma][j(a,t) - k(a,t)] \qquad a,\ t \geq 0 \qquad a \neq t$$
$$j(0,t) - k(0,t) = 0 \qquad t > 0$$
$$j(a,0) - k(a,0) = 0 \qquad a \geq 0$$

By the uniqueness claim of Lemma 4.1, $j(a,t) - k(a,t) = 0$ for all a, $t \geq 0$. □

LEMMA 4.3 Let (4.123), (4.124), (4.125), (4.127), (4.136), (4.137), (4.138) hold, let $\gamma \in R$, let C be a bounded linear operator in $L^1$, let h: $[0,\infty) \to L^1$ such that h is continuous, and let w: $[0,\infty) \to L^1$ such that w is continuous and satisfies

$$w(t) = S_\gamma(t)\phi + L_\gamma(t)f + \int_0^t S_\gamma(t - s)[h(s) + (\gamma + C)w(s)]\, ds \qquad t \geq 0 \tag{4.141}$$

Then,

$$w(t) = \hat{S}(t)\phi + L_\gamma(t)f + \int_0^t \hat{S}(t - s)[h(s) + (\gamma + C)L_\gamma(s)f]\, ds \qquad t \geq 0 \tag{4.142}$$

where

(4.143) $\hat{S}(t)$, $t \geq 0$, is the strongly continuous semigroup of bounded linear operators in $L^1$ with infinitesimal generator $B_0 + C$ (see Proposition 4.14).

*Proof.* For each positive integer n let $\phi_n \in D(B_0)$, let $h_n$: $[0,\infty) \to L^1$ such that $h_n$ is continuously differentiable, let

$q_n$: $[0,\infty) \to L^1$ such that $q_n$ is continuously differentiable, let $\phi_n \to \phi$, and let $h_n \to h$, $q_n \to L_\gamma(t)f$ uniformly on bounded intervals of $[0,\infty)$. For each positive integer n, define

$$(4.144) \quad w_n(t) \stackrel{\text{def}}{=} \hat{S}(t)\phi_n + q_n(t) + \int_0^t \hat{S}(t - s)[h_n(s) + (\gamma + C)q_n(s)]\, ds \qquad t \geq 0$$

From (4.97) we have that for $n = 1, 2, \ldots, t \geq 0$,

$$\begin{aligned} \frac{d}{dt}[w_n(t) - q_n(t)] &= (B_0 + C)[w_n(t) - q_n(t)] + h_n(t) \\ &\quad + (\gamma + C)q_n(t) \\ &= (B_0 - \gamma)[w_n(t) - q_n(t)] + h_n(t) \\ &\quad + (\gamma + C)w_n(t) \end{aligned}$$

Again from (4.97) we have that for $n = 1, 2, \ldots, t \geq 0$,

$$w_n(t) - q_n(t) = S_\gamma(t)\phi_n + \int_0^t S_\gamma(t - s)[h_n(s) + (\gamma + C)w_n(s)]\, ds$$

Let N and $\sigma$ be constants such that $|S_\gamma(t)| \leq Ne^{\sigma t}$ for $t \geq 0$. Then, for $n = 1, 2, \ldots, t \geq 0$,

$$\begin{aligned} \|w(t) - w_n(t)\|_{L^1} &\leq Ne^{\sigma t}\|\phi - \phi_n\|_{L^1} + \|q_n(t) - L_\gamma(t)f\|_{L^1} \\ &\quad + \int_0^t Ne^{\sigma(t-s)}[\|h(s) - h_n(s)\|_{L^1} \\ &\quad + |\gamma + C|\,\|w(s) - w_n(s)\|_{L^1}]\, ds \end{aligned}$$

Now apply Gronwall's lemma, let $n \to \infty$, and use the continuity properties of $\hat{S}(t)$, $t \geq 0$, to argue that (4.142) holds. □

We require the following hypotheses on the birth function F and the aging function G:

(4.145) $F$: $L^1 \to R^n$, $F$ is continuously Fréchet differentiable, and for $i = 1, 2, \ldots, n$ there exists $\beta_i$: $[0,\infty) \times [0,\infty) \to [0,\infty)$ such that $F(\phi)_i = \int_0^\infty \beta_i(a, K_i\phi)\phi_i(a)\, da$, $\phi \in L^1$, where $K_i$: $L^1 \to [0,\infty)$, $K_i\phi = \int_0^\infty k_i(a)\phi(a)\, da$, $\phi \in L^1$, $k_i$: $[0,\infty) \to B(R^n, [0,\infty))$, $\beta_i$ is continuous and bounded, $\beta_i$ has a continuous and bounded partial derivative in its second place, denoted by $\beta_i'$, $k_i$ is continuous and bounded, and $F'(\hat{\phi})\phi$ satisfies (3.73) for $\phi$, $\hat{\phi} \in L^1$.

(4.146) $G$: $L^1 \to L^1$, $G$ is continuously Fréchet differentiable, and for $i = 1, 2, \ldots, n$ there exists $\mu_i$: $[0,\infty) \times [0,\infty) \to [0,\infty)$ such that $G(\phi)_i(a) = -\mu_i(a, J_i\phi)\phi_i(a)$, $\phi \in L^1$, almost all $a > 0$, where $J_i$: $L^1 \to [0,\infty)$, $J_i\phi = \int_0^\infty j_i(a)\phi(a)\, da$, $\phi \in L^1$, $j_i$: $[0,\infty) \to B(R^n, [0,\infty))$, $\mu_i$ is continuous and bounded, $\mu_i$ has a continuous and bounded partial derivative in its second place, denoted by $\mu_i'$, $j_i$ is continuous and bounded, and for all $\hat{\phi} \in L^1$, the mapping $G_{\hat{\phi}}$: $L^1 \to L^1$ defined by $G_{\hat{\phi}}(\phi)_i(a) \stackrel{\text{def}}{=} -\mu_i(a, J_i\hat{\phi})\phi_i(a)$, $\phi \in L^1$, almost all $a > 0$, $i = 1, 2, \ldots, n$, satisfies (2.49) and (3.67).

REMARK 4.14 Notice that if $F$ satisfies (4.145) then $F$ satisfies (2.1), (2.22), and if $G$ satisfies (4.146), then $G$ satisfies (2.2), (2.23).

THEOREM 4.13 Let (4.145), (4.146) hold, let $T_\phi = +\infty$ for all $\phi \in L^1_+$, and let $S(t)$, $t \geq 0$, be the strongly continuous nonlinear semigroup in $L^1_+$ associated with (ADP) (as in Theorem 3.1) with infinitesimal generator $B$ (as in Theorem 3.2). Let $\hat{\phi} \in L^1_+$ be an equilibrium solution of (ADP) and let $\hat{B}$: $L^1 \to L^1$ be defined by

$$\hat{B}\phi \stackrel{\text{def}}{=} -\phi' + G'(\hat{\phi})\phi,\quad D(\hat{B}) \stackrel{\text{def}}{=} \{\phi \in L^1:\ \phi \text{ is absolutely continuous on } [0,\infty),\ \phi' \in L^1, \text{ and } \phi(0) = F'(\hat{\phi})\phi\}.$$

The following hold:

(4.147) If $\sup_{\lambda \in \sigma(\hat{B}) - E\sigma(\hat{B})} \operatorname{Re} \lambda < 0$, then $\hat{\phi}$ is a locally exponentially asymptotically stable equilibrium of (ADP).

(4.148) If there exists $\lambda_1 \in \sigma(\hat{B})$ such that $\text{Re}\,\lambda_1 > 0$ and $\sup_{\lambda \in \sigma(\hat{B}) - E\sigma(\hat{B}), \lambda \neq \lambda_1} \text{Re}\,\lambda < \text{Re}\,\lambda_1$, then $\hat{\phi}$ is an unstable equilibrium of (ADP).

*Proof.* Let $F(\phi) = F(\hat{\phi}) + F'(\hat{\phi})(\phi - \hat{\phi}) + o_F(\phi - \hat{\phi})$ for all $\phi \in L^1$, where $o_F\colon\ L^1 \to R^n$, $|o_F(\phi)| \leq b_F(r)\|\phi\|_{L^1}$ for $\|\phi\|_{L^1} \leq r$, and $b_F\colon\ [0,\infty) \to [0,\infty)$ is an increasing continuous function such that $b_F(0) = 0$, and let $G(\phi) = G(\hat{\phi}) + G'(\hat{\phi})(\phi - \hat{\phi}) + o_G(\phi - \hat{\phi})$ for all $\phi \in L^1$, where $o_G\colon\ L^1 \to L^1$, $\|o_G(\phi)\|_{L^1} \leq b_G(r)\|\phi\|_{L^1}$, for $\|\phi\|_{L^1} \leq r$, and $b_G\colon\ [0,\infty) \to [0,\infty)$ is an increasing continuous function such that $b_G(0) = 0$ (see Definition 2.4).

Notice that for $\phi \in L^1$, $i = 1, 2, \ldots, n$,

$$(F'(\hat{\phi})\phi)_i = \int_0^\infty [\beta_i'(a,K_i\hat{\phi})K_i\phi\hat{\phi}_i(a) + \beta_i(a,K_i\hat{\phi})\phi_i(a)]\,da$$

$$= \left(\int_0^\infty \beta_i'(b,K_i\hat{\phi})\hat{\phi}_i(b)\,db\right)\int_0^\infty k_i(a)\phi(a)\,da$$

$$+ \int_0^\infty \beta_i(a,K_i\hat{\phi})\phi_i(a)\,da$$

so that there exists $\hat{\beta}\colon\ [0,\infty) \to B(R^n,R^n)$ such that $\hat{\beta}$ satisfies (4.123) and

$$F'(\hat{\phi})\phi = \int_0^\infty \hat{\beta}(a)\phi(a)\,da \qquad \phi \in L^1$$

Notice also that for $\phi \in L^1$, almost all $a > 0$, $i = 1, 2, \ldots, n$

$$(G'(\hat{\phi})\phi)_i(a) = -\mu_i'(a,J_i\hat{\phi})J_i\phi\hat{\phi}_i(a) - \mu_i(a,J_i\hat{\phi})\phi_i(a)$$

so that if we define $\hat{\mu}\colon\ [0,\infty) \to B(R^n,R^n)$ by

$$\hat{\mu}_{ij}(a) = \begin{cases} \mu_i(a,J_i\hat{\phi}) & a \geq 0,\ i = j \\ 0 & a \geq 0,\ i \neq j \end{cases}$$

and $C\colon\ L^1 \to L^1$ by

$$(C\phi)_i(a) \overset{\text{def}}{=} -\mu_i'(a,J\hat{\phi})J\phi\hat{\phi}_i(a), \ \phi \in L^1, \text{ almost all } a > 0,$$
$$i = 1, 2, \ldots, n$$

then $\hat{\mu}$ satisfies (4.124) and

$$G'(\hat{\phi})\phi = -\hat{\mu}\phi + C\phi \qquad \phi \in L^1$$

Let $\phi \in D(B)$, let $\phi_0 \overset{\text{def}}{=} \phi - \hat{\phi}$, let $k(a,t) \overset{\text{def}}{=} (S(t)\phi)(a) - \hat{\phi}(a)$, $a, t \geq 0$, let $f(t) \overset{\text{def}}{=} o_F(k(\cdot,t))$, $t \geq 0$, $h(t) \overset{\text{def}}{=} o_G(k(\cdot,t))$, $t \geq 0$, and $g(\cdot,t) \overset{\text{def}}{=} Ck(\cdot,t) + h(t)$, $t \geq 0$. Observe that f satisfies (4.125), g satisfies (4.126), and $\hat{B}$ has the form $\hat{B} = B_0 + C$, where $B_0$ is defined as in (4.136).

From (2.69) we obtain

$$\begin{aligned} Dk(a,t) &= G(S(t)\phi)(a) - G(\hat{\phi})(a) \\ &= G'(\hat{\phi})k(\cdot,t)(a) + o_G(k(\cdot,t))(a) \\ &= -\hat{\mu}(a)k(a,t) + g(a,t) \qquad a, t \geq 0 \end{aligned}$$

$$\begin{aligned} k(0,t) &= F(S(t)\phi) - F(\hat{\phi}) \\ &= F'(\hat{\phi})k(\cdot,t) + o_F(k(\cdot,t)) \\ &= \int_0^\infty \hat{\beta}(a)k(a,t)\, da + f(t) \qquad t \geq 0 \end{aligned}$$

$$k(a,0) = \phi_0(a) \qquad a \geq 0$$

Let $\gamma > \beta_0 + \mu_0$, where $\beta_0$, $\mu_0$ are as in (4.123), (4.124), respectively, let $\alpha \overset{\text{def}}{=} \beta_0 + \mu_0 - \gamma$, let $S_\gamma(t)$, $t \geq 0$, be as in (4.136), let $L_\gamma(t)$ be as in (4.138), and let $\hat{S}(t)$, $t \geq 0$, be as in (4.143). From Lemma 4.2 we obtain

$$\begin{aligned} k(\cdot,t) &= S_\gamma(t)\phi_0 + L_\gamma(t)f \\ &\quad + \int_0^t S_\gamma(t-s)[g(\cdot,s) + \gamma k(\cdot,s)]\, ds \\ &= S_\gamma(t)\phi_0 + L_\gamma(t)f \\ &\quad + \int_0^t S_\gamma(t-s)[h(s) + (\gamma + C)k(\cdot,s)]\, ds \qquad t \geq 0 \end{aligned}$$

From Lemma 4.3 we obtain

$$(4.149)\quad k(\cdot,t) = S(t)\phi - \hat{\phi} = L_\gamma(t)o_F(S(\cdot)\phi - \hat{\phi}) + \hat{S}(t)(\phi - \hat{\phi}) + \int_0^t \hat{S}(t-s)[o_G(S(s)\phi - \hat{\phi}) + (\gamma + C)L_\gamma(s)o_F(S(\cdot)\phi - \hat{\phi})]\,ds \qquad t \geq 0$$

From Definition 2.4 it is seen that there exist increasing continuous functions $\hat{b}_F$, $\hat{b}_G$: $[0,\infty) \to [0,\infty)$ such that $\hat{b}_F(0) = \hat{b}_G(0) = 0$ and $|o_F(\phi_1) - o_F(\phi_2)| \leq \hat{b}_F(r)\|\phi_1 - \phi_2\|_{L^1}$, $\|o_G(\phi_1) - o_G(\phi_2)\|_{L^1} \leq \hat{b}_G(r)\|\phi_1 - \phi_2\|_{L^1}$ for $\|\phi_1\|_{L^1}$, $\|\phi_2\|_{L^1} \leq r$. From the deneseness of $D(B)$ in $L^1_+$ (Theorem 3.4) and the continuity properties of $S(t)$, $t \geq 0$ (Theorem 3.1), $\hat{S}(t)$, $t \geq 0$, $L_\gamma(t)$, $t \in R$, $o_F$, $o_G$, we then see that (4.149) holds for all $\phi \in L^1_+$.

Furthermore, the solution of (4.149) is unique. That is, suppose that for $i = 1, 2$, $w_i$: $[0,\infty) \to L^1$ is a continuous function satisfying

$$w_i(t) = L_\gamma(t)o_F(w_i(\cdot)) + \hat{S}(t)(\phi - \hat{\phi}) + \int_0^t \hat{S}(t-s)[o_G(w_i(s)) + (\gamma + C)L_\gamma(s)o_F(w_i(\cdot))]\,ds \qquad t \geq 0$$

Let $r > 0$ such that $\|w_i(t)\|_{L^1} \leq r$ for $0 \leq t \leq t_0$, $i = 1, 2$, and let $M$, $\omega$ be constants such that $|\hat{S}(t)| \leq Me^{\omega t}$, $t \geq 0$. Then from (4.139) we have for $0 \leq t \leq t_0$,

$$\|w_1(t) - w_2(t)\|_{L^1} \leq 2\int_0^t e^{\alpha(t-s)}\hat{b}_F(r)\|w_1(s) - w_2(s)\|_{L^1}\,ds + \int_0^t Me^{\omega(t-s)}\Big[\hat{b}_G(r)\|w_1(s) - w_2(s)\|_{L^1} + 2|\gamma + C|\hat{b}_F(r)\int_0^s e^{\alpha(s-\tau)} \times \|w_1(\tau) - w_2(\tau)\|_{L^1}\,d\tau\Big]\,ds$$

which implies

$$e^{-\omega t}\|w_1(t) - w_2(t)\|_{L^1}$$
$$\leq [2\hat{b}_F(r) + M\hat{b}_G(r)] \int_0^t e^{-\omega s}\|w_1(s) - w_2(s)\|_{L^1} \, ds$$
$$+ 2M|\gamma + C|\hat{b}_F(r) \int_0^t e^{-\omega\tau}\|w_1(\tau) - w_2(\tau)\|_{L^1} \, d\tau$$

The claimed uniqueness now follows from Gronwall's lemma.

Since C is a compact linear operator in $L^1$, $\omega_1(\hat{B}) = \omega_1(B_0 + C) = \omega_1(B_0)$ by Proposition 4.14. Since (4.146) holds, there exists a constant $\underline{\mu} > 0$ as in (3.67) for $G_{\hat{\phi}}$. By (4.72) of Theorem 4.6 $\omega_1(B_0) \leq -\underline{\mu}$ and so

$$(4.150) \qquad \omega_1(\hat{B}) \leq -\underline{\mu}$$

Suppose that (4.147) holds. By (4.57) of Proposition 4.13 $\omega_0(\hat{B}) < 0$. By (4.47) of Proposition 4.12 there exist constants $M \geq 1$ and $\omega \in (\alpha,0)$ such that $|\hat{S}(t)| \leq Me^{\omega t}$, $t \geq 0$. Let $r > 0$ such that

$$2b_F(r) + M\left[1 + 2\,\frac{\gamma + |C|}{\omega - \alpha}\right] b_G(r) \stackrel{\text{def}}{=} \sigma < -\frac{\omega}{2}$$

Let $\varepsilon = r/M$, let $\phi \in L^1_+$ such that $\|\phi - \hat{\phi}\|_{L^1} < \varepsilon$, and let $t_1 \leq \infty$ be the largest extended real number such that $\|S(t)\phi - \hat{\phi}\|_{L^1} \leq r$, $0 \leq t < t_1$. Set $w(t) \stackrel{\text{def}}{=} \|S(t)\phi - \hat{\phi}\|_{L^1}$, $t \geq 0$.

From (4.149) and (4.139) we obtain for $0 \leq t < t_1$,

$$(4.151) \qquad e^{-\omega t}w(t) \leq M\|\phi - \hat{\phi}\|_{L^1}$$
$$+ 2e^{(\alpha-\omega)t}b_F(r) \int_0^t e^{-\alpha s}w(s) \, ds$$
$$+ Mb_G(r) \int_0^t e^{-\omega s}w(s) \, ds$$

$$+ \ 2Mb_G(r)(\gamma + |C|) \int_0^t e^{-\omega s} \left[ \int_0^s e^{\alpha(s-\tau)} w(\tau) \, d\tau \right] ds$$

$$\leq M\|\phi - \hat{\phi}\|_{L^1} + 2b_F(r) \int_0^t e^{(\alpha-\omega)(t-s)} e^{-\omega s} w(s) \, ds$$

$$+ \ Mb_G(r) \left[ 1 + 2 \frac{\gamma + |C|}{\omega - \alpha} \right] \int_0^t e^{-\omega\tau} w(\tau) \, d\tau$$

$$< M\|\phi - \hat{\phi}\|_{L^1} + \sigma \int_0^t e^{-\omega s} w(s) \, ds$$

By Gronwall's lemma we have that

$$e^{-\omega t} w(t) \leq M\|\phi - \hat{\phi}\| e^{\sigma t} \qquad 0 \leq t < t_1$$

which implies that

$$\|S(t)\phi - \hat{\phi}\|_{L^1} \leq M\|\phi - \hat{\phi}\| e^{\omega t/2} \qquad 0 \leq t < t_1$$

Consequently, $t_1 = +\infty$, and $\hat{\phi}$ is locally exponentially asymptotically stable.

Suppose that (4.148) holds. Let $\Lambda \overset{\text{def}}{=} \{\lambda_1\}$ in Proposition 4.15. By (4.148) and (4.150) we may choose $\omega > 0$ such that

$$\max\{\omega_1(\hat{B}), \sup_{\lambda \in \sigma(\hat{B}) - E\sigma(\hat{B}) - \Lambda} \operatorname{Re} \lambda\} < \omega < \operatorname{Re} \lambda_1$$

By Proposition 4.15 there exists a decomposition $X = X_1 \oplus X_2$, $X_2 \neq \{0\}$, associated projection operators $P_1$, $P_2$ such that $P_1 X = X_1$, $P_2 X = X_2$, and $\delta > 0$ such that

(4.152) $\hat{S}(t)(X_i) \subset X_i$ for $t \geq 0$, $i = 1, 2$, $\hat{B}$ is bounded on $X_2$, $\|\hat{S}(t)P_1\phi\|_{L^1} \leq Me^{(\omega-\delta)t}\|P_1\phi\|_{L^1}$ for $\phi \in L^1$, $t \geq 0$, and $\|\hat{S}(t)P_2\phi\|_{L^1} \leq Me^{(\omega+\delta)t}\|P_2\phi\|_{L^1}$ for $\phi \in L^1$, $t \leq 0$.

Define

$$\sigma \overset{\text{def}}{=} \frac{2}{\omega - \alpha} + \left[ \frac{M(|P_1| + |P_2|)}{\delta} \right] \left[ 1 + 2 \left( \frac{\gamma + |C|}{\omega - \alpha} \right) \right]$$

and choose $r > 0$ such that

$$[b_F(r) + b_G(r)]\sigma < \frac{1}{2M} \tag{4.153}$$

$$[\hat{b}_F(r) + \hat{b}_G(r)]\sigma < 1 \tag{4.154}$$

Using (4.153) we may choose $\phi_0 \in X_2$ such that

$$\frac{2b_F(r)r}{\omega - \alpha} + \frac{M|P_1|b_G(r)r}{\delta} + 2M|P_1|(\gamma + |C|)\frac{b_F(r)r}{(\omega - \alpha)\delta} < \|\phi_0\|_{L^1} < \frac{r}{2M} \tag{4.155}$$

We claim that for each positive integer $m = 1, 2, \ldots,$ there exists a unique function $v_m\colon\ (-\infty,m] \to L^1$ such that

$$\begin{aligned} v_m(t) &= \hat{S}(t - m)\phi_0 + L_\gamma(t)o_F(v_m(\cdot)) \\ &\quad - \int_t^m \hat{S}(t - s)P_2[o_G(v_m(s)) \\ &\quad + (\gamma + C)L_\gamma(s)o_F(v_m(\cdot))]\, ds \\ &\quad + \int_{-\infty}^t \hat{S}(t - s)P_1[o_G(v_m(s)) \\ &\quad + (\gamma + C)L_\gamma(s)o_F(v_m(\cdot))]\, ds \end{aligned} \tag{4.156}$$

satisfying

$$\|v_m(t)\|_{L^1} \le re^{\omega(t-m)} \qquad t \le m \tag{4.157}$$

To establish this claim fix m and define the complete metric space Z, with metric $\rho$, by

$$Z \overset{\text{def}}{=} \{v\colon\ v \text{ is a continuous function from } (-\infty,m] \text{ to } L^1 \text{ such that } \sup_{-\infty<t\le m}\|v(t)\|_{L^1} \le re^{\omega(t-m)}\}$$

$$\rho(u,v) \overset{\text{def}}{=} \sup_{t\le m} e^{-\omega(t-m)}\|u(t) - v(t)\|_{L^1} \qquad u, v \in Z$$

Define a mapping H on Z by

$$(Hv)(t) \overset{\text{def}}{=} \hat{S}(t - m)\phi_0 + L_\gamma(t)o_F(v(\cdot)) - \int_t^m \hat{S}(t - s)P_2[o_G(v(s)) + (\gamma + C)L_\gamma(s)o_F(v(\cdot))]\, ds + \int_{-\infty}^t \hat{S}(t - s)P_1[o_G(v(s)) + (\gamma + C)L_\gamma(s)o_F(v(\cdot))]\, ds \qquad t \le m$$

Observe that H maps Z into Z, since for $v \in Z$, $t \le m$, $(Hv)(t)$ is continuous in t, and by (4.139), (4.153), and (4.155)

$$\begin{aligned}
\|(Hv)(t)\|_{L^1} &\le Me^{(\omega+\delta)(t-m)}\|\phi_0\|_{L^1} \\
&+ \int_0^t 2e^{\alpha(t-s)}b_F(r)re^{\omega(s-m)}\, ds \\
&+ \int_t^m Me^{(\omega+\delta)(t-s)}|P_2|\Big[b_G(r)re^{\omega(s-m)} \\
&+ (\gamma + |C|)2\int_0^s e^{\alpha(s-\tau)}b_F(r)re^{\omega(\tau-m)}\, d\tau\Big]\, ds \\
&+ \int_{-\infty}^t Me^{(\omega-\delta)(t-s)}|P_1|\Big[b_G(r)re^{\omega(s-m)} \\
&+ (\gamma + |C|)2\int_0^s e^{\alpha(s-\tau)}b_F(r)re^{\omega(\tau-m)}\, d\tau\Big]\, ds \\
&\le \Bigg[M\|\phi_0\|_{L^1} + \frac{2b_F(r)r}{\omega - \alpha} + M|P_2|\left(\frac{b_G(r)r}{\delta} + 2(\gamma + |C|)\frac{b_F(r)r}{(\omega - \alpha)\delta}\right) \\
&+ M|P_1|\left(\frac{b_G(r)r}{\delta} + 2(\gamma + |C|)\frac{b_F(r)r}{(\omega - \alpha)\delta}\right)\Bigg]e^{\omega(t-m)} \\
&\le [M\|\phi_0\| + (b_F(r) + b_G(r))r\sigma]e^{\omega(t-m)} \\
&\le re^{\omega(t-m)} \qquad t \le m
\end{aligned}$$

A similar estimate uses (4.154) to show that for $v, \hat{v} \in Z$, $t \le m$,

$$\|(Hv)(t) - (H\hat{v})(t)\|_{L^1} \le c\rho(v,\hat{v})$$

where c is a constant < 1. By the contraction mapping theorem there exists a unique fixed point $v_m \in Z$ of H, which is the unique solution of (4.156) satisfying (4.157).

We next claim that for m = 1, 2, ..., $0 \le t \le m$,

$$(4.158) \quad v_m(t) = \hat{S}(t)v_m(0) + L_\gamma(t)o_F(v_m(\cdot)) + \int_0^t \hat{S}(t-s)[o_G(v_m(s)) + (\gamma + C)L_\gamma(s)o_F(v_m)(\cdot))]\, ds$$

To establish this claim let m be fixed, for each positive integer k let $\phi_k \in D(\hat{B})$, let $p_k, q_k$: $(-\infty,m] \to L^1$ such that $p_k, q_k$ are continuously differentiable and have compact support, and let $\phi_k \to \phi_0$ and let $p_k, q_k \to L_\gamma(t)o_F(v_m(\cdot)), o_G(v_m(t)) + (\gamma + C)L_\gamma(t)o_F(v_m(\cdot))$, respectively, uniformly on bounded intervals of $(-\infty,m]$. Define for $k = 1, 2, \ldots, t \le m$,

$$v_{m,k}(t) \overset{\text{def}}{=} \hat{S}(t-m)\phi_k + p_k(t) - \int_t^m \hat{S}(t-s)P_2 q_k(s)\, ds + \int_{-\infty}^t \hat{S}(t-s)P_1 q_k(s)\, ds$$

Observe that $v_{m,k} \to v_m$ uniformly on bounded intervals of $(-\infty,m]$ as $k \to \infty$. Arguing as in the proof of (4.104) we see that

$$\frac{d}{dt}[v_{m,k}(t) - p_k(t)] = \hat{B}[v_{m,k}(t) - p_k(t)] + q_k(t) \qquad t \le m,\ k = 1, 2, \ldots$$

which by (4.97) implies that

$$v_{m,k}(t) = \hat{S}(t)[v_{m,k}(0) + p_k(0)] + p_k(t) + \int_0^t \hat{S}(t-s)q_k(s)\, ds \qquad 0 \le t \le m,\ k = 1, 2, \ldots$$

Now let $k \to \infty$ to obtain (4.158).

Define $\phi_m \overset{\text{def}}{=} v_m(0) + \hat{\phi}$, m = 1, 2, ... . From (4.158) and the uniqueness of solutions to (4.149) we obtain $v_m(t) = S(t)\phi_m - \hat{\phi}$ for $0 \le t \le m$, m = 1, 2, ... . From (4.157) we obtain $\|\phi_m - \hat{\phi}\|_{L^1} \le e^{-\omega m}$

for m = 1, 2, ... . From (4.156) we obtain

$$\|S(m)\phi_m - \hat{\phi}\|_{L^1} = \|v_m(m)\|_{L^1} \geq \|\phi_0\|_{L^1} - \|L_\gamma(m)o_F(v_m(\cdot))\|_{L^1}$$

$$- \int_{-\infty}^{m} \|\hat{S}(m - s)P_1[o_G(v_m(s))$$

$$+ (\gamma + C)L_\gamma(s)o_F(v_m(\cdot))]\|_{L^1} \, ds$$

$$\geq \|\phi_0\|_{L^1} - 2\int_0^m e^{\alpha(m-s)} b_F(r) r e^{\omega(s-m)} \, ds$$

$$- \int_{-\infty}^{m} Me^{(\omega-\delta)(m-s)} |P_1| [b_G(r) r e^{\omega(s-m)}$$

$$+ (\gamma + |C|) 2\int_0^s e^{\alpha(s-\tau)} b_F(r) r e^{\omega(\tau-m)} \, d\tau] \, ds$$

$$\geq \|\phi_0\|_{L^1} - 2b_F(r) r/(\omega - \alpha)$$

$$- M|P_1| b_G(r) r/\delta - 2M|P_1|(\gamma + |C|) b_F(r) r/(\omega - \alpha)\delta$$

$$m = 1, 2, \ldots$$

From (4.155) we then obtain that $\hat{\phi}$ is unstable. □

The next theorem provides sufficient conditions for the extinction of the ith species of a multispecies population. As with Theorem 4.2 and Theorem 4.5 the net reproduction rate of the ith species is required to be less than 1.

THEOREM 4.14 (J. Prüss [242]) Let (4.145), (4.146) hold, Let $T_\phi = +\infty$ for all $\phi \in L^1_+$, and let $S(t)$, $t \geq 0$, be the strongly continuous nonlinear semigroup in $L^1_+$ associated with (ADP) (as in Theorem 3.1) with infinitesimal generator B (as in Theorem 3.2). Let i be a fixed integer in [1,n] and let there exist continuous and bounded functions $\beta_0$: $[0,\infty) \to [0,\infty)$ and $\mu_0$: $[0,\infty) \to [0,\infty)$ such that

$$(4.159) \qquad \int_0^\infty \beta_0(a) \exp\left[-\int_0^a \mu_0(b) \, db\right] da < 1$$

(4.160) $\beta_i(a,K\phi) \le \beta_0(a)$ and $\mu_i(a,J\phi) \ge \mu_0(a)$ for all $\phi \in L^1_+$, $a \ge 0$

Then, there exist constants $M \ge 1$ and $\omega > 0$ such that

(4.161) $$\|(S(t)\phi)_i\|_{L^1(0,\infty;R)} \le Me^{-\omega t}\|\phi_i\|_{L^1(0,\infty;R)} \quad \text{for } \phi \in L^1_+,\ t \ge 0$$

*Proof.* Let $\phi \in D(B)$ and define $k(a,t) \stackrel{\text{def}}{=} (S_i(t)\phi)(a)$, a, $t \ge 0$. Notice that $k(a,t) \ge 0$ for a, $t \ge 0$, since $S(t)\phi \in L^1_+$. From (2.69) we have that

$$Dk(a,t) = -\mu_0(a)k(a,t) + g(a,t) \qquad a,\ t \ge 0$$

$$k(0,t) = \int_0^\infty \beta_0(a)k(a,t)\,da + f(t) \qquad t \ge 0$$

$$k(a,0) = \phi_i(a) \qquad a \ge 0$$

where

$$g(a,t) \stackrel{\text{def}}{=} [\mu_i(a,J_iS(t)\phi) - \mu_0(a)]k(a,t) \qquad a,\ t \ge 0$$

$$f(t) \stackrel{\text{def}}{=} \int_0^\infty [\beta_i(a,K_iS(t)\phi) - \beta_0(a)]k(a,t)\,da \qquad t \ge 0$$

Let $\Pi_0(b,a) \stackrel{\text{def}}{=} \exp[-\int_a^b \mu_0(s)\,ds]$, $0 \le a \le b$, let $R_0\colon [0,\infty) \to [0,\infty)$ satisfy $R_0 = \beta_0 + \beta_0 * R_0$, and let $L_0(t)$, $t \ge 0$, be defined by (4.138). By Lemma 4.2 we have that

(4.162) $$k(\cdot,t) = S_0(t)\phi_i + L_0(t)f + \int_0^t S_0(t-s)g(\cdot,s)\,ds \qquad t \ge 0$$

where $S_0(t)$, $t \ge 0$, is the strongly continuous semigroup of bounded linear operators in $L^1(0,\infty;R)$ with infinitesimal generator $B_0\colon L^1(0,\infty;R) \to L^1(0,\infty;R)$ defined by

$B_0\psi \stackrel{\text{def}}{=} -\psi' - \mu_0\psi$, $\psi \in D(B_0)$, $D(B_0) \stackrel{\text{def}}{=} \{\psi \in L^1(0,\infty;R)\colon\ \psi$ is absolutely continuous on $[0,\infty)$, $\psi' \in L^1(0,\infty;R)$ and $\psi(0) = \int_0^\infty \beta_0(a)\psi(a)\,da\}$

By (4.160) $g(a,t) \le 0$ for $a$, $t \ge 0$ and $f(t) \le 0$ for $t \ge 0$. From Theorem 2.4 we see that $S_0(t)(L^1_+(0,\infty;R)) \subset L^1_+(0,\infty;R)$ for $t \ge 0$. From (4.138) we see that $(L_0(t)f)(a) \le 0$ for $a$, $t \ge 0$. By (4.162) we then see that $k(a,t) \le (S_0(t)\phi_i)(a)$ for $t \ge 0$, almost all $a \ge 0$, which implies that

$$\|(S(t)\phi)_i\|_{L^1(0,\infty;R)} \le \|S_0(t)\phi_i\|_{L^1(0,\infty;R)} \qquad t \ge 0 \tag{4.163}$$

By Theorem 3.1 and Theorem 3.4, (4.163) holds for all $\phi \in L^1_+$. By (4.159) and Theorem 4.10, 0 is a globally exponentially asymptotically stable equilibrium of $S_0(t)$, $t \ge 0$. Hence, (4.162) holds. □

Applications of Theorem 4.13 are given for the following single species models with nonlinear birth and mortality functions:

EXAMPLE 4.14 Let $R^n = R$. Let $F$ have the form (2.17) and let $G$ have the form (2.18) as in Examples 2.1, 2.4, 3.1, and 4.1, where $\beta$ and $\mu$ are twice continuously differentiable, $\beta$, $\beta'$, $\mu$, and $\mu'$ are all bounded, and $\mu$ is bounded below by a positive constant. Recall from Remark 2.4 that $F$ and $G$ are continuously Fréchet differentiable (as in Definition 2.4) with

$$F'(\hat{\phi})\phi = \beta'(P\hat{\phi})P\phi \int_0^\infty e^{-\alpha a}\hat{\phi}(a)\, da + \beta(P\hat{\phi}) \int_0^\infty e^{-\alpha a}(a)\, da \qquad \phi,\ \hat{\phi} \in L^1 \tag{4.164}$$

$$(G'(\hat{\phi})\phi)(a) = -\mu'(P\hat{\phi})P\phi\hat{\phi}(a) - \mu(P\hat{\phi})\phi(a) \qquad \phi,\ \hat{\phi} \in L^1,\ \text{a.e. } a > 0 \tag{4.165}$$

Further, observe that $F$ satisfies (4.145) and $G$ satisfies (4.146).

Let $\hat{\phi} \in L^1_+$ such that $\hat{\phi}$ is an equilibrium solution of (ADP). From Example 4.1, $\hat{\phi}(a) = \mu(\hat{P})\hat{P}\exp[-\mu(\hat{P})a]$, $a \ge 0$, where $P\hat{\phi} = \hat{P}$, and either $\hat{P} = 0$ or $\hat{P} > 0$ and satisfies $\alpha + \mu(\hat{P}) = \beta(\hat{P})$. Define the linear operator $\hat{B}$ in $L^1$ by

(4.166) $\quad \hat{B}\phi \overset{\text{def}}{=} -\phi' + G'(\hat{\phi})\phi, \ D(\hat{B}) \overset{\text{def}}{=} \{\phi \in L^1\colon\ \phi$ is absolutely continuous on $[0,\infty)$, $\phi' \in L^1$, and $\phi(0) = F'(\hat{\phi})\phi\}$

The stability of $\hat{\phi}$ will be determined by $\sigma(\hat{B}) - E\sigma(\hat{B})$ according to Theorem 4.13. Let $\lambda \in C$ and let $\hat{B}\phi = \lambda\phi$ for some $\phi \in L^1$, $\phi \neq 0$. As in Example 4.11 the definition of $\hat{B}$ yields that $\phi$ satisfies (4.114) and

$$\phi(0) = \beta'(\hat{P})P\phi\mu(\hat{P})\hat{P}\beta(\hat{P})^{-1} + \beta(\hat{P}) \int_0^\infty e^{-\alpha a}\phi(a)\, da \tag{4.167}$$

If $\hat{\phi} = 0$, then (4.114) and (4.167) yield that $\phi(0) = \beta(0)(\alpha + \lambda + \mu(0))^{-1}\phi(0)$, so that $\lambda = \beta(0) - \mu(0) - \alpha$. Consequently, (4.147) implies that 0 is locally exponentially asymptotically stable if $\beta(0) < \mu(0) + \alpha$, and (4.148) implies that 0 is unstable if $\beta(0) > \mu(0) + \alpha$.

If $\hat{\phi} \neq 0$ and $\lambda = 0$, then (4.114) and (4.167) yield (4.115) and

$$\begin{aligned}\phi(0) = {}& \beta'(\hat{P})P\phi\mu(\hat{P})\hat{P}\beta(\hat{P})^{-1} + \beta(\hat{P})[\phi(0)(\alpha + \mu(\hat{P}))^{-1} \\ & - \mu'(\hat{P})P\phi\mu(\hat{P})\hat{P}(\alpha + \mu(\hat{P}))^{-2}]\end{aligned}$$

Since $\beta(\hat{P}) = \alpha + \mu(\hat{P})$, this last equation implies that $\beta'(\hat{P}) = \mu'(\hat{P})$.

If $\hat{\phi} \neq 0$, $\lambda \neq 0$, and Re $\lambda > -\mu(\hat{P})$, then (4.117) and (4.118) hold. Then, (4.117), (4.118), (4.167), and the fact that $\beta(\hat{P}) = \alpha + \mu(\hat{P})$ imply that

$$\begin{aligned}(\lambda + \mu(\hat{P}) + \mu'(\hat{P})\hat{P})P\phi = {}& \beta'(\hat{P})P\phi\mu(\hat{P})\hat{P}\beta(\hat{P})^{-1} \\ & + \beta(\hat{P})[(\alpha + \mu(\hat{P}))\phi(0) - \mu'(\hat{P})P\phi\mu(\hat{P})\hat{P}] \\ & \times (\alpha + \lambda + \mu(\hat{P}))^{-1}(\alpha + \mu(\hat{P}))^{-1} \\ = {}& \beta'(\hat{P})P\phi\mu(\hat{P})\hat{P}\beta(\hat{P})^{-1} + [\beta(\hat{P})(\lambda + \mu(\hat{P}) \\ & + \mu'(\hat{P})\hat{P})P\phi - \mu'(\hat{P})P\phi\mu(\hat{P})\hat{P}](\beta(\hat{P}) + \lambda)^{-1}\end{aligned}$$

which implies that

$$\begin{aligned}& \lambda^2 + \lambda[\mu(\hat{P}) + \mu'(\hat{P})\hat{P} - \beta'(\hat{P})\mu(\hat{P})\hat{P}\beta(\hat{P})^{-1} \\ & + \mu(\hat{P})\hat{P}(\mu'(\hat{P}) - \beta'(\hat{P})) = 0\end{aligned}$$

If $\mu'(\hat{P}) > \beta'(\hat{P})$, then the coefficients of this quadratic equation are all positive (since $0 < \mu(\hat{P}) \leq \mu(\hat{P}) + \alpha = \beta(\hat{P})$), and hence $\text{Re}\,\lambda < 0$. If $\mu'(\hat{P}) < \beta'(\hat{P})$, then this quadratic equation has a real positive root.

We conclude the following for the case that $\hat{\phi} \neq 0$: if $\mu'(\hat{P}) > \beta'(\hat{P})$, then (4.147) implies that $\hat{\phi}$ is locally exponentially asymptotically stable, and if $\mu'(\hat{P}) < \beta'(\hat{P})$, then (4.148) implies that $\hat{\phi}$ is unstable.

EXAMPLE 4.15 Let $R^n = R$. Let $F$ have the form (2.19) and let $G$ have the form (2.18) as in Examples 2.2, 2.5, 3.2, and 4.2, where $\beta$ and $\mu$ are twice continuously differentiable, $\beta$, $\beta'$, $\mu$, and $\mu'$ are all bounded, and $\mu$ is bounded below by a positive constant. Observe that $F$ satisfies (4.145), $G$ satisfies (4.146) and $F$ and $G$ are continuously Fréchet differentiable with $G'$ given by (4.166) and $F'$ given by

$$(4.168) \qquad F'(\hat{\phi})\phi = \beta'(P\hat{\phi})P\phi \int_0^\infty (1 - e^{-\alpha a})\hat{\phi}(a)\, da + \beta(P\hat{\phi}) \int_0^\infty (1 - e^{-\alpha a})\phi(a)\, da \qquad \phi,\ \hat{\phi} \in L^1$$

Let $\hat{\phi} \in L^1_+$ such that $\hat{\phi}$ is an equilibrium solution of (ADP). From Example 4.2, $\hat{\phi}(a) = \mu(\hat{P})\hat{P}\exp[-\mu(\hat{P})a]$, $a \geq 0$, where $P\hat{\phi} = \hat{P}$, and either $\hat{P} = 0$ or $\hat{P} > 0$ and satisfies $\mu(\hat{P})(\alpha + \mu(\hat{P})) = \alpha\beta(\hat{P})$. Define $\hat{B}$ as in (4.166) with $F'$ as in (4.168). Let $\lambda \in C$ and let $\hat{B}\phi = \lambda\phi$ for some $\phi \in L^1$, $\phi \neq 0$. As in Example 4.11 the definition of $\hat{B}$ yields that $\hat{\phi}$ satisfies (4.114) and

$$(4.169) \qquad \phi(0) = \beta'(\hat{P})P\phi\mu(\hat{P})\hat{P}\beta(\hat{P})^{-1} + \beta(\hat{P}) \int_0^\infty (1 - e^{-\alpha a})\phi(a)\, da$$

If $\hat{\phi} = 0$, then (4.114) and (4.169) yield that $\alpha\beta(0) = (\lambda + \mu(0))(\lambda + \alpha + \mu(0))$. As in Example 4.12, $\alpha\beta(0) < \mu(0)(\alpha + \mu(0))$ implies that 0 is locally exponentially asymptotically stable by (4.147), and $\alpha\beta(0) > \mu(0)(\alpha + \mu(0))$ implies that 0 is unstable by (4.148).

If $\hat{\phi} \neq 0$ and $\lambda = 0$, then (4.114) and (4.169) yield (4.115) and

$$\phi(0) = \beta'(\hat{P})P\phi\mu(\hat{P})\hat{P}\beta(\hat{P})^{-1} + \beta(\hat{P})[\alpha\phi(0)\mu(\hat{P})^{-1}(\alpha + \mu(\hat{P}))^{-1} - \mu'(\hat{P})P\phi\mu(\hat{P})\hat{P}\alpha(\alpha + 2\mu(\hat{P}))\mu(\hat{P})^{-2}(\alpha + \mu(\hat{P}))^{-2}]$$

Since $\mu(\hat{P})(\alpha + \mu(\hat{P})) = \alpha\beta(\hat{P})$, this last equation implies that $\beta'(\hat{P})\alpha = \mu'(\hat{P})(\alpha + 2\mu(\hat{P}))$.

If $\hat{\phi} \neq 0$, $\lambda \neq 0$, and $\mathrm{Re}\,\lambda > -\mu(\hat{P})$, then (4.117), (4.118), (4.169), and the fact that $\mu(\hat{P})(\alpha + \mu(\hat{P})) = \alpha\beta(\hat{P})$ imply that

$$\begin{aligned}
(\lambda + \mu(\hat{P}) + \mu'(\hat{P})\hat{P})P\phi &= \beta'(\hat{P})P\phi\mu(\hat{P})\hat{P}\beta(\hat{P})^{-1} \\
&\quad + \beta(\hat{P})\alpha(\lambda + \mu(\hat{P}))^{-1}(\alpha + \lambda + \mu(\hat{P}))^{-1} \\
&\quad \times \{\phi(0) - \mu'(\hat{P})P\phi\mu(\hat{P})\hat{P}(\alpha + \lambda + 2\mu(\hat{P})) \\
&\quad \times \mu(\hat{P})^{-1}(\alpha + \mu(\hat{P}))^{-1}\} \\
&= \beta'(\hat{P})P\phi\mu(\hat{P})\hat{P}\beta(\hat{P})^{-1} + \beta(\hat{P})\alpha P\phi(\lambda + \mu(\hat{P}))^{-1} \\
&\quad \times (\alpha + \lambda + \mu(\hat{P}))^{-1}\{\lambda + \mu(\hat{P}) + \mu'(\hat{P})\hat{P} \\
&\quad - \mu'(\hat{P})\hat{P}(\alpha + \lambda + 2\mu(\hat{P}))(\alpha + \mu(\hat{P}))^{-1}\} \\
&= \beta'(\hat{P})P\phi\mu(\hat{P})\hat{P}\beta(\hat{P})^{-1} \\
&\quad + \beta(\hat{P})\alpha P\phi(\alpha + \lambda + \mu(\hat{P}))^{-1} \\
&\quad \times \{1 - \mu'(\hat{P})\hat{P}(\alpha + \mu(\hat{P}))^{-1}\}
\end{aligned}$$

which implies that $\lambda^2 + b\lambda + c = 0$, where $b \overset{\text{def}}{=} \alpha + 2\mu(\hat{P}) + \mu'(\hat{P})\hat{P} - \beta'(\hat{P})\mu(\hat{P})\hat{P}\beta(\hat{P})^{-1}$ and $c \overset{\text{def}}{=} \hat{P}\mu'(\hat{P})(2\mu(\hat{P}) + \alpha) - \hat{P}\alpha\beta'(\hat{P})$.

We conclude the following for the case that $\hat{\phi} \neq 0$: if $b > 0$ and $c > 0$, then (4.147) implies that $\hat{\phi}$ is locally exponentially asymptotically stable, and if $c > 0$ and $b < 0$ or if $c < 0$, then (4.148) implies that $\hat{\phi}$ is unstable.

EXAMPLE 4.16 Let $R^n = R$. Let F have the form (2.20) and let G have the form (2.18) as in Examples 2.3, 2.6, 3.3, and 4.3, where $\beta$ and $\mu$ are twice continuously differentiable, $\beta$, $\beta'$, $\mu$, and $\mu'$ are all bounded, and $\mu$ is bounded below by a positive constant. Observe that F satisfies (4.145), G satisfies (4.146), and F and G are continuously Fréchet differentiable with G' given by (4.166) and F' given by

$$(4.170)\quad F'(\hat{\phi})\phi = \beta'(P\hat{\phi})P\phi \int_0^\infty ae^{-\alpha a}\hat{\phi}(a)\,da + \beta(P\hat{\phi}) \int_0^\infty ae^{-\alpha a}\phi(a)\,da \qquad \phi,\ \hat{\phi} \in L^1$$

Let $\hat{\phi} \in L^1_+$ such that $\hat{\phi}$ is an equilibrium solution of (ADP). From Example 4.3, $\hat{\phi}(a) = \mu(\hat{P})\hat{P}\exp[-\mu(\hat{P})a]$, $a \geq 0$, where $P\hat{\phi} = \hat{P}$, and either $\hat{P} = 0$ or $\hat{P} > 0$ and satisfies $\beta(\hat{P}) = (\alpha + \mu(\hat{P}))^2$. Define $\hat{B}$ as in (4.166) with F' as in (4.170). Let $\lambda \in C$ and let $\hat{B}\phi = \lambda\phi$ for some $\phi \in L^1$, $\phi \neq 0$. As in Example 4.11, $\hat{\phi}$ satisfies (4.114) and

$$(4.171)\quad \phi(0) = \beta'(\hat{P})P\phi\mu(\hat{P})\hat{P}\beta(\hat{P})^{-1} + \beta(\hat{P}) \int_0^\infty ae^{-\alpha a}\phi(a)\,da$$

If $\hat{\phi} = 0$, then (4.114) and (4.171) yield that $(\lambda + \mu(0) + \alpha)^2 = \beta(0)$. As in Example 4.13, $\beta(0) < (\alpha + \mu(0))^2$ implies that 0 is locally exponentially asymptotically stable by (4.147), and $\beta(0) > (\alpha + \mu(0))^2$ implies that 0 is unstable by (4.148).

If $\hat{\phi} \neq 0$ and $\lambda = 0$, then (4.114) and (4.171) yield (4.115) and

$$\phi(0) = \beta'(\hat{P})P\phi\mu(\hat{P})\hat{P}\beta(\hat{P})^{-1} + \beta(\hat{P})[\phi(0)\beta(\hat{P})^{-1} - 2\mu'(\hat{P})P\phi\mu(\hat{P})\hat{P}\beta(\hat{P})^{-3/2}]$$

This last equation then implies that $\beta'(\hat{P}) = 2\mu'(\hat{P})\beta(\hat{P})^{1/2}$.

If $\hat{\phi} \neq 0$, $\lambda \neq 0$, and $\operatorname{Re}\lambda > -\mu(\hat{P})$, then (4.117), (4.118), (4.171), and the fact that $\beta(\hat{P}) = (\alpha + \mu(\hat{P}))^2$ imply that

$$\begin{aligned}
&(\lambda + \mu(\hat{P}) + \mu'(\hat{P})\hat{P})P\phi \\
&= \beta'(\hat{P})P\phi\mu(\hat{P})\hat{P}\beta(\hat{P})^{-1} + \beta(\hat{P})(\alpha + \lambda + \mu(\hat{P}))^{-2}[\phi(0) - \mu'(\hat{P})P\phi\mu(\hat{P})\hat{P} \\
&\quad \times (2(\alpha + \mu(\hat{P})) + \lambda)(\alpha + \mu(\hat{P}))^{-2}] \\
&= \beta'(\hat{P})P\phi\mu(\hat{P})\hat{P}\beta(\hat{P})^{-1} + \beta(\hat{P})(\lambda + \beta(\hat{P})^{1/2})^{-2}P\phi[\lambda + \mu(\hat{P}) \\
&\quad + \mu'(\hat{P})\hat{P} - \mu'(\hat{P})\mu(\hat{P})\hat{P}(2\beta(\hat{P})^{1/2} + \lambda)\beta(\hat{P})^{-1}]
\end{aligned}$$

which implies that

$$\lambda^3 + [\mu(\hat{P}) + \mu'(\hat{P})\hat{P} + 2\beta(\hat{P})^{1/2} - \mu(\hat{P})\hat{P}\beta'(\hat{P})\beta(\hat{P})^{-1}]\lambda^2$$
$$+ [2\beta(\hat{P})^{1/2}\mu(\hat{P}) + 2\beta(\hat{P})^{1/2}\mu'(\hat{P})\hat{P} + \mu(\hat{P})\hat{P}\mu'(\hat{P})$$
$$- 2\mu(\hat{P})\hat{P}\beta'(\hat{P})\beta(\hat{P})^{1/2}]\lambda + [2\beta(\hat{P})^{1/2}\mu(\hat{P})\hat{P}\mu'(\hat{P}) - \beta'(\hat{P})\mu(\hat{P})\hat{P}] = 0$$

We conclude the following for the case that $\hat{\phi} \neq 0$: if $\beta'(\hat{P}) \neq 2\mu'(\hat{P})\beta(\hat{P})^{1/2}$ and the roots of this cubic equation all have negative real part, then $\hat{\phi}$ is locally exponentially asymptotically stable by (4.147), and if this cubic equation has a root with positive real part, then $\hat{\phi}$ is unstable by (4.148).

## 4.6 NOTES

The use of Liapunov functions, omega limit sets, and the invariance principle involves a topological approach to the stability theory of nonlinear age-dependent population dynamics. As with other applications of this approach, the primary difficulty is to find suitable Liapunov functions. Since the setting of the problem (ADP) is $L^1(0,\infty;R^n)$, these Liapunov functions are defined on an infinite dimensional space. The use of Liapunov functions in the investigation of stability theory for the related subjects of functional differential equations and Volterra integral equations has been used by many authors, and we refer the reader to our references for some of the relevant articles.

The fundamental theorem of Lotka (Theorem 4.10 in Section 4.3) was first proved by W. Feller in [104] using Laplace transform techniques. In [104] it was shown that in certain cases the solutions converge to the stable age distribution in an oscillatory fashion.

The method of linearization has been used by many authors in the study of the asymptotic behavior of nonlinear age-dependent population dynamics. Among these we mention J. Cushing [69], J. Cushing and M. Saleem [70], K. Gopalsamy [117], M. Gurtin and R. C. MacCamy [130], M. Gyllenberg [137], J. Prüss [240], [241], [242], and F. J.

S. Wang [293]. In [241] J. Prüss uses the method of Hopf bifurcation to study the existence and stability of periodic solutions of nonlinear age-dependent population models. The principal difficulty in applying the method of linearization in a stability analysis of a nonlinear age-dependent population model is the problem of locating the spectrum of the infinitesimal generator of the linearized problem.

# 5
# Biological Population Models

## 5.1 SPECIES INTERACTION

In this section we apply the results of the preceding chapters to an age-dependent population model of two interacting biological species. The form of the model will allow the interaction to be interpreted as either two species in competition or two species as predator and prey.

Let $L^1 \overset{\text{def}}{=} L^1(0,\infty;R^2)$ and define the birth function $F\colon\ L^1 \to R^2$ and the aging function $G\colon\ L^1 \to L^1$ by

$$(5.1)\qquad F(\phi) = [F(\phi)_1, F(\phi)_2]^T \qquad \phi = [\phi_1,\phi_2]^T \in L^1$$

$$F(\phi)_i \overset{\text{def}}{=} \int_0^\infty \beta_i(1 - e^{-\alpha_i a})\phi_i(a)\, da \qquad i = 1,\ 2$$

$$(5.2)\qquad G(\phi) = [G(\phi)_1, G(\phi)_2]^T \qquad \phi = [\phi_1,\phi_2]^T \in L^1$$

$$G(\phi)_i \overset{\text{def}}{=} -[\mu_{i1}(P\phi_1) + \mu_{i2}(P\phi_2)]\phi_i \qquad i = 1,\ 2$$

where $P\phi_i \overset{\text{def}}{=} \int_0^\infty \phi_i(a)\, da$, $\phi = [\phi_1,\phi_2]^T$, $i = 1, 2$, $\alpha_1$, $\beta_1$, $\alpha_2$, $\beta_2$ are all positive constants, $\mu_{11}$, $\mu_{12}$, $\mu_{21}$, $\mu_{22}$ are all bounded and twice continuously differentiable functions from R to $[0,\infty)$, and for some constant $\underline{\mu} > 0$, $\Sigma_{j=1}^2\, \mu_{ij}(P) \geq \underline{\mu}$ for all $P \geq 0$, $i = 1, 2$. Consequently, the model has the form

$$D\ell_i(a,t) = -[\mu_{i1}(P\ell_1(\cdot,t)) + \mu_{i2}(P\ell_2(\cdot,t))]\ell_i(a,t) \qquad i = 1,\ 2$$

$$\ell_i(0,t) = \int_0^\infty \beta_i(1 - e^{-\alpha_i a})\ell_i(a,t)\, da \qquad i = 1,\ 2$$

Each component of the density function $\phi = [\phi_1,\phi_2]^T$ corresponds to one of two species present in the same environment. Each species has a linear birth process independent of the other species. The form of the birth process means that members of each species reproduce as long as they survive. Each species has a nonlinear aging process dependent on both species. The aging process is a pure mortality process in which the mortality modulus has both an intraspecies and an interspecies contribution. The form of the mortality process corresponds to a harsh environment, since the mortality modulus is independent of age.

As in Examples 2.2 and 2.5, the hypotheses (5.1) and (5.2) imply that (2.1), (2.2), (2.22), (2.23), (2.28), and (2.49) hold. Hence, we may apply Theorems 2.1, 2.2, 2.4, 2.5, 2.8, and 2.9 to obtain the existence of a unique nonnegative solution to (ADP) for each $[\phi_1,\phi_2]^T \in L^1_+$ defined for all time. As in Example 3.2, Theorems 3.1, 3.2, 3.3, and 3.4 imply that these solutions correspond to a strongly continuous nonlinear semigroup $S(t)$, $t \geq 0$, in $L^1_+$ with infinitesimal generator $B\colon L^1_+ \to L^1$ defined by

$$B\phi = [(B\phi)_1,(B\phi)_2]^T \qquad (B\phi)_i \overset{\text{def}}{=} -\phi_i' + G(\phi)_i \qquad i = 1, 2$$

$D(B) \overset{\text{def}}{=} \{\phi = [\phi_1,\phi_2]^T \in L^1_+\colon\ \phi_i$ is absolutely continuous on $[0,\infty)$, $\phi_i' \in L^1(0,\infty;R)$ and $\phi_i(0) = \int_0^\infty \beta_i(1 - e^{-\alpha_i a})\phi_i(a)\,da$, $i = 1, 2\}$

An equilibrium solution $\hat{\phi} \in L^1_+$ of (ADP) will satisfy $B\hat{\phi} = 0$, that is,

$$\hat{\phi}_i'(a) = -\left[\sum_{j=1}^{2} \mu_{ij}(P\hat{\phi}_j)\right]\hat{\phi}_i(a) \qquad a \geq 0 \qquad i = 1, 2 \tag{5.3}$$

$$\hat{\phi}_i(0) = \beta_i \int_0^\infty (1 - e^{-\alpha_i a})\hat{\phi}_i(a)\,da \qquad i = 1, 2 \tag{5.4}$$

Define $\hat{P}_i \overset{\text{def}}{=} P\hat{\phi}_i$, $\hat{\mu}_i = \Sigma_{j=1}^2 \mu_{ij}(\hat{P}_j)$, $i = 1, 2$. The solution of (5.3) is

(5.5) $\hat{\phi}_i(a) = \exp[-\hat{\mu}_i a]\hat{\phi}_i(0) \qquad a \geq 0 \qquad i = 1, 2$

The boundary condition (5.4) then implies that

(5.6) $\alpha_i \beta_i = \hat{\mu}_i(\alpha_i + \hat{\mu}_i) \qquad i = 1, 2$

Thus, an equilibrium solution of (ADP) in $L^1_+$ is given by (5.5), where $\hat{P}_i \geq 0$, i = 1, 2, satisfies (5.6), and

(5.7) $\hat{\phi}_i(0) = \hat{\mu}_i \hat{P}_i \qquad i = 1, 2$

We consider first the extinction of one of the species. Let i be fixed and let there exist a positive constant $\mu_0$ such that $\mu_{i1}(P\phi_1) + \mu_{i2}(P\phi_2) \geq \mu_0$ for all $[\phi_1,\phi_2]^T \in L^1_+$ and $\alpha_i \beta_i < \mu_0(\alpha_i + \mu_0)$. Define $\beta_0, \mu_0$: $[0,\infty) \to [0,\infty)$ by $\beta_0(a) \overset{def}{=} \beta_i(1 - e^{-\alpha_i a})$, $\mu_0(a) \overset{def}{=} \mu_0$, $a \geq 0$. Then, the hypothesis of Theorem 4.5 is satisfied, and so $\lim_{t\to\infty} \| (S(t)\phi)_i \|_{L^1(0,\infty;R)} = 0$.

We next consider the stability of nontrivial equilibrium solutions of (ADP). Suppose that $\hat{\phi}$ is such a nontrivial equilibrium, that is, $\hat{\phi}$ is given by (5.5), where $\hat{P}_1$, $\hat{P}_2$ satisfy (5.6), and both $\hat{P}_1$, $\hat{P}_2$ are positive. Let $\hat{B}$: $L^1 \to L^1$ be defined by

(5.8) $\hat{B}\phi \overset{def}{=} -\phi' + G'(\hat{\phi})\phi$, $\phi \in D(\hat{B})$, $D(\hat{B}) \overset{def}{=} \{\phi = [\phi_1,\phi_2]^T \in L^1$: $\phi_i$ is absolutely continuous on $[0,\infty)$, $\phi_i' \in L^1(0,\infty; \ )$ and $\phi_i(0) = \int_0^\infty \beta_i(1 - e^{-\alpha_i a})\phi_i(a)\,da$, $i = 1, 2\}$

where $G'(\hat{\phi})$: $L^1 \to L^1$ is the bounded linear operator given by

$$(G'(\hat{\phi})\phi)_i = -\left[\left(\sum_{j=1}^{2} \mu_{ij}(\hat{P}_j)\right)\phi_i + \left(\sum_{j=1}^{2} \mu_{ij}'(\hat{P}_j)P\phi_j\right)\hat{\phi}_i\right]$$

$$\phi = [\phi_1,\phi_2]^T \in L^1 \qquad i = 1, 2$$

Observe that for $x = [x_1,x_2]^T \in R^2$,

$$\sum_{i=1}^{2} (\operatorname{sgn} x_i)\left(-\sum_{j=1}^{2} \mu_{ij}(\hat{P}_j)x_i\right) = -\sum_{i=1}^{2}\left(\sum_{j=1}^{2} \mu_{ij}(\hat{P}_j)\right)|x_i|$$

$$\le - \sum_{i=1}^{2} \underline{\mu}|x_i| = -\underline{\mu}|x|$$

Also, the mapping $\phi_i \to (\Sigma_{j=1}^2 \mu'_{ij}(\hat{P}_j)P\phi_j)\hat{\phi}_i$, $i = 1, 2$, is compact from $L^1(0,\infty;R)$ into $L^1(0,\infty;R)$. Thus, by Proposition 4.14 and Theorem 4.6, $\omega_1(\hat{B}) \le -\underline{\mu}$. According to Theorem 4.12, the stability of $\hat{\phi}$ is determined by $\sigma(\hat{B}) - E\sigma(\hat{B})$.

Let $\lambda \in C$ such that for some $\phi \in L^1$, $\phi \ne 0$, $\hat{B}\phi = \lambda\phi$. Define $\hat{\mu}'_i \overset{\text{def}}{=} \Sigma_{j=1}^2 \mu'_{ij}(\hat{P}_j)P\psi_j$, $i = 1, 2$. From (5.5), (5.7), and (5.8) we obtain

$$\text{(5.9)} \qquad \phi_i(a) = \exp[-(\lambda + \hat{\mu}_i)a]\phi_i(0) - \int_0^a \exp[-(\lambda + \hat{\mu}_i)(a - b)]\hat{\mu}'_i\hat{\mu}_i\hat{P}_i \exp[-\hat{\mu}_i b]\, db \qquad a \ge 0 \qquad i = 1, 2$$

Suppose that $\lambda = 0$. From (5.9) we have that

$$\phi_i(a) = \exp[-\hat{\mu}_i a](\phi_i(0) - \hat{\mu}'_i\hat{\mu}_i\hat{P}_i a) \qquad a \ge 0 \qquad i = 1, 2$$

Also, from (5.6) we have that

$$\phi_i(0) = \int_0^\infty \beta_i(1 - e^{-\alpha_i a}) \exp[-\hat{\mu}_i a](\phi_i(0) - \hat{\mu}'_i\hat{\mu}_i\hat{P}_i a)\, da$$
$$= \phi_i(0) - \hat{\mu}'_i\hat{P}_i(\alpha_i + 2\hat{\mu}_i)(\alpha + \hat{\mu}_i)^{-1} \qquad i = 1, 2$$

which implies that $\hat{\mu}'_i = 0$, $i = 1, 2$ (since $\hat{P}_i$, $\alpha_i$, $\hat{\mu}_i$ are all positive). Thus,

$$\mu'_{11}(\hat{P}_1)P\phi_1 + \mu'_{12}(\hat{P}_2)P\phi_2 = 0$$
$$\mu'_{21}(\hat{P}_1)P\phi_1 + \mu'_{22}(\hat{P}_2)P\phi_2 = 0$$

and since $P\phi_1$, $P\phi_2$ cannot both be 0, $\mu'_{11}(\hat{P}_1)\mu'_{22}(\hat{P}) = \mu'_{21}(\hat{P}_1)\mu'_{12}(\hat{P}_2)$.

Suppose that $\lambda \ne 0$ and $\operatorname{Re} \lambda > -\hat{\mu}_i$, $i = 1, 2$. From (5.9) we have that

$$\text{(5.10)} \qquad \phi_i(a) = \exp[-(\lambda + \hat{\mu}_i)a]\phi_i(0) - \lambda^{-1}\hat{\mu}'_i\hat{\mu}_i\hat{P}_i \times \{\exp[-\hat{\mu}_i a] - \exp[-(\lambda + \hat{\mu}_i)a]\} \qquad a \ge 0 \qquad i = 1, 2$$

From (5.10) we obtain

$$\phi_i(0) = \int_0^\infty \beta_i(1 - e^{-\alpha_i a}) \{\exp[-(\lambda + \hat{\mu}_i)a]\phi_i(0)$$

$$- \lambda^{-1}\hat{\mu}_i'\hat{\mu}_i\hat{P}_i (\exp[-\hat{\mu}_i a] - \exp[-(\lambda + \hat{\mu}_i)a])\} \quad da$$

$$= \beta_i\{\phi_i(0)\alpha_i(\lambda + \hat{\mu}_i)^{-1}(\alpha_i + \lambda + \hat{\mu}_i)^{-1}$$

$$- \lambda^{-1}\hat{\mu}_i'\hat{\mu}_i\hat{P}_i[\alpha_i\hat{\mu}_i^{-1}(\alpha_i + \hat{\mu}_i)^{-1}$$

$$- \alpha_i(\lambda + \hat{\mu}_i)^{-1}(\alpha_i + \lambda + \hat{\mu}_i)^{-1}]\}$$

$$= \alpha_i\beta_i(\lambda + \hat{\mu}_i)^{-1}(\alpha + \lambda + \hat{\mu}_i)^{-1}\{\phi_i(0)$$

$$- \hat{\mu}_i'\hat{P}_i(\alpha_i + \lambda + 2\hat{\mu}_i)(\alpha_i + \hat{\mu}_i)^{-1}\} \qquad i = 1, 2$$

Now use (5.6) to obtain

$$\phi_i(0) = \alpha_i\beta_i(\lambda + \hat{\mu}_i)^{-1}(\alpha_i + \lambda + \hat{\mu}_i)^{-1}\{\phi_i(0)$$

$$- \mu_i'\hat{P}_i(\alpha_i + \lambda + 2\hat{\mu}_i)\alpha_i^{-1}\beta_i^{-1}\hat{\mu}_i\} \qquad i = 1, 2$$

which implies that

$$\text{(5.11)} \qquad \lambda\phi_i(0) = -\hat{\mu}_i'\hat{\mu}_i\hat{P}_i \qquad i = 1, 2$$

From (5.10) we also obtain

$$P\phi_i = \int_0^\infty \phi_i(a)\ da$$

$$= [\phi_i(0) - \hat{\mu}_i'\hat{P}_i](\lambda + \hat{\mu}_i)^{-1} \qquad i = 1, 2$$

which implies that

$$[\lambda + \hat{\mu}_1 + \mu_{11}'(\hat{P}_1)\hat{P}_1]P\phi_1 + \mu_{12}'(\hat{P}_2)\hat{P}_1P\phi_2 = \phi_1(0)$$

$$\mu_{21}'(\hat{P}_1)\hat{P}_2P\phi_1 + [\lambda + \hat{\mu}_2 + \mu_{22}'(\hat{P}_2)\hat{P}_2]P\phi_2 = \phi_2(0)$$

Now substitute (5.11) into this system of equations to obtain

$$(\lambda[\lambda + \hat{\mu}_1 + \mu'_{11}(\hat{P}_1)\hat{P}_1] + \mu'_{11}(\hat{P}_1)\hat{P}_1\hat{\mu}_1)P\phi_1 + (\lambda\mu'_{12}(\hat{P}_2)\hat{P}_1 + \mu'_{12}(\hat{P}_2)\hat{P}_1\hat{\mu}_1)P\phi_2 = 0$$

$$(\lambda\mu'_{21}(\hat{P}_1)\hat{P}_2 + \mu'_{21}(\hat{P}_1)\hat{P}_2\hat{\mu}_2)P\phi_1 + (\lambda[\lambda + \hat{\mu}_2 + \mu'_{22}(\hat{P}_2)\hat{P}_2] + \mu'_{22}(\hat{P}_2)\hat{P}_2\hat{\mu}_2)P\phi_2 = 0$$

Since $P\phi_1$, $P\phi_2$ cannot both be 0, the determinant of this system must be 0. Thus

$$\begin{aligned}&(\lambda[\lambda + \hat{\mu}_1 + \mu'_{11}(\hat{P}_1)\hat{P}_1] + \mu'_{11}(\hat{P}_1)\hat{P}_1\hat{\mu}_1)(\lambda[\lambda + \hat{\mu}_2 + \mu'_{22}(\hat{P}_2)\hat{P}_2] \\ &+ \mu'_{22}(\hat{P}_2)\hat{P}_2\hat{\mu}_2) - \mu'_{12}(\hat{P}_2)\mu'_{21}(\hat{P}_1)\hat{P}_1\hat{P}_2(\lambda + \hat{\mu}_1)(\lambda + \hat{\mu}_2) \\ &= (\lambda + \hat{\mu}_1)(\lambda + \hat{\mu}_2)[\lambda^2 + \lambda(\mu'_{11}(\hat{P}_1)\hat{P}_1 + \mu'_{22}(\hat{P}_2)\hat{P}_2) \\ &\quad + (\mu'_{11}(\hat{P}_1)\mu'_{22}(\hat{P}_2) - \mu'_{12}(\hat{P}_2)\mu'_{21}(\hat{P}_1))\hat{P}_1\hat{P}_2] = 0\end{aligned}$$

Set $b \overset{\text{def}}{=} \mu'_{11}(\hat{P}_1)\hat{P}_1 + \mu'_{22}(\hat{P}_2)\hat{P}_2$ and $c \overset{\text{def}}{=} (\mu'_{11}(\hat{P}_1)\mu'_{22}(\hat{P}_2) - \mu'_{12}(\hat{P}_2)\mu'_{21}(\hat{P}_1))\hat{P}_1\hat{P}_2$. We have that if $\lambda \neq 0$ and $\operatorname{Re} \lambda > -\hat{\mu}_i$, $i = 1, 2$, then $\lambda^2 + b\lambda + c = 0$.

From Theorem 4.12 we conclude the following: if $b > 0$ and $c > 0$, then (4.109) implies that $\hat{\phi}$ is locally exponentially asymptotically stable, and if $c > 0$ and $b < 0$ or if $c < 0$, then (4.110) implies that $\hat{\phi}$ is unstable. If $\mu_{12}(P)$ and $\mu_{21}(P)$ are increasing functions of P, then this system corresponds to a model of two competing species. If $\mu_{12}(P)$ is decreasing in P and $\mu_{21}(P)$ is increasing in P, then this system corresponds to a predator-prey model in which the first component represents the predator and the second component represents the prey. If we assume that $\mu'_{11}(\hat{P}_1)\hat{P}_1 + \mu'_{22}(\hat{P}_2)\hat{P}_2$ is positive, then the stability of $\hat{\phi}$ is determined according to whether $\mu'_{11}(\hat{P}_1)\mu'_{22}(\hat{P}_2)$ is greater than or less than $\mu'_{12}(\hat{P}_2)\mu'_{21}(\hat{P}_1)$, that is, whether the intra-species mortality dominates or is dominated by the interspecies mortality.

## 5.2 POPULATION GENETICS

In [153] F. Hoppensteadt formulated a mendelian genetics population model with age structure. This model considers the case of one locus on the chromosome at which the gene has two alleles (or variants). In this section we will analyze a similar model and determine conditions under which such a population tends toward extinction. We place the following assumptions on the genetics population:

Assumption 1 The population consists of three classes of genotypes AA, Aa, aa (we do not distinguish between Aa and aA genotypes) determined according to Mendel's laws for two alleles A and a for one gene at one locus on the chromosome.

Assumption 2 Selection may occur through both births and deaths, so that each genotype class has its own fertility and mortality moduli.

Assumption 3 The birth process is not effected by crowding, but crowding effects are present in the mortality process, so that the fertility moduli are independent of the densities, but the mortality moduli are density-dependent.

Assumption 4 The model is based on the female population only, so that fertility and mortality rates are independent of sex.

Assumption 5 The initial age distribution of each of the three genotypes is known.

To fit this model to the formulation of age-dependent population dynamics in the preceding chapters, let $R^n = R^3$, let $L^1 = L^1(0,\infty;R^3)$, and let $\ell(\cdot,t) = [\ell_1(\cdot,t), \ell_2(\cdot,t), \ell_3(\cdot,t)]^T \in L^1$, where the components $\ell_1(a,t)$, $\ell_2(a,t)$, $\ell_3(a,t)$ correspond to the population densities with respect to age a at time t of the genotype classes AA, Aa, aa, respectively. We suppose that the aging function G for this population corresponds to a pure mortality process satisfying

(5.12) $G\colon L^1 \to L^1$, $G$ satisfies (4.2), $G([\phi_1,\phi_2,\phi_3]^T(a) \overset{\text{def}}{=} [-\mu_1(a,\phi)\phi_1(a), -\mu_2(a,\phi)\phi_2(a), -\mu_3(a,\phi)\phi_3(a)]^T$ for all $\phi = [\phi_1,\phi_2,\phi_3]^T \in L^1$ and almost all $a > 0$, where $\mu_i\colon [0,\infty) \times L^1 \to [0,\infty)$ for $i = 1, 2, 3$, as in (4.2) (i), (ii), (iii), and (iv).

We suppose that the birth function $F$ satisfies

(5.13) $F\colon L^1 \to R^3$, $F([\phi_1,\phi_2,\phi_3]^T) \overset{\text{def}}{=} \int_0^\infty [\beta_1\phi_1(\phi_1 + \phi_2) + \beta_2\phi_2(\phi_1 + \phi_2), (1/2)\{\beta_1\phi_1(\phi_2 + \phi_3) + \beta_2\phi_2(\phi_1 + 2\phi_2 + \phi_3) + \beta_3\phi_3(\phi_1 + \phi_2)\}, \beta_2\phi_2(\phi_2 + \phi_3) + \beta_3\phi_3(\phi_2 + \phi_3)]^T/\{\phi_1 + 2\phi_2 + \phi_3\}\, da$ (we have suppressed the variable of integration in the integral) for all $\phi = [\phi_1,\phi_2,\phi_3]^T \in L^1_+$, $[\phi_1,\phi_2,\phi_3]^T \neq [0,0,0]^T$, where $\beta_i\colon [0,\infty) \to [0,\infty)$ is continuous and bounded for each $i = 1, 2, 3$, and $F([\phi_1,\phi_2,\phi_3]^T) \overset{\text{def}}{=} [0,0,0]^T$ for either $[\phi_1,\phi_2,\phi_3]^T \notin L^1_+$ or $[\phi_1,\phi_2,\phi_3]^T = [0,0,0]^T$.

The explanation for this birth function arises as follows: Define $P(a,t) = \ell_1(a,t) + 2\ell_2(a,t) + \ell_3(a,t)$, so that the total population at time $t$ is

$$\int_0^\infty P(a,t)\, da$$

[the factor 2 times $\ell_2(a,t)$ occurs becuase we do not distinguish between Aa and aA genotypes]. The form of the birth function in (5.13) results from the various mating possibilities between males and females belonging to the different genotype classes, as determined by Mendel's laws (see Figure 5.1). For example, consider the ways in which the genotype classes can mate to produce AA genotypes. The birth rate of AA offspring from AA females and AA males or Aa males is given by

$$\text{(5.14)} \qquad \int_0^\infty \beta_1(a)\left\{\ell_1(a,t)\left[\frac{\ell_1(a,t)}{P(a,t)}\right] + \ell_1(a,t)\left[\frac{\ell_2(a,t)}{P(a,t)}\right]\right\} da$$

| Male / Female | A | a |
|---|---|---|
| A | AA | Aa |
| a | aA | aa |

*Figure 5.1*

where $\beta_1(a)$ is the birth modulus of the AA genotype class, $\ell_1(a,t)/P(a,t)$ is the proportion of the population which is of AA genotype, and $\ell_2(a,t)/P(a,t)$ is the proportion of the population which is of Aa genotype. (Note that all the matings between AA females and AA males result in AA offspring and half the matings between AA females and Aa males result in AA offspring, but we must multiply by 2 to account for the AA offspring of AA females and aA males, as well.) The birth rate of AA offspring from Aa females and AA males or Aa males is given by

$$(5.15) \qquad \int_0^\infty \beta_2(a)\left\{\ell_2(a,t)\left[\frac{\ell_1(a,t)}{P(a,t)}\right] + \ell_2(a,t)\left[\frac{\ell_2(a,t)}{P(a,t)}\right]\right\} da$$

where $\beta_2(a)$ is the birth modulus of the Aa genotype class. (Note that half of the matings between Aa females and AA males result in AA offspring, but we must multiply by 2 to account for the AA offspring of aA females and AA males, as well. Also, a fourth of the matings between Aa females and Aa males are AA offspring, but we must multiply by 4 to account for the AA offspring of Aa females and aA males, aA females and Aa males, and aA females and aA males.) The total birth rate of AA offspring is then given by the sum of (5.14) and (5.15), which explains the first component of the formula for the birth function F in (5.13). The other two components of F are obtained by similar reasoning.

For convenience define $\tau_1 = 1$, $\tau_2 = 2$, and $\tau_3 = 1$. The total birth rate of females at time t is

$$(5.16) \qquad \sum_{i=1}^{3} \tau_i \ell_i(0,t) = \sum_{i=1}^{3} \tau_i F(\ell(\cdot,t))_i = \int_0^\infty \sum_{i=1}^{3} \tau_i \beta_i(a) \ell_i(a,t)\, da$$

which may be seen by direct verification from (5.13). (For example,

$$\int_0^\infty \tau_1 \beta_1(a) \ell_1(a,t)\, da$$
$$= \int_0^\infty \left\{ \tau_1 \beta_1(a) \ell_1(a,t) \frac{[\ell_1(a,t) + \ell_2(a,t)}{P(a,t)} + \tau_2 \frac{1}{2} \beta_1(a) \ell_1(a,t) \frac{[\ell_1(a,t) + \ell_3(a,t)]}{P(a,t)} \right\} da$$

and similar calculations hold for the terms involving $\beta_2(a)$ and $\beta_3(a)$.)

We claim that F satisfies (2.1) and (2.22), and in fact, F is globally Lipschitz continuous from $L^1$ to $R^3$. We give an indication of the verification of this fact as follows: Define f: $R^3 \to R$ by $f(u,v,w) \overset{\text{def}}{=} u^2/(u + 2v + w)$, $[u,v,w]^T \in R^3_+$, $[u,v,w]^T \neq [0,0,0]^T$, and $f(u,v,w) = 0$ otherwise. Observe that if $[u,v,w]^T \in R^3_+$, $[u,v,w]^T \neq [0,0,0]^T$, then $(\partial/\partial u) f(u,v,w) = u(u + 4v + 2w)/(u + 2v + w)^2 \leq 2$. Thus, if $u_1 > 0$, $u_2 > 0$, $v \geq 0$, $w \geq 0$, then

$$|f(u_1,v,w) - f(u_2,v,w)|$$
$$= \left| \int_{u_1}^{u_2} \frac{\partial}{\partial u} f(u,v,w)\, du \right| \leq 2|u_1 - u_2|$$

If $u \geq 0$, $v \geq 0$, $w \geq 0$, then

$$|f(u,v,w) - f(0,v,w)| \leq |u - 0|$$

Thus, f is Lipschitz continuous in its first place with Lipschitz constant $\leq 2$. Similar estimates can be used to show that f is Lipschitz continuous in its second and third places, and thus, that f is Lipschitz continuous from $R^3$ to R. Similar estimates can also be made for each term of F in (5.13) to establish the existence of a constant C such that

(5.17) $$|F(\phi) - F(\hat{\phi})| \le C \sup_{a\ge 0, i=1,2,3} \beta_i(a) \|\phi - \hat{\phi}\|_{L^1} \quad \text{for all } \phi, \hat{\phi} \in L^1$$

We see, therefore, that F satisfies (2.1) and (2.22). Notice that F does not satisfy (4.1), that is, F does not have the form

$$F(\phi)_i = \int_0^\infty \beta_i(a,\phi)\phi_i(a)\, da \qquad \phi \in L^1 \qquad i = 1, 2, 3$$

Since G satisfies (4.2), we obtain automatically that G satisfies (2.2), (2.23), (2.49), and (3.64) (see Remark 3.4 and Remark 4.1). From (5.17) and the fact that $F(0) = 0$ we obtain that (2.28) is satisfied with $\omega = C \sup_{a\ge 0, i=1,2,3} \beta_i(a)$. Thus, we may apply Theorems 2.1, 2.2, 2.4, 2.5, 2.8, and 2.9 in Chapter 2, as well as Theorems 3.1, 3.2, 3.3, and 3.4 in Chapter 3. Consequently, the solutions corresponding to nonnegative initial age distributions exist for all time, are nonnegative, are unique, depend continuously upon their initial age distributions, and form a strongly continuous nonlinear semigroup in $L^1_+$.

We now establish a sufficient condition for the extinction of the genetics population. We suppose the following conditions:

(5.18) There exists $\delta > 0$ such that $\int_0^\infty e^{\delta a}\beta_i(a) \exp[-\int_0^a \mu_i(b,0)\, db]\, da < 1$ for $i = 1, 2, 3$.

(5.19) $\mu_i(a,\phi) \ge \mu_i(a,0)$ for all $\phi \in L^1_+$, $a \ge 0$, $i = 1, 2, 3$.

Under the assumptions (5.18) and (5.19) we claim that there exists $\omega > 0$ such that if $\phi \in L^1_+$ and $\phi$ has compact support in $[0,\infty)$, then there exists a constant $K = K(\phi)$ (depending on $\phi$) such that

(5.20) $$\int_0^\infty \sum_{i=1}^3 \tau_i \ell_i(a,t)\, da \le K(\phi) e^{-\omega t} \int_0^\infty \sum_{i=1}^3 \tau_i \phi_i(a)\, da \qquad t \ge 0$$

The proof of (5.20) is very similar to the proof of Theorem 4.2. By (5.18) we may choose $\omega$ such that $0 < \omega < \underline{\mu}$ and

$$(5.21) \qquad \int_0^\infty e^{\omega a}\beta_i(a) \exp\left[-\int_0^a \mu_i(b,0)\, db\right] da < 1 \qquad i = 1, 2, 3$$

where $\underline{\mu}$ is the constant in (4.2) (iv). Define the functions $p_i$: $[0,\infty) \to R$, $i = 1, 2, 3$, by

$$p_i(a) \stackrel{\text{def}}{=} \exp\left[-\omega a + \int_0^a \mu_i(b,0)\, db\right]$$
$$\times \left\{1 - \int_0^a e^{\omega b}\beta_i(b) \exp\left[-\int_0^b \mu_i(\tau)\, d\tau\right] db\right\}$$

Let $\phi = [\phi_1,\phi_2,\phi_3]^T \in L_+^1$ such that $\phi$ has compact support in $[0,\infty)$. For $t \geq 0$, define

$$V(t) \stackrel{\text{def}}{=} \int_0^\infty \sum_{i=1}^3 \tau_i \ell_i(a,t) p_i(a)\, da$$

Using (5.16) and an argument very similar to the one used to prove Theorem 4.2, one can show that

$$\limsup_{h\to 0^+} h^{-1}[V(t + h) - V(t)]$$
$$\leq \sum_{i=1}^3 \tau_i \ell_i(0,t) p_i(0) + \int_0^\infty \sum_{i=1}^3 \tau_i[-\omega p_i(a) - \beta_i(a)]\ell_i(a,t)\, da$$
$$= -\omega V(t)$$

The claim (5.20) then follows using the same argument as in Theorem 4.2.

The condition (5.18) assures the extinction of the genetics population. This condition involves an age dependence in the fertility and mortality moduli for each of the three genotype classes separately. This condition demonstrates the role of age structure in determining whether or not the population becomes extinct.

## 5.3 EPIDEMIC POPULATIONS

Among the most important mathematical population models are those which describe the evolution of infectious diseases. It is evident that for many diseases the chronological age structure of the population is an important consideration in the design of such models. For childhood diseases this age structure is especially important, since such factors as exposure, susceptibility, immunity, inoculation, and quarantine are frequently influenced by age considerations. Among the first continuous deterministic epidemic models incorporating chronological age effects were those of F. Hoppensteadt in [153] and K. Dietz in [92]. In this section we will analyze a similar model of age-dependent epidemic population dynamics.

One of the principal values of epidemic population models is that the relative importance of various parameters can be analyzed. Many mathematical epidemic models possess so-called threshold conditions, that is, certain critical values involving the various parameters which must be exceeded in order for the epidemic to occur. In the epidemic model we discuss below such a threshold phenomenon is present, and as we shall see, this threshold phenomenon exhibits an age dependency.

We divide the epidemic population into three subclasses: the susceptible class (S) (members of the population not infected, but capable of becoming infected); the infective class (I) (members of the population capable of transmitting the disease to susceptibles); and the removed class (R) (members of the population neither susceptible nor infective). Let $s(a,t)$, $i(a,t)$, and $r(a,t)$ be the density functions (with respect to age a) at time t of the susceptible, infective, and removed classes, respectively. Let $S(t) \stackrel{\text{def}}{=} \int_0^\infty s(a,t)\, da$, $I(t) \stackrel{\text{def}}{=} \int_0^\infty i(a,t)\, da$, and $R(t) \stackrel{\text{def}}{=} \int_0^\infty r(a,t)\, da$ be the corresponding total population at time t of the susceptible, infective, and removed classes, respectively. We place the following assumptions on the model:

Assumption 1 The initial age distributions of the susceptible class, $s_0(a)$, the infective class, $i_0(a)$, and the removed class, $r_0(a)$, are all known.

Assumption 2 The probability that a susceptible of age a survives as a susceptible to age b without entering the removed class is $\exp[-\int_a^b \mu(c)\, dc]$, $0 \le a \le b$, where $\mu$: $[0,\infty) \to [0,\infty)$ such that $\mu$ is bounded and Lipschitz continuous on $[0,\infty)$ and there exists $\underline{\mu} > 0$ such that $\mu(a) \ge \underline{\mu}$ for $a \ge 0$ (the function $\mu$ is called the removal modulus of the susceptible class and accounts for removal of susceptibles due to such factors as mortality or immunization).

Assumption 3 The probability that an infective of age a survives as an infective to age b is $\exp[-\int_a^b \gamma(c)\, dc]$, $0 \le a \le b$, where $\gamma$: $[0,\infty) \to [0,\infty)$ such that $\gamma$ is bounded and Lipschitz continuous and there exists a constant $\underline{\gamma} > 0$ such that $\gamma(a) \ge \underline{\gamma}$ for $a \ge 0$ (the function $\gamma$ is called the removal modulus of the infective class and accounts for removal of infectives due to such factors as mortality, quarantine, or recovery with permanent immunity).

Assumption 4 The probability that at time t a susceptible of age a survives as a susceptible to age b without becoming infective is $\exp[-I(t) \int_a^b \eta(c)\, dc]$, $0 \le a \le b$, where $\eta$: $[0,\infty) \to [0,\infty)$ such that $\eta$ is Lipschitz continuous and there exists a constant $a_0 > 0$ such that $\eta(a) = 0$ for $a \ge a_0$ (the function $\eta$ is called the contact modulus and measures the influence of the infective class upon the susceptible class).

Assumption 5 The birth rate of the susceptible class is constant, so that $s(0,t) = \beta$, $t \ge 0$, for some positive constant $\beta$, and the birth rate of the infective class and removed class is 0 (the assumption that the birth rate of susceptibles is constant is reasonable for infectious diseases which do not affect birth rates and for populations which are demographically stable in the time interval under consideration).

As in Section 1.4, Assumptions 1-5 lead to the following equations for this model:

(5.22) $Ds(a,t) = -\eta(a)I(t)s(a,t) - \mu(a)s(a,t)$

(5.23) $Di(a,t) = \eta(a)I(t)s(a,t) - \gamma(a)i(a,t)$

(5.24) $Dr(a,t) = \mu(a)s(a,t) + \gamma(a)i(a,t)$

(5.25) $s(0,t) = \beta,\ i(0,t) = 0,\ r(0,t) = 0$

(5.26) $s(a,0) = s_0(a),\ i(a,0) = i_0(a),\ r(a,0) = r_0(a)$

We note that it suffices to consider the model in terms of only $s(a,t)$ and $i(a,t)$, since once these densities are known, $r(a,t)$ can be obtained from (5.24), (5.25), and (5.26) as in Section 1.2:

$$r(a,t) = \begin{cases} \int_0^a [\mu(c)s(c,\ c + t - a) + \gamma(c)i(c,\ c + t - a)]\ dc & a < t \\ r_0(a - t) + \int_{a-t}^a [\mu(c)s(c,\ c + t - a) + \gamma(c)i(c,\ c + t - a)]\ dc & a > t \end{cases}$$

In order to fit this model to the formulation (ADP) of the preceding chapters we take $R^n = R^2$, $L^1 \overset{\text{def}}{=} L^1(0,\infty;R^2)$, and define the birth function F and the aging function G as follows:

(5.27) $F\colon\ L^1 \to R^2,\ F([\phi_1,\phi_2]^T) \overset{\text{def}}{=} [\beta,0]^T,\ [\phi_1,\phi_2]^T \in L^1$, where $\beta$ is the positive constant in Assumption 5.

(5.28) $G\colon\ L^1 \to L^1,\ G([\phi_1,\phi_2]^T)(a) \overset{\text{def}}{=} [-\eta(a)P\phi_2\phi_1(a) - \mu(a)\phi_1(a),\ \eta(a)P\phi_2\phi_1(a) - \gamma(a)\phi_2(a)]^T,\ [\phi_1,\phi_2]^T \in L^1$, a.e. $a > 0$, where $P\phi \overset{\text{def}}{=} \int_0^\infty \phi(a)\ da$ for $\phi \in L^1(0,\infty;R)$ and $\mu$, $\gamma$, $\eta$ are as in Assumptions 2-4.

Notice that this G is an example of an aging function which is not a pure mortality function. Notice also that F satisfies (2.1), (2.22), and G satisfies (2.2), (2.23), (2.49), and (3.64). Thus, we may apply Theorems 2.1, 2.2, 2.3, 2.4, 2.8, and 2.9 in Chapter 2 to obtain the existence of a unique nonnegative solution $[s(\cdot,t),i(\cdot,t)]^T$ to (ADP) for each $[\phi_1,\phi_2]^T \in L^1_+$ defined on a maximal interval of existence.

We claim next that this maximal interval of existence is $[0,\infty)$. By Theorem 2.3 it suffices to show that the solution $[s(\cdot,t),i(\cdot,t)]^T$

is bounded on bounded intervals of t. Let $[s_0, i_0]^T \in L^1_+$ and let $t > 0$. From (2.47) we obtain that for almost all $c \in (-t, 0)$,

$$\begin{aligned}0 &\le s(t + c, t) + i(t + c, t) \\ &= \beta - \int_{-c}^{t} [\mu(\tau + c)s(\tau + c, \tau) + \gamma(\tau + c)i(\tau + c, \tau)]\, d\tau \\ &\le \beta\end{aligned}$$

and from (2.48) we obtain that for almost all $c \in (0, \infty)$,

$$\begin{aligned}0 &\le s(t + c, t) + i(t + c, t) \\ &= s_0(c) + i_0(c) - \int_0^t [\mu(\tau + c)s(\tau + c, \tau) \\ &\quad + \gamma(\tau + c)i(\tau + c, \tau)]\, d\tau \\ &\le s_0(c) + i_0(c)\end{aligned}$$

Thus, we have that

$$\begin{aligned}&\|s(\cdot,t)\|_{L^1} + \|i(\cdot,t)\|_{L^1} \\ &= \int_0^\infty [s(a,t) + i(a,t)]\, da \\ &= \int_{-t}^\infty [s(t + c, t) + i(t + c, t)]\, dc \\ &\le \int_0^\infty \beta\, dc + \int_0^\infty [s_0(c) + i_0(c)]\, dc \\ &= \beta t + \|s_0\|_{L^1} + \|i_0\|_{L^1}\end{aligned}$$

and hence the solution exists globally in time. Thus, we may apply Theorems 3.1, 3.2, 3.3, and 3.4 in Chapter 3 to obtain a nonlinear semigroup associated with the solutions of (ADP).

We next consider the existence of equilibrium solutions for this model. Suppose that $[\hat{s}, \hat{i}]^T \in L^1_+$ is an equilibrium solution of (ADP) and $\hat{i}_0 \neq 0$. From Proposition 4.1 we obtain

$$\frac{d}{da}\hat{s}(a) = -\eta(a)\hat{I}\hat{s}(a) - \mu(a)\hat{s}(a) \qquad a \ge 0,\ \hat{s}(0) = \beta$$

$$\frac{d}{da}\hat{i}(a) = \eta(a)\hat{I}\hat{s}(a) - \gamma(a)\hat{i}(a) \qquad a \geq 0,\ \hat{i}(0) = 0$$

where $\hat{I} \stackrel{\text{def}}{=} \int_0^\infty \hat{i}(a)\,da$. By direct integration we obtain

$$(5.29) \qquad \hat{s}(a) = \beta \exp\left[-\int_0^a (\eta(b)\hat{I} + \mu(b))db\right] \qquad a \geq 0$$

$$(5.30) \qquad \hat{i}(a) = \int_0^a \exp\left[-\int_b^a \gamma(c)\ dc\right]\eta(b)\hat{I}\hat{s}(b)\ db \qquad a \geq 0$$

Substitute the formula for $\hat{s}$ in (5.29) into (5.30) and integrate from 0 to $\infty$ to obtain

$$\begin{aligned}(5.31) \qquad \hat{I} &= \int_0^\infty\left[\int_0^a \exp\left[-\int_b^a \gamma(c)\ dc\right]\eta(b)\hat{I}\ \beta \right.\\ &\qquad \left.\times \exp\left[-\int_0^b (\eta(c)\hat{I} + \mu(c))\ dc\right]\ db\right]\ da\\ &= \int_0^\infty\left[\int_b^\infty \exp\left[-\int_b^a \gamma(c)\ dc\right]\eta(b)\hat{I}\ \beta \right.\\ &\qquad \left.\times \exp\left[-\int_0^b (\eta(c)\hat{I} + \mu(c))\ dc\right]\ da\right]\ db\\ &= \int_0^\infty\left(\int_0^\infty \exp\left[-\int_b^{b+\tau} \gamma(c)\ dc\right]\ d\tau\right)\eta(b)\hat{I}\ \beta\\ &\qquad \times \exp\left[-\int_0^b (\eta(c)\hat{I} + \mu(c))\ dc\right]\ db\end{aligned}$$

If we divide (5.31) by $\hat{I}$ we see that $\hat{I}$ must satisfy the equation $\alpha(\hat{I}) = 1$, where

$$\begin{aligned}(5.32) \qquad \alpha(I) &\stackrel{\text{def}}{=} \int_0^{a_0}\left(\int_0^\infty \exp\left[-\int_b^{b+\tau} \gamma(c)\ dc\right]\ d\tau\right)\eta(b)\ \beta\\ &\qquad \times \exp\left[-\int_0^b (\eta(c)I + \mu(c))\ dc\right]\ db \qquad I \geq 0\end{aligned}$$

We claim now that a necessary and sufficient condition for the existence of a necessarily unique equilibrium solution $[\hat{s},\hat{i}]^T \in L^1_+$

such that $\hat{i} \neq 0$ is that $\alpha(0) > 1$. This claim follows immediately from the computations above provided that we can show the $\alpha$ is continuous, strictly decreasing, and $\lim_{I\to\infty}\alpha(I) = 0$. (Notice that if $\alpha(0) \leq 1$, then $\alpha(I) = 1$ has no nontrivial solution.) That $\alpha$ is continuous and strictly decreasing is obvious from (5.32). To see that $\lim_{I\to\infty}\alpha(I) = 0$, observe that for $I > 0$,

$$\alpha(I) \leq \int_0^{a_0}\left(\int_0^\infty \exp\left[-\int_b^{b+\tau} \gamma(c)\ dc\right]\ d\tau\right)\eta(b)\beta \exp\left[-\int_0^b \eta(c)I\ dc\right]\ db$$

$$= \int_0^{a_0}\left(\int_b^\infty \exp\left[-\int_b^a \gamma(c)\ dc\right]\ da\right)\left(-\beta I^{-1}\frac{d}{db}\exp\left[-\int_0^b \eta(c)I\ dc\right]\right)\ db$$

$$= \beta I^{-1}\left[\int_0^{a_0}\left(\frac{d}{db}\int_b^\infty \exp\left[-\int_b^a \gamma(c)\ dc\right]\ da\right)\left(\exp\left[-\int_0^b \eta(c)I\ dc\right]\right)\ db\right.$$

$$-\left(\int_{a_0}^\infty \exp\left[-\int_{a_0}^a \gamma(c)\ dc\right]\ da\right)\left(\exp\left[-\int_0^{a_0} \eta(c)I\ dc\right]\right)$$

$$\left.+\int_0^\infty \exp\left[-\int_0^a \gamma(c)\ dc\right]\ da\right]$$

$$\leq \beta I^{-1}\left[\int_0^{a_0}\left(\int_b^\infty \gamma(b)\exp\left[-\int_b^a \gamma(c)\ dc\right]\ da\right)\ db\right.$$

$$\left.+\int_0^\infty \exp\left[-\int_0^a \gamma(c)\ dc\right]\ da\right]$$

The claim that $\lim_{I\to\infty}\alpha(I) = 0$ now follows.

We next claim that if $\alpha(0) < 1$, then for all $[s_0, i_0]^T \in L^1_+$,

$$(5.33)\qquad \lim_{t\to\infty}\int_0^\infty i(a,t)\ da = 0$$

To establish this claim observe that $s(a,t)$ has the explicit formula

$$(5.34)\qquad s(a,t) = \begin{cases} \beta \exp[-\int_{t-a}^t(\eta(\tau + a - t)I(\tau) + \mu(\tau + a - t))\ d\tau] & a < t \\ s_0(a - t)\exp[-\int_0^t (\eta(\tau + a - t)I(\tau) + \mu(\tau + a - t))\ d\tau] & \text{a.e. } a > t \end{cases}$$

(which may be verified as in Section 1.2). Let $c \in R$ and define $w_c(t) \overset{\text{def}}{=} i(t + c, t)$ for $t \geq t_c \overset{\text{def}}{=} \max\{0,-c\}$. By (2.67) and (2.68) we have that for almost all $c \in R$ and for all $t \geq t_c$,

$$\frac{d}{dt} w_c(t) = \eta(t + c)I(t)s(t + c, t) - \gamma(t + c)w_c(t)$$

which implies that for almost all $c \in R$ and for all $t \geq t_c$,

$$\begin{aligned} w_c(t) &= \exp\left[-\int_{t_c}^{t} \gamma(\tau + c)\, d\tau\right] w_c(t_c) \\ &\quad + \int_{t_c}^{t} \exp\left[-\int_{\tau}^{t} \gamma(\sigma + c)\, d\sigma\right]\eta(\tau + c)I(\tau)s(\tau + c, \tau)\, d\tau \end{aligned}$$

Thus, for $t > 0$, (5.25) and (5.26) yield that

$$\begin{aligned} \int_{-t}^{\infty} w_c(t)\, dc &= \int_{-t}^{0} w_c(t)\, dc + \int_{0}^{\infty} w_c(t)\, dc \\ &= \int_{-t}^{0} \left\{\exp\left[-\int_{-c}^{t} \gamma(\tau + c)\, d\tau\right] w_c(-c)\right. \\ &\quad + \int_{-c}^{t} \exp\left[-\int_{\tau}^{t} \gamma(\sigma + c)\, d\sigma\right]\eta(\tau + c) \\ &\quad \left.\times I(\tau)s(\tau + c, \tau)\, d\tau\right\} dc \\ &\quad + \int_{0}^{\infty} \left\{\exp\left[-\int_{0}^{t} \gamma(\tau + c)\, d\tau\right] w_c(0)\right. \\ &\quad + \int_{0}^{t} \exp\left[-\int_{\tau}^{t} \gamma(\sigma + c)\, d\sigma\right]\eta(\tau + c) \\ &\quad \left.\times I(\tau)s(\tau + c, \tau)\, d\tau\right\} dc \\ &= \int_{-t}^{0} \exp\left[-\int_{-c}^{t} \gamma(\tau + c)\, d\tau\right] i(0,-c)\, dc \\ &\quad + \int_{0}^{t} \left[\int_{-\tau}^{0} \exp\left[-\int_{\tau}^{t} \gamma(\sigma + c)\, d\sigma\right]\eta(\tau + c)\right. \\ &\quad \left.\times I(\tau)s(\tau + c, \tau)\, dc\right] d\tau \\ &\quad + \int_{0}^{\infty} \exp\left[-\int_{0}^{t} \gamma(\tau + c)\, d\tau\right] i(c,0)\, dc \end{aligned}$$

$$+ \int_0^t \left[ \int_0^\infty \exp\left[ -\int_\tau^t \gamma(\sigma + c)\, d\sigma \right] \eta(\tau + c) \times I(\tau) s(\tau + c, \tau)\, dc \right] d\tau$$

$$= \int_0^\infty \exp\left[ -\int_0^t \gamma(\tau + c)\, d\tau \right] i_0(c)\, dc$$

$$+ \int_0^t \left[ \int_{-\tau}^\infty \exp\left[ -\int_\tau^t \gamma(\sigma + c)\, d\sigma \right] \eta(\tau + c) \times I(\tau) s(\tau + c, \tau)\, dc \right] d\tau$$

Then, for $t \ge a_0$

$$\int_0^\infty i(a,t)\, da \le \int_0^\infty e^{-\underline{\gamma} t} i_0(c)\, dc$$

$$+ \int_0^t \left[ \int_0^\infty \exp\left[ -\int_\tau^t \gamma(\sigma + b - \tau) d\sigma \right] \eta(b) \times I(\tau) s(b,\tau)\, db \right] d\tau$$

$$\le e^{-\underline{\gamma} t} \int_0^\infty i_0(c)\, dc$$

$$+ e^{-\underline{\gamma} t} \int_0^{a_0} \left[ \int_0^{a_0} e^{\underline{\gamma}\tau} \eta(b) I(\tau) s(b,\tau)\, db \right] d\tau$$

$$+ \int_{a_0}^t \left[ \int_0^{a_0} \exp\left[ -\int_b^{b+t-\tau} \gamma(c)\, dc \right] \eta(b) \times I(\tau) s(b,\tau)\, db \right] d\tau$$

Define $C = C(s_0, i_0)$ by

$$C \overset{\text{def}}{=} \int_0^\infty i_0(c)\, dc + \int_0^{a_0} \left[ \int_0^{a_0} e^{\underline{\gamma}\tau} \eta(b) I(\tau) s(b,\tau)\, db \right] d\tau$$

and use (5.34) to obtain that for $t \ge a_0$,

$$I(t) \le e^{-\underline{\gamma}\tau} C + \int_{a_0}^t k(t - \tau) I(\tau)\, d\tau \tag{5.35}$$

where

$$k(\sigma) \stackrel{\text{def}}{=} \int_0^{a_0} \exp\left[-\int_b^{b+\sigma} \gamma(c)\, dc\right] \eta(b)\beta \exp\left[-\int_0^b \mu(c)\, dc\right] db \qquad \sigma \geq 0$$

Notice that

$$k(\sigma) \leq e^{-\underline{\gamma}\sigma} \int_0^{a_0} \eta(b)\, db \qquad \sigma \geq 0$$

and consequently, we may define $q$: $[0,\underline{\gamma}) \to [0,\infty)$ by

$$q(\delta) \stackrel{\text{def}}{=} \int_0^{\infty} e^{\delta\sigma} k(\sigma)\, d\sigma \qquad 0 \leq \delta < \underline{\gamma}$$

Suppose now that $\alpha(0) < 1$, which means that $q(0) < 1$. Since $q$ is continuous, there exists $\varepsilon \in (0,1)$ and $\delta \in (0,\underline{\gamma})$ such that $q(\delta) \leq \varepsilon$. Set $u(t) \stackrel{\text{def}}{=} \sup_{0\leq\tau\leq t} e^{\delta\tau} I(\tau)$, $t \geq 0$. From (5.35) we obtain that for $t \geq a_0$,

$$\begin{aligned} e^{\delta t} I(t) &\leq e^{-(\underline{\gamma}-\delta)t} C + \int_{a_0}^{t} e^{\delta(t-\tau)} k(t-\tau) e^{\delta t} I(\tau)\, d\tau \\ &\leq C + \left(\int_{a_0}^{t} e^{\delta(t-\tau)} k(t-\tau)\, d\tau\right) u(t) \\ &\leq C + \varepsilon u(t) \end{aligned}$$

which implies that for $t \geq a_0$

$$u(t) \leq \frac{C}{1-\varepsilon}$$

Thus, for $t \geq a_0$

$$I(t) \leq C(s_0,i_0)e^{-\delta t}/(1-\varepsilon)$$

and so (5.33) is established.

We summarize our conclusions for this model as follows: A necessary and sufficient condition for the existence of a necessarily unique equilibrium solution $[\hat{s},\hat{i}]^T \in L_+^1$ of (ADP) such that $\hat{i} \neq 0$ is that $\alpha(0) > 1$. If $\alpha(0) < 1$, then every solution $[s(\cdot,t),i(\cdot,t)]^T$ of (ADP) corresponding to an initial age distribution $[s_0,i_0]^T \in L^1_+$ has the property that $\int_0^\infty i(a,t)\,da \le Me^{-\delta t}$, $t \ge 0$, where M is a positive constant dependent on $[s_0,i_0]^T$, but $\delta$ is positive constant independent of $[s_0,i_0]^T$. These conditions on $\alpha$ are threshold conditions in the sense that if $\alpha(0) < 1$, then the infective population decays exponentially to 0, and if $\alpha(0) > 1$, then an equilibrium state exists in which the infective population is not 0. Since $\alpha$ involves the age-dependent removal moduli $\mu$ and $\gamma$ and the age-dependent contact modulus $\eta$, these threshold conditions illustrate the effects of chronological age structure upon the behavior of the infectious disease within the epidemic population.

## 5.4 LOGISTIC POPULATIONS

In Section 1.1 we provided a discussion of the logistic equation as a model of age-independent population growth. The age-independent logistic equation possesses a nonlinear term as a mechanism to depress growth as the population becomes large. The effects of crowding are thus incorporated into the logistic model of population growth. In this section we provide a model of age-dependent logistic population growth. There have been a number of studies of age-dependent logistic populations, including those of F. Hoppensteadt [153], P. Marcati [211], S. Busenberg and M. Iannelli [33], and M. Gurtin [127]. The discussion we present here is based upon the results in [33], [211], and [127].

We suppose a single species age-dependent population, possessing a self-limiting mechanism in its mortality process, but not its birth process. Accordingly, we formulate the model in $L^1 = L^1(0,\infty;R)$ and define the birth function F and the aging function G by

(5.36) $F\colon L^1 \to R$, $F(\phi) \overset{\text{def}}{=} \int_0^\infty \beta(a)\phi(a)\,da$, $\phi \in L^1$, where $\beta\colon [0,\infty) \to [0,\infty)$ is bounded and continuous.

(5.37) $G: L^1 \to L^1$, $G(\phi)(a) \overset{\text{def}}{=} -[\mu(a) + \eta(P\phi)]\phi(a)$, $\phi \in L^1$, a.e. $a > 0$, where $\mu$: $[0,\infty) \to [0,\infty)$ is bounded, Lipschitz continuous, and satisfies $\mu(a) \geq \underline{\mu}$ for all $a \geq 0$ and some positive constant $\underline{\mu}$, $\eta$: $R \to [0,\infty)$ is Lipschitz continuous on bounded sets, and $P\phi \overset{\text{def}}{=} \int_0^\infty \phi(a)\, da$ for all $\phi \in L^1$.

Thus, the model has the form

$$D\ell(a,t) = -[\mu(a) + \eta(P\ell(\cdot,t))]\ell(a,t)$$

$$\ell(0,t) = \int_0^\infty \beta(a)\ell(a,t)\, da$$

The birth function F is linear, whereas the aging function G is nonlinear. The form of the aging function, which involves only mortality, assumes a separability of the mortality modulus into a term $\mu$ depending only on age and a term $\eta$ depending only on total population. In [127] Gurtin describes this type of mortality modulus as one which separates the mortality due to inherent species aging and the mortality due to environmental limitations. Notice that if F satisfies (5.36), then F satisfies (2.1), (2.22), (3.73), and (4.67), and if G satisfies (5.37), then G satisfies (2.2), (2.23), and (2.49). Further, (2.28) holds. By Theorems 2.1 and 2.4, (ADP) has a unique nonnegative solution for every initial age distribution $\phi \in L^1_+$ defined on a maximal interval of existence, which must be $[0,\infty)$ by Theorem 2.5.

Let $S(t)$, $t \geq 0$, be the strongly continuous nonlinear semigroup in $L^1_+$ associated with (ADP) as in Theorem 3.1. By Theorem 3.2 the infinitesimal generator B of $S(t)$, $t \geq 0$, is

$$B\phi \overset{\text{def}}{=} -\phi' + G(\phi) \qquad \phi \in D(B)$$

$D(B) \overset{\text{def}}{=} \{\phi \in L^1_+$: $\phi$ is absolutely continuous on $[0,\infty)$, $\phi' \in L^1$, and $\phi(0) = F(\phi)\}$

From Proposition 3.6 we have that

$$\frac{d^+}{dt} S(t)\phi = BS(t)\phi \qquad \phi \in D(B),\ t \geq 0 \tag{5.38}$$

Now let $\phi \in D(B)$ and define $\ell(a,t) \overset{\text{def}}{=} (S(t)\phi)(a)$, $a, t \geq 0$. From Theorem 2.9 we have that

$$D\ell(a,t) = -[\mu(a) + \eta(P\ell(\cdot,t))]\ell(a,t) \qquad a, t \geq 0 \tag{5.39}$$

$$\ell(0,t) = \int_0^\infty \beta(a)\ell(a,t)\, da \qquad t \geq 0 \tag{5.40}$$

$$\ell(a,0) = \phi(a) \qquad a \geq 0 \tag{5.41}$$

Define $P(t) \overset{\text{def}}{=} \int_0^\infty \ell(a,t)\, da$, $t \geq 0$, and use (5.38) to obtain that for $t \geq 0$,

$$\begin{aligned} P'^{+}(t) &= \int_0^\infty \frac{d^+}{dt} (S(t)\phi)(a)\, da \\ &= \int_0^\infty \left[- \frac{d}{da} (\ell(\cdot,t))(a) - (\mu(a) + \eta(P(t))\ell(a,t)\right] da \\ &= \int_0^\infty [\beta(a) - \mu(a)]\ell(a,t)\, da - \eta(P(t))P(t) \end{aligned} \tag{5.42}$$

Suppose that $P(t) > 0$ for all $t \geq 0$ and define $w(a,t) \overset{\text{def}}{=} \ell(a,t)/P(t)$, $a, t \geq 0$. From (5.39), (5.40), (5.41), and (5.42) we obtain

$$\begin{aligned} Dw(a,t) &= D\ell(a,t)/P(t) + \ell(a,t)P'^{+}(t)/(P(t))^2 \\ &= -\left\{\mu(a) + \int_0^\infty [\beta(b) - \mu(b)]w(b,t)\, db\right\}w(a,t) \\ &\qquad\qquad a, t \geq 0 \end{aligned} \tag{5.43}$$

$$w(0,t) = \int_0^\infty \beta(a)w(a,t)\, da \qquad t \geq 0 \tag{5.44}$$

$$w(a,0) = \phi(a)/P(0) \qquad a \geq 0 \tag{5.45}$$

$$\int_0^\infty w(a,t)\, da = 1 \qquad t \geq 0 \tag{5.46}$$

Next, define

$$k(t) \overset{\text{def}}{=} \int_0^\infty [\mu(a) - \beta(a)]w(a,t)\, da \qquad t \geq 0$$

$$v(a,t) \overset{\text{def}}{=} \exp\left[-\int_0^t k(s)\, ds\right] w(a,t) \qquad a, t \geq 0$$

From (5.43), (5.44), (5.45) we obtain

$$\begin{aligned} (5.47) \qquad Dv(a,t) &= Dw(a,t)k(t) + w(a,t)k'(t) \\ &= -\mu(a)v(a,t) \qquad a, t \geq 0 \end{aligned}$$

$$(5.48) \qquad v(0,t) = \int_0^\infty \beta(a)v(a,t)\, da \qquad t \geq 0$$

$$(5.49) \qquad v(a,0) = \phi(a)/P(0) \qquad a \geq 0$$

The problem (5.47), (5.48), (5.49) corresponds to classical linear age-dependent population dynamics as discussed in Section 1.2. The solution v of (5.47), (5.48), (5.49) is given uniquely by (1.12) and (1.14). The asymptotic behavior of this solution is given by Theorem 4.10. We consider two cases:

*Case 1* $\quad \int_0^\infty \beta(a)\Pi(a,0)\, da < 1$

*Case 2* $\quad \int_0^\infty \beta(a)\Pi(a,0)\, da > 1$

where $\Pi(a,b) \overset{\text{def}}{=} \exp[-\int_b^a \mu(c)\, dc]$, $0 \leq b \leq a$. In Case 1 Theorem 4.10 yields

$$(5.50) \qquad \lim_{t\to\infty} \int_0^\infty v(a,t)\, da = 0$$

In Case 2 Theorem 4.10 and Remark 4.12 yield

$$(5.51) \qquad \lim_{t\to\infty} e^{-\lambda_1 t} v(a,t) = V_{\lambda_1}(\phi/P(0)e^{-\lambda_1 a}\Pi(a,0)/M_{\lambda_1} \text{ uniformly in bounded intervals of } a$$

where $\lambda_1$ is the unique positive real number such that

$$\int_0^\infty e^{-\lambda_1 a} \beta(a)\Pi(a,0)\; da = 1$$

$V_{\lambda_1}(\phi/P(0))$ is the natural reproductive value of the initial age distribution $\phi$, and $M_{\lambda_1}$ is the mean age of childbirth for the stable age distribution.

If Case 1 holds, then (5.46) and (5.50) imply that

$$\text{(5.52)} \qquad \lim_{t\to\infty} \exp\left[-\int_0^t k(s)\; ds\right] = \lim_{t\to\infty} \int_0^\infty \exp\left[-\int_0^t k(s)\; ds\right] w(a,t)\; da$$

$$= \lim_{t\to\infty} \int_0^\infty v(a,t)\; da = 0$$

If Case 2 holds, then (5.46) and (5.51) imply that

$$\lim_{t\to\infty} \exp\left[-\int_0^t k(s)\; ds\right] e^{-\lambda_1 t}$$

$$= \lim_{t\to\infty} \int_0^\infty w(a,t)\; \exp\left[-\int_0^t k(s)\; ds\right] e^{\lambda_1 t}\; da$$

$$= \lim_{t\to\infty} \int_0^\infty v(a,t) e^{-\lambda_1 t}\; da = (V_{\lambda_1}(\phi)/M_{\lambda_1}) \int_0^\infty e^{-\lambda_1 a} \Pi(a,0)\; da$$

Therefore, in Case 2 we must have that

$$\text{(5.53)} \qquad \lim_{t\to\infty} \exp\left[-\int_0^t k(s)\; ds\right] = \infty$$

$$\text{(5.54)} \qquad \lim_{t\to\infty} w(a,t) = \lim_{t\to\infty} (e^{-\lambda_1 t} v(a,t)) \left(e^{\lambda_1 t} \exp\left[\int_0^t k(s)\; ds\right]\right)$$

$$= e^{-\lambda_1 a} \Pi(a,0) \Big/ \int_0^\infty e^{-\lambda_1 b} \Pi(b,0)\; db$$

$$\overset{\text{def}}{=} w_\infty(a) \text{ uniformly in bounded intervals of } a \geq 0$$

$$\text{(5.55)} \qquad \lim_{t\to\infty} k(t) = \int_0^\infty [\mu(a) - \beta(a)] w_\infty(a)\; da$$

$$= \int_0^\infty \left[ - \frac{d}{da} (\Pi(a,0)) e^{-\lambda_1 a} \right] da \Big/ \int_0^\infty e^{-\lambda_1 b} \Pi(b,0)\, db$$

$$- \int_0^\infty \beta(a) w_\infty(a)\, da$$

$$= \left( \int_0^\infty \Pi(a,0)(-\lambda_1 e^{-\lambda_1 a})\, da + 1 \right) \Big/ \int_0^\infty e^{-\lambda_1 b} \Pi(b,0)\, db - w_\infty(0)$$

$$= -\lambda_1$$

From (5.42) we obtain

$$(5.56) \qquad P'^{+}(t) = -k(t)P(t) - \eta(P(t))P(t) \qquad t \geq 0$$

Consequently, the asymptotic behavior of the solutions to (ADP) is determined by (5.52), (5.53), (5.54), (5.55), and the asymptotic behavior of the solutions to (5.56). We suppose now that $\eta$ has the logistic form

$$(5.57) \qquad \eta(P) \overset{\text{def}}{=} |\lambda_1| P/K \qquad P \in R$$

where K is a positive constant called the *environmental carrying capacity*. Then, (5.56) is a Bernoulli equation and the solution is given explicitly by

$$P(t) = \frac{\exp[-\int_0^t k(s)\, ds] P(0)}{1 + \lambda_1 K^{-1} P(0) \int_0^t \exp[-\int_0^s k(\tau)\, d\tau]\, ds}$$

If Case 1 holds, then (5.52) implies that $\lim_{t\to\infty} P(t) = 0$. If Case 2 holds, then (5.53) and (5.55) imply that $\lim_{t\to\infty} P(t) = K$, and (5.54) then implies that

$$(5.58) \qquad \lim_{t\to\infty} \ell(a,t) = K e^{-\lambda_1 a} \Pi(a,0) \Big/ \int_0^\infty e^{-\lambda_1 b} \Pi(b,0)\, db$$

uniformly in bounded intervals of a.

For this logistic model of age-dependent population dynamics we now see an analogy with the logistic model of age-independent population dynamics in Section 1.1. The intrinsic growth constant $\lambda_1$ in the age-independent model (1.2) is analogous to the intrinsic growth constant $\lambda_1$ in the age-dependent model. The environmental carrying capacity K in (1.2) is analogous to the environmental carrying capacity in (5.57). If these environmental carrying capacities are $\infty$, then crowding effects are not present. In this case the models are linear and fail to have nontrivial equilibrium states. If the environmental carrying capacities are positive, then crowding effects are present. In this case the models are nonlinear and possess unique nontrivial equilibrium states to which the populations must stabilize. For the age-dependent logistic population (5.58) provides an explicit formula for this equilibrium state as a function of age.

## 5.5 NOTES

The literature of nonlinear age-dependent population dynamics includes a number of studies of multiple species interaction. A discrete version of an age-dependent population model for competing species is given in [274] by C. Travis, W. Post, D. DeAngelis, and J. Perkowski, wherein a stability analysis is made of the Leslie matrix model for two competing species, each divided into two age groups. Age-dependent models of predator-prey interaction have been investigated by J. Cushing in [69], J. Cushing and M. Saleem in [70], K. Gopalsamy in [118], M. Gurtin and D. Levine in [129], and A. Haimovici in [141]. In [240], [241], and [242], J. Prüss treats general models of interacting species.

Nonlinear models of age-dependent population genetics were first proposed by F. Hoppensteadt in [153]. The results of Section 5.2 appeared in [304] and [306]. Nonlinear models of age-dependent epidemic populations were first formulated by K. Dietz in [92] and F. Hoppensteadt in [153]. Other treatments of age-dependent epidemic populations may be found in the articles of G. Di Blasio [85], L. Lamberti and P. Vernole [186], P. Waltman [292], and the author [302], [307]. In the model in [92] the age corresponds to chronological age, whereas in [85], [186], [292], [302], and [307] the age corresponds to disease age. In [153] both chronological and disease age are incorporated into the epidemic model studied.

There have been several treatments of age-dependent logistic populations in the literature of nonlinear age-dependent population dynamics. In [211] P. Marcati uses Laplace transform techniques to study the asymptotic behavior of the model in Section 5.4. In [293] F. J. S. Wang studies another age-dependent logistic model using integral equations techniques. In [127] M. Gurtin studies an age-dependent logistic population from the point of view of convergence to stable age (or persistent age) distributions. The proof used in the result of Section 5.4 is based on the work of S. Busenberg and M. Iannelli in [33].

Some other applications of nonlinear age-dependent population dynamics to biological populations are found in the articles of G. L. Curry, R. M. Feldman, B. L. Deuermeyer, and M. S. Keener [65] (cotton fruiting), R. Elderkin [99] (seed dispersal), R. Elderkin, et al. [101] (malarial populations), M. Gurtin and L. Murphy [135], [136] (renewable resource harvesting), M. Gyllenberg [137] (bacterial cultures), and M. Witten [310], [311] (cell growth).

# References


1. H. Amann, Fixed point equations and nonlinear eigenvalue problems in ordered B-spaces, *SIAM Review 18* (1976), 620-709.

2. A. Ambrosetti, Proprietá spetralli di certi operatori lineari non compatti, *Rend. Sem. Mat. Univ. Padova 42* (1969), 189-200.

3. V. G. Angelov and V. G. Bainov, On the existence and uniqueness of bounded solutions to functional-differential equations of neutral type in Banach spaces, *Scripta Fac. Sci. Nat. Univ. Purk. Brun. 10* (1980), No. 8 (*Matematica*), 367-376.

4. A. Ardito and P. Ricciardi, Existence and regularity for linear delay partial differential equations (to appear).

5. D. M. Auslander, G. F. Oster, C. B. Huffaker, Dynamics of interacting populations, *J. Franklin Inst. 297* (1974), 345-376.

6. M. Badii and G. F. Webb, Representation of solutions of functional differential equations with time-dependent delays, *Houston J. Math. 6* (1980), 577-594.

7. M. Badii and G. F. Webb, Nonlinear nonautonomous functional differential equations in $L^p$ spaces, *Nonl. Anal. 5* (1981), 203-223.

8. N. T. J. Bailey, *The Mathematical Theory of Infectious Diseases*, Griffin, London, 1975.

9. J. Ball, On the asymptotic behavior of generalized processes with applications to nonlinear evolution equations, *J. Differential Equations 27* (1978), 224-265.

10. H. T. Banks and J. A. Burns, An abstract framework for approximation of solutions to optimal control problems governed by hereditary systems, *Int. Conf. Differential Equations*, Academic Press, 1975, pp. 10-25.

11. H. T. Banks and J. A. Burns, Hereditary control problems: numerical methods based on averaging approximations, *SIAM J. Control Optimization 16* (1978), 169-208.

12. H. T. Banks and F. Kappel, Spline approximations for functional differential equations, *J. Differential Equations 34* (1979), 496-522.

13. V. Barbu, *Nonlinear Semigroups and Differential Equations in Banach Spaces*, Noordhoff International Publishing Co., Leyden, The Netherlands, 1976.

14. V. Barbu and S. I. Grossman, Asymptotic behavior of linear integro-differential systems, *Trans. Amer. Math. Soc. 173* (1972), 277-288.

15. M. Bardi, An equation of growth of a single specie with realistic dependence on crowding and seasonal factors (to appear).

16. R. G. Bartle, *The Elements of Real Analysis*, Second Edition, John Wiley, New York, 1964.

17. G. I. Bell and E. C. Anderson, Cell growth and division. I. A mathematical model with applications to cell volume distributions in mammalian suspension cultures, *Biophys. J. 7* (1967), 329-351.

18. A. Belleni-Morante, *Applied Semigroups and Evolution Equations*, Oxford University Press, Oxford, 1979.

19. R. Bellman and K. L. Cooke, *Differential-Difference Equations*, Academic Press, 1963.

20. H. Bernardelli, Population waves, *J. Burma Res. Soc. 31* (1941), 1-18.

21. C. Bernier and A. Manitius, On semigroups in $R^n \times L^p$ corresponding to differential equations with delays, *Can. J. Math.*, Vol. XXX (1978), 897-914.

22. J. G. Borisovic and A. S. Turbabin, On the Cauchy-problem for linear nonhomogeneous differential equation with retarded argument, *Soviet Math. Dokl. 10* (1969), 401-405.

23. D. W. Brewer, A nonlinear semigroup for a functional differential equation, *Trans. Amer. Math. Soc. 236* (1978), 173-191.

24. D. W. Brewer, A nonlinear contraction semigroup for a functional differential equation, *Proc. Helsinki Symp. Integral Equations*, Springer-Verlag, 1978.

25. H. Brezis, Opérateurs maximaux monotones et semigroupes de contractions dans les espaces de Hilbert, *Math. Studies 5*, North Holland, 1973.

26. H. Brezis and A. Pazy, Accretive sets and differential equations in Banach spaces, *Israel J. Math. 8* (1970), 367-383.

27. H. Brezis and A. Pazy, Convergence and approximation of semigroups of nonlinear operators in Banach spaces, *J. Funct. Anal. 7* (1972), 63-74.

28. F. E. Browder, On the spectral theory of elliptic differential operators, *Math. Ann. 142* (1961), 22-130.

29. J. A. Burns and E. M. Cliff, Methods for approximating solutions to linear hereditary quadratic optimal control problems, *IEEE Trans. Automatic Control 23* (1978), 21-36.

30. J. A. Burns and T. L. Herdman, Adjoint semigroup theory for a Volterra integrodifferential system, *Bull. Amer. Math. Soc. 81* (1975), 1099-1102.

31. J. A. Burns, T. L. Herdman, and H. W. Stech, The Cauchy problem for linear functional differential equations, *Integral and Functional Differential Equations*, Marcel Dekker, 1981, pp. 137-147.

32. J. A. Burns, T. L. Herdman, and H. W. Stech, Differential-boundary operators and associated neutral functional differential equations (to appear).

33. S. Busenberg and M. Iannelli, A class of nonlinear diffusion problems in age-dependent population dynamics, *J. Nonlinear Analysis T.M.A.* (to appear)

34. S. Busenberg and M. Iannelli, Nonlinear diffusion problems in age-structured population dynamics (to appear).

35. S. Busenberg and M. Iannelli, A degenerate nonlinear diffusion problem in age-structured population dynamics (to appear).

36. G. Chen and R. Grimmer, Semigroups and integral equations, *J. Integral Equations 2* (1980), 133-154.

37. M. Chipot, On the equations of age-dependent population dynamics, Lefschetz Center for Dynamical Systems, Report #82-1.

38. M. Chipot and L. Edelstein, A mathematical theory of size distributions in tissue culture (to appear).

39. C. W. Clark, Mathematical Bioeconomics: *The Optimal Management of Renewable Resources*, Wiley, 1976.

40. A. J. Coale, A new method for calculating Lotka's r - The intrinsic rate of growth in a stable population, *Population Studies 11* (1957), 92-94.

41. A. J. Coale, Age patterns of marriage, *Population Studies 25* (1971), 193-214.

42. A. J. Coale, *The Growth and Structure of Human Populations*, Princeton University Press, Princeton, 1972.

43. E. A. Coddington and N. Levinson, *Theory of Ordinary Differential Equations*, McGraw-Hill, New York, 1955.

44. C. V. Coffman and B. D. Coleman, On the growth of populations with narrow spread in reproductive age: II. Conditions of convexity, *J. Math. Biol. 6* (1978), 285-303.

45. C. V. Coffman and B. D. Coleman, On the growth of populations with narrow spread in reproductive age: III, *J. Math. Biol. 7* (1979), 281-301.

46. C. V. Coffman and J. J. Schäffer, Linear differential equations with delays: Existence, uniqueness, growth, and compactness under natural Caratheodory conditions, *J. Differential Equations 16* (1974), 26-44.

47. B. D. Coleman, Thermodynamics of materials with memory, *Arch. Rat. Mech. Anal. 17* (1964), 1-46.

48. B. D. Coleman, Thermodynamics of discrete mechanical systems with memory, *Adv. Chem. Phys. 24* (1973), 95-154.

49. B. D. Coleman, On the growth of populations with narrow spread in reproductive age, *J. Math. Biol. 6* (1978), 1-19.

50. B. D. Coleman, Nonautonomous logistic equations as models of the adjustment of populations to environmental change, *Math. Biosci. 45* (1979), 159-173.

51. B. D. Coleman and V. J. Mizel, Norms and semigroups in the theory of fading memory, *Arch. Rat. Mech. Anal. 23* (1966), 87-123.

52. B. D. Coleman and V. J. Mizel, On the general theory of fading memory, *Arch. Rat. Mech. Anal. 29* (1968), 18-31.

53. B. D. Coleman and W. Noll, An approximation theorem for functionals, with applications in continuum mechanics, *Arch. Rat. Mech. Anal. 6* (1960), 355-370.

54. B. D. Coleman and D. R. Owen, On the initial value problem for a class of functional-differential equations, *Arch. Rat. Mech. Anal. 55* (1975), 277-299.

55. K. L. Cooke, Delay differential equations, *Mathematics of Biology*, Centro Internazionale Mathematico Estivo, Napoli, 1979, pp. 5-80.

56. M. G. Crandall, An introduction to evolution governed by accretive operators, *Dynamical Systems*, Vol. 1, Academic Press, 1976.

57. M. G. Crandall and L. C. Evans, On the relation of the operator $\partial/\partial s + \partial/\partial\tau$ to evolution governed by accretive operators, *Israel J. Math. 21* (1975), 261-278.

58. M. G. Crandall and T. M. Liggett, Generation of semigroups of nonlinear transformations on general Banach spaces, *Amer. J. Math. 93* (1971), 265-298.

59. M. G. Crandall, S. O. Londen, and J. A. Nohel, An abstract nonlinear Volterra integrodifferential equation, *J. Math. Anal. Appl. 64* (1978), 701-735.

60. M. G. Crandall and A. Pazy, Semigroups of nonlinear operators and dissipative sets, *J. Funct. Anal. 3* (1969), 376-418.

61. M. G. Crandall and A. Pazy, On accretive sets in Banach spaces, *J. Funct. Anal. 5* (1970), 204-217.

62. M. G. Crandall and A. Pazy, Nonlinear evolution equations in Banach spaces, *Israel J. Math. 11* (1972), 57-94.

63. J. F. Crow and M. Kimura, *An Introduction to Population Genetics Theory*, Harper and Row, New York, 1970.

64. C. W. Cryer and L. Tavernini, The numerical solution of Volterra functional differential equations by Euler's method, *SIAM J. Numer. Anal. 9* (1972), 105-129.

65. G. L. Curry, R. M. Feldman, B. L. Deuermeyer, and M. E. Keener, Approximating a closed-form solution for cotton fruiting dynamics, *Math. Biosci. 54* (1981), 91-113.

66. J. Cushing, Integrodifferential equations and delay models in population dynamics, *Lecture Notes in Biomathematics 29*, Springer-Verlag, 1977.

67. J. M. Cushing, Volterra integrodifferential equations in population dynamics, *Mathematics of Biology*, Centro Internazionale Matematics Estivo, Napoli, 1979, pp. 81-148.

68. J. M. Cushing, Model stability and instability in age structured populations, *J. Theoret. Biol. 86* (1980), 709-730.

69. J. M. Cushing, Stability and maturation periods in age structured populations, *Proc. Conf. Differential Equations and Applications*, Claremont College, Academic Press, 1981.

70. J. M. Cushing and M. Saleem, A predator-prey model with age structure (to appear).

71. C. M. Dafermos, Contraction semigroups and trend to equilibrium in continuum mechanics, *Lecture in Mathematics*, No. 503, Springer-Verlag, 1976, pp. 295-306.

72. C. M. Dafermos and M. Slemrod, Asymptotic behavior of nonlinear contraction semigroups, *J. Funct. Anal. 13* (1973), 97-106.

73. J. K. Daleckii and M. G. Krein, Stability of solutions of differential equations in Banach spaces, *Transl. Math. Monographs*, Vol. 43, Amer. Math. Soc., Providence, 1974.

74. G. Da Prato and M. Iannelli, Linear abstract integro-differential equations of hyperbolic type in Hilbert spaces, *Rend. Sem. Mat. Univ. Padova 62* (1980), 191-206.

75. R. Datko, Lyapunov functionals for certain linear delay differential equations in a Hilbert space, *J. Math. Anal. Appl. 25* (1971), 258-274.

76. E. B. Davies, *One-Parameter Semigroups*, Academic Press, London, 1980.

77. K. Deimling, Ordinary differential equations in Banach spaces, *Lecture Notes in Mathematics 596*, Springer-Verlag, 1977.

78. K. Deimling, Equilibria of an age-dependent population model, *Nonlinear Differential Equations*, Proceedings International Conference, Trento, 1980, Academic Press, 1981, pp. 129-132.

79. M. G. Delfour and S. K. Mitter, Controlability, observability and optimal feedback control of hereditary differential systems, *SIAM J. Control 10* (1972), 298-328.

80. M. C. Delfour and S. K. Mitter, Hereditary differential systems with constant delays, I. General case, *J. Differential Equations 12* (1972), 213-235.

81. R. Derndinger, Über das Spektrum positiver Generatoren, *Math. Zeit. 172* (1980), 281-293.

82. G. W. Desch, The state space $L^\infty$ for functional differential equations, *Evolution Equations and Their Applications*, F. Kappel and W. Schappacher (Editors), Pitman, Boston, 1982.

83. G. Di Blasio, Nonlinear age-dependent population diffusion, *J. Math. Biol. 8* (1979), 265-284.

84. G. Di Blasio, Nonlinear age-dependent population growth with history-dependent birth rate, *Math. Biosci. 46* (1979), 279-291.

85. G. Di Blasio, A problem arising in the mathematical theory of epidemics (to appear).

86. G. Di Blasio, M. Iannelli, and E. Sinestrari, An abstract partial differential equation with a boundary condition of renewal type, *Boll. U.M.I. Serie V*, Vol. XVIII-C, N. 1 (1981), 260-274.

87. G. Di Blasio, M. Iannelli, and E. Sinestrari, Approach to equilibrium in age structured populations with an increasing recruitment process, *J. Math. Biol. 13* (1982), 371-382.

88. G. Di Blasio and L. Lamberti, An initial-boundary value problem for age-dependent population diffusion, *SIAM J. Appl. Math. 35* (1978), 593-615.

89. O. Diekmann, Limiting behavior in an epidemic model, *Nonl. Anal. Theory, Methods, Appl. 1* (1977), 459-470.

90. O. Diekmann, Thresholds and travelling waves for the geographical spread of infection, *J. Math. Biol. 6* (1978), 109-130.

91. O. Diekmann, The stable size distribution: An example in structured population dynamics, *Mathematisch Centrum*, Report TW 231, Amsterdam.

92. K. Dietz, Transmission and control of arbovirus diseases, Proc. SIMS Conference on Epidemics, Alta, Utah, 1974.

93. J. Dieudonné, *Foundations of Modern Analysis, Pure and Applied Mathematics Series*, Academic Press, 1969.

94. N. Dunford and J. Schwartz, *Linear Operators, Part I: General Theory*, Interscience, 1958.

95. J. Dyson and R. Villella-Bressan, Functional differential equations and nonlinear evolution operators, Proc. Roy. Soc. Edin. 75A, 20 (1975/76), 223-234.

96. J. Dyson and R. Villella-Bressan, On an abstract functional differential equation with infinite delay, Comm. Conf. Ordinary and Partial Differential Equations, Dundee, Scotland, 1976.

97. J. Dyson and R. Villella-Bressan, Nonlinear functional differential equations in $L^1$-spaces, *Nonl. Anal. 1* (1977), 383-395.

98. J. Dyson and R. Villella-Bressan, Semigroup of translations associated with functional and functional differential equations, Proc. Roy. Soc. Edin., 82A (1979), 171-181.

99. R. H. Elderkin, Seed dispersal in a patchy environment with global age dependence, *J. Math. Biol. 13* (1982), 283-303.

100. R. H. Elderkin, Nonlinear, globally age-dependent population models: Some basic theory (to appear).

101. R. H. Elderkin, D. P. Berkowitz, F. A. Farris, C. F. Gunn, F. J. Hickernell, S. N. Kass, F. I. Mansfield, and R. G. Taranto, On the steady state of an age dependent model for malaria (to appear).

102. L. E. El'sgol'ts, *Introduction to the Theory of Differential Equations with Deviating Arguments*, Holden-Day, 1966.

103. H. Engler, Functional differential equations in Banach spaces: Growth and decay of solutions (to appear).

104. W. Feller, On the integral equation of renewal theory, *Ann. Math. Stat. 12* (1941), 243-267.

105. R. A. Fisher, *The Genetical Theory of Natural Selection*, 2nd Ed., Dover, New York, 1958.

106. W. Fitzgibbon, Nonlinear Volterra equations with infinite delay, *Monatsh. Math. 84* (1977), 275-288.

107. W. Fitzgibbon, Stability for abstract nonlinear Volterra equations involving finite delay, *J. Math. Anal. Appl. 60* (1977), 429-434.

108. W. E. Fitzgibbon, Abstract Volterra equations with infinite delay, *Nonlinear Systems and Applications* (Int. Conf.), V. Lakshmikantham, Ed., 1977, 513-524.

109. W. E. Fitzgibbon, Semilinear functional differential equations in Banach spaces, *J. Differential Equations 29* (1978), 1-14.

110. W. Fitzgibbon, Representation and approximation of solutions to semilinear Volterra equations with delay, *J. Differential Equations 32* (1979), 233-249.

111. W. Fitzgibbon, Semilinear integrodifferential equations in Banach space (to appear).

112. H. Flaschka and M. Leitman, On semigroups of nonlinear operators and the solution of the functional differential equation $\dot{x}(t) = F(x_t)$, *J. Math. Anal. Appl. 49* (1975), 649-658.

113. J. C. Frauenthal, *Introduction to Population Modeling*, Birkhäuser, Boston, 1980.

114. J. C. Frauenthal, Mathematical modeling in epidemiology, *Lecture Notes in Mathematics*, Springer-Verlag, Berlin-Heidelberg-New York, 1980.

115. M. G. Garroni and M. Langlais, Age-dependent population diffusion with external constraint, *J. Math. Biol. 14* (1982), 77-94.

116. J. A. Goldstein, Semigroups of operators and applications (to appear).

117. K. Gopalsamy, On the asymptotic age distributions in dispersive population, *Math. Biosci. 31* (1976), 191-205.

118. K. Gopalsamy, Time lags and density dependence in age dependent two species competition, *Bull. Australian Math. Soc.* (to appear).

119. G. Greiner and R Nagel, On the stability of strongly continuous semigroups of positive operators on $L^2(\mu)$ (to appear).

120. G. Greiner, J. Voigt, M. Wolff, On the spectral bound of the generator semigroups of positive operators, *J. Operator Theory 5* (1981), 245-256.

121. D. H. Griffel, Age-dependent population growth, *J. Inst. Math. Appl. 17* (1976), 141-152.

122. R. C. Grimmer, Resolvent operators for integral equations in a Banach space (to appear).

123. R. Grimmer and G. Seifert, Stability properties of Volterra integrodifferential equations, *J. Differential Equations 19* (1975), 142-166.

124. R. Grimmer and M. Zeman, Nonlinear Volterra integrodifferential equations in a Banach space (to appear).

125. U. Groh and F. Neubrander, Stabilität startstetiger, positiver Operator-halbgruppen auf C*-Algebren, *Math. Ann. 256* (1981), 509-516.

126. M. E. Gurtin, A system of equations for age-dependent population diffusion, *J. Theoret. Biol. 40* (1973), 389-392.

127. M. E. Gurtin, The mathematical theory of age-structured populations (to appear).

128. M. E. Gurtin, Some questions and open problems in continuum mechanics and population dynamics (to appear).

129. M. E. Gurtin and D. S. Levine, On predator-prey interaction with predation dependent on age of prey, *Math. Biosci. 47* (1979), 207-219.

130. M. E. Gurtin and R. C. MacCamy, Nonlinear age-dependent population dynamics, *Arch. Rat. Mech. Anal. 54* (1974), 281-300.

131. M. E. Gurtin and R. C. MacCamy, On the diffusion of biological population, *Math. Biosci. 38* (1977), 35-49.

132. M. E. Gurtin and R. C. MacCamy, Population dynamics with age dependence, *Nonlinear Analysis and Mechanics*: Heriot-Watt Symposium III, Pitman, Boston-London-Melbourne, 1979.

133. M. E. Gurtin and R. C. MacCamy, Some simple models for nonlinear age-dependent population dynamics, (1979), 199-211.

134. M. E. Gurtin and R. C. MacCamy, Product solutions and asymptotic behavior for age-dependent, dispersing populations (to appear).

135. M. E. Gurtin and L. F. Murphy, On the optimal harvesting of age-structured population: Some simple models, *Math. Biosci.* *55* (1981), 115-136.

136. M. E. Gurtin and L. F. Murphy, On the optimal harvesting of persistent age-structured populations, *J. Math. Biol.* (to appear).

137. Mats Gyllenberg, Nonlinear age-dependent population dynamics in continuously propagated bacterial cultures, *Math. Bios. 62* (1982), 45-74.

138. Mats Gyllenberg, Stability of a nonlinear age-dependent population model containing a control variable (to appear).

139. J. R. Haddock, Asymptotic behavior of solutions of nonlinear functional differential equations in Banach space, *Trans. Amer. Math. Soc. 231* (1977), 83-92.

140. A. Haimovici, On the growth of a population dependent on ages and involving resources and pollution, *Math. Biosci. 43* (1979), 213-237.

141. A. Haimovici, On the age dependent growth of two interacting populations, *Boll. Un. Mat. Ital. 15* (1979), 405-429.

142. J. K. Hale, Dynamical systems and stability, *J. Math. Anal. Appl. 26* (1969), 39-59.

143. J. K. Hale, *Ordinary Differential Equations*, Interscience Series on Pure and Applied Mathematics, Vol. 21, Wiley-Interscience, New York, 1969.

144. J. Hale, *Functional Differential Equations*, Appl. Math. Series, Vol. 3, Springer-Verlag, Berlin-Heidelberg-New York, 1971.

145. J. K. Hale, Functional differential equations with infinite delays, *J. Math. Anal. Appl. 48* (1974), 276-283.

146. A. Haraux, Nonlinear evolution equations - Global behavior of solutions, *Lecture Notes in Mathematics*, Springer-Verlag, Berlin-Heidelberg-New York, 1981.

147. M. Heard, An abstract parabolic Volterra integrodifferential equation, *SIAM J. Math. Anal. 13* (1982), 81-105.

148. H. J. A. M. Heijmans, An eigenvalue problem related to cell growth, Mathematisch Centrum Report TW 229, Amsterdam.

149. D. Henry, Linear autonomous neutral functional differential equations, *J. Differential Equations 15* (1974), 106-127.

150. D. Henry, Geometric theory of semilinear parabolic equations, *Lecture Notes in Mathematics*, Vol. 840, Springer-Verlag, Berlin-Heidelberg-New York, 1981.

151. E. Hille and R. S. Phillips, *Functional Analysis and Semigroups*, Amer. Math. Soc., Providence, 1959.

152. F. Hoppensteadt, An age dependent epidemic model, *J. Franklin Inst. 297* (1974), 325-333.

153. F. Hoppensteadt, Mathematical theories of populations: Demographics, genetics, and epidemics, *SIAM Reg. Conf. Series in Appl. Math.*, 1975.

154. F. Hoppensteadt, A nonlinear renewal equation with periodic and chaotic solutions, *SIAM-AMS Proceedings 10* (1976), 51-60.

155. F. C. Hoppensteadt, Perturbation methods in biology, *Mathematics of Biology*, Centro Internazionale Matematics Estivo, Napoli, 1979, pp. 264-322.

156. F. Iacob and N. H. Pavel, Invariant sets for a class of perturbed differential equations of retarded type, *Israel J. Math. 28* (1977), 254-264.

157. J. Kaplan and J. Yorke, Toward a unification of ordinary differential equations with nonlinear semigroup theory, Int. Conf. Ordinary Differential Equations, USC, 1974, Academic Press, 1975.

158. F. Kappel, Approximation of neutral functional differential equations in the state space $R^n \times L^2$ (to appear).

159. F. Kappel and K. Kunisch, Spline approximations for neutral functional differential equations, *SIAM J. Numer. Anal. 18* (1981), 1058-1080.

160. F. Kappel and K. Kunisch, Approximation of the state of infinite delay and Volterra-type equations (to appear).

161. F. Kappel and W. Schappacher, Autonomous nonlinear functional differential equations and averaging approximations, *Nonl. Anal. 2* (1978), 391-421.

162. A. G. Kartsatos and M. Parrott, Global solutions of functional evolution equations involving locally defined Lipschitzean perturbations (to appear).

163. A. G. Kartsatos and M. Parrott, A method of lines for a nonlinear abstract functional evolution equation (to appear).

164. A. G. Kartsatos and M. Parrott, Existence of solutions of a nonlinear abstract functional differential equation (to appear).

165. A. G. Kartsatos and M. Parrott, Functional evolution equations involving time dependent maximal monotone operators in Banach spaces (to appear).

166. A. G. Kartsatos and M. Parrott, Convergence of the Kato approximants for evolution equations involving functional perturbations (to appear)

167. T. Kato, *Perturbation Theory for Linear Operators*. Springer-Verlag, 1966.

168. N. Keyfitz, *Introduction to the Mathematics of Population*, Addison-Wesley, Reading, 1968.

169. S. G. Krein, Linear differential equations in Banach space, *Translations of Mathematical Monographs*, Amer. Math. Soc., Providence, 1971.

170. R. R. Kuczynski, *The Balance of Births and Deaths, Volume I, Western and Northern Europe*, Macmillan Company for the Brookings Institution, New York, 1928.

171. R. R. Kuczynski, *The Balance of Births and Deaths, Volume II, Eastern and Southern Europe*, The Brookings Institution, Washington, D.C., 1931.

172. R. R. Kuczynski, *Fertility and Reproduction*, Falcon Press, New York, 1931/32.

173. R. R. Kuczynski, *The Measurement of Population Growth*, Sedgewich and Jackson, Std., London, 1935 (reproduced by Gordon and Breach Science Publishers, New York, 1969).

174. K. Kunisch, Abstract Cauchy problem and abstract integral equations for neutral functional differential equations, *Archiv der Mathematik 31* (1978), 580-588.

175. K. Kunisch, Neutral functional differential equations in $L^p$-spaces and averaging approximation, *Nonl. Anal. 3* (1979), 419-448.

176. K. Kunisch, A semigroup approach to partial differential equations with delay (to appear).

177. K. Kunisch and W. Schappacher, Order-preserving evolution operators of functional differential equations, *Boll. Un. Mat. Ital. 16* (1979), 480-500.

178. K. Kunisch and W. Schappacher, Mild and strong solutions for partial differential equations with delay, *Ann. Mat. Pura Appl. CXXV* (1980), 193-219.

179. K. Kunisch and W. Schappacher, Variation of constants formulas for partial differential equations with delay, *Nonl. Anal. 5* (1981), 123-142.

180. K. Kunisch, W. Schappacher, and G. Webb, Nonlinear age-dependent population dynamics with random diffusion (to appear).

181. K. Kuratowski, *Topology*, Vol. I, Academic Press, New York, 1966.

182. G. Ladas and V. Lakshmikantham, *Differential Equations in Abstract Spaces*, Academic Press, 1972.

183. V. Lakshmikantham and S. Leela, *Differential and Integral Inequalities, Theory and Applications*, Volume I, Academic Press, New York, 1969.

184. V. Lakshmikantham and S. Leela, *Differential and Integral Inequalities*, Volume II, Academic Press, New York, 1969.

185. L. Lamberti and P. Vernole, Existence and asymptotic behavior of solutions of an age structured population model, *Boll. UMI, Anal. Funz. Appl.*, Serie Vol. XVIII-C (1981).

186. L. Lamberti and P. Vernole, An age structured epidemic model-asymptotic behavior (to appear).

187. H. L. Langhaar, General population theory in the age-time continuum, *J. Franklin Inst. 293* (1972), 199-214.

188. M. Langlais, A nonlinear problem in age dependent population diffusion (to appear).

189. M. Langlais, On some linear age dependent population diffusion models (to appear).

190. J. P. LaSalle, The stability of dynamical systems, *SIAM Reg. Conf. Series in Appl. Math. 25*, 1976.

191. J. P. LaSalle, Stability theory and invariance principles, *Dynamical Systems, An International Symposium*, Vol. 1, Academic Press, 1976.

192. S. Leela and V. Moauro, Existence of solutions in a closed set for delay differential equations in Banach spaces, *Nonl. Anal. 2* (1978), 47-58.

193. P. H. Leslie, On the use of matrices in certain population mathematics, *Biometrika 33* (1945), 183-212.

194. P. H. Leslie, Some further notes on the use of matrices in population mathematics, *Biometrika 35* (1948), 213-245.

195. D. Levine and M. Gurtin, Models of predation and cannibalism in age-structured populations, *Differential Equations and Applications in Ecology, Epidemics, and Population Problems*, Academic Press, New York, 1981, pp. 145-161.

196. E. G. Lewis, On the generation and growth of a population, *Sankhya 6* (1942), 93-96.

197. J. Lighthouse, Function space flow invariance for functional differential equations of retarded type, *Proc. Amer. Math. Soc. 77* (1979), 91-98.

198. J. Lighthouse and S. Rankin, A partial functional differential equation of Sobolev type (to appear).

199. S. O. Londen, On an integro-differential Volterra equation with a maximal monotone mapping, *J. Differential Equations 27* (1978), 405-420.

200. A. Lopez, *Problems in Stable Population Theory*, Office of Population Research, Princeton, 1961.

201. L. Lopez and Donato Trigiante, A hybrid scheme for solving a model of population dynamics (to appear).

202. A. J. Lotka, The stability of the normal age distribution, *Proc. Nat. Acad. Science 8* (1922), 339-345.

203. A. J. Lotka, On an integral equation in population analysis, *Ann. Math. Stat. 10* (1939), 1-25.

204. A. J. Lotka, *Elements of Mathematical Biology*, Dover, 1956.

205. R. C. MacCamy, A population model with nonlinear diffusion, *J. Differential Equations 39* (1981), 52-72.

206. R. C. MacCamy, Simple population models with diffusion, *Comp. Math. Appl. 8* (1982).

207. T. R. Malthus, *An Essay on the Principle of Population* (first edition), printed for J. Johnson in St. Paul's Churchyard, London, 1798.

208. A. Manitius and R. Triggiani, Function space controllability of linear retarded systems: A derivation from abstract operator conditions, *SIAM J. Control and Optimization 16* (1978), 599-645.

209. P. Marcati, Asymptotic behavior of the renewal equation arising in the Gurtin population model, Proc. Int. Conf. Arlington, 1980 (to appear).

210. P. Marcati, Asymptotic behavior in age-dependent population dynamics with hereditary renewal law, *SIAM J. Math. Anal. 12* (1981), 904-916.

211. P. Marcati, On the global stability of the logistic age dependent population growth (to appear).

212. P. Marcati, Some considerations on the mathematical approach to nonlinear age dependent population dynamics (to appear).

213. P. Marcati and R. Serafini, Asymptotic behavior in age dependent population dynamics with spatial spread, *Boll. Un. Mat. Ital. 16-B* (1979), 734-753.

214. M. Marcus and V. Mizel, Semilinear hereditary hyperbolic systems with nonlocal boundary conditions, A, *J. Math. Anal. Appl. 76* (1980), 440-475.

215. M. Marcus and V. Mizel, Semilinear hereditary hyperbolic systems with nonlocal boundary conditions, B, *J. Math. Anal. Appl. 77* (1980), 1-19.

216. M. Marcus and V. J. Mizel, Limiting equations for problems involving long range memory (to appear).

217. R. H. Martin, *Nonlinear Operators and Differential Equations in Banach Spaces*, John Wiley and Sons, 1976.

218. A. G. McKendrick, Applications of mathematics to medical problems, *Proc. Edin. Math. Soc. 44* (1926), 98-130.

219. R. M. Melvin, Topologies for neutral functional differential equations, *J. Differential Equations 13* (1973), 24-31.

220. R. K. Miller, *Nonlinear Volterra Integral Equations*, Benjamin, 1971.

221. R. K. Miller, Linear Volterra integrodifferential equations as semigroups, *Funkcial. Ekvac. 17* (1974), 39-55.

222. R. K. Miller and R. K. Wheeler, Well posedness and stability of linear Volterra integrodifferential equations in abstract spaces, *Funkcial. Ekvac. 2* (1978), 279-305.

223. J. Murray, *Lectures on Nonlinear-Differential Equation Models in Biology,* Clarendon Press, Oxford, 1977.

224. T. Naito, On autonomous linear functional differential equations with infinite retardations, *J. Differential Equations 21* (1976), 297-315.

225. R. D. Nussbaum, The radius of the essential spectrum, *Duke Math. J. 38* (1970), 473-478.

226. G. Oster, Internal variables in population dynamics, *Some Mathematical Questions in Biology, VII*, Amer. Math. Soc., Providence, 1976.

227. G. Oster, Lectures in population dynamics, *Lectures in Applied Mathematics 16*, Amer. Math. Soc., 1977, pp. 149-170.

228. M. E. Parrott, Representation and approximation of generalized solutions of a nonlinear functional differential equation (to appear).

229. N. H. Pavel, *Analysis of Some Nonlinear Problems in Banach Spaces and Applications*, Universitatea "Al. I. Cuza" Iasi, Facultatea de Matematică, 1982.

230. A. Pazy, Semigroups of linear operators and applications to partial differential equations, Univ. Maryland Lecture Notes, No. 10, 1974.

231. A. Pazy, A class of semi-linear equations of evolution, *Israel J. Math. 20* (1975), 26-36.

232. A. F. Plant, Nonlinear semigroups of translations in Banach space generated by functional differential equations, *J. Math. Anal. Appl. 60* (1977), 67-74.

233. A. T. Plant, Stability of nonlinear functional differential equations using weighted norms, *Houston J. Math. 3* (1977), 99-108.

234. A. T. Plant, On the asymptotic stability of solutions of Volterra integrodifferential equations, *J. Differential Equations 39* (1981), 39-51.

235. J. H. Pollard, *Mathematical Models for the Growth of Human Populations*, Cambridge University Press, 1973.

236. M. A. Pozio, Behavior of solutions of some abstract functional differential equations and applications to predator-prey dynamics, *Nonl. Anal. 4* (1980), 917-938.

237. M. A. Pozio, Decay estimates for partial functional differential equations (to appear).

238. M. A. Pozio, Some conditions for global asymptotic stability of equilibria of integrodifferential equations (to appear).

239. A. J. Pritchard and J. Zabczyk, Stability and stabilizability of infinite dimensional systems, *SIAM Review 23* (1981), 25-52.

240. J. Prüss, Equilibrium solutions of age-specific population dynamics of several species, *J. Math. Biol. 11* (1981), 65-84.

241. J. Prüss, On the qualitative behavior of populations with age-specific interactions (to appear).

242. J. Prüss, Stability analysis for equilibria in age-specific population dynamics (to appear).

243. S. Rankin, Existence and asymptotic behavior of a functional differential equation in Banach space (to appear).

244. G. Reddien and G. F. Webb, Numerical approximation of nonlinear functional differential equations with $L^2$ initial functions, *SIAM J. Math. Anal. 9* (1978), 1151-1171.

245. C. Rorres, Stability of an age specific population with density dependent fertility, *Theoret. Population Biol. 10* (1976), 26-46.

246. C. Rorres, Local stability of a population with density-dependent fertility, *Theoret. Population Biol. 16* (1979), 283-300.

247. C. Rorres, A nonlinear model of population growth in which fertility is dependent on birth rate, *SIAM J. Appl. Math. 37* (1979), 423-432.

248. M. Rotenberg, Equilibrium and stability in populations whose interactions are age specific, *J. Theoret. Biol. 54* (1975), 207-224.

249. H. L. Royden, *Real Analysis*, Second Edition, Macmillan, New York, 1968.

250. S. Rubinov, Age-structured equations in the theory of cell populations, *Studies in Mathematical Biology*, Part II, Math. Assoc. Amer., Washington, 1978.

251. W. Rudin, *Real and Complex Analysis*, McGraw-Hill, New York, 1966.

252. D. Salamon, On finite dimensional perturbation of semigroups corresponding to differential delay systems (to appear).

253. P. A. Samuelson, Resolving a historical confusion in population analysis, *Human Biol. 48* (1976), 559-580.

254. S. H. Saperstone, *Semidynamical Systems in Infinite Dimensional Spaces*, Appl. Math. Sci. Series 37, Springer-Verlag, Berlin, 1981.

255. M. Schechter, *Principles of Functional Analysis*, Academic Press, New York, 1971.

256. O. Scherbaum and G. Rasch, Cell size distribution and single cell growth in Tetrahymena Pyriformis, *G. L. Acta Pathol. Microbiol. Scandinav. 41* (1957), 161-182.

257. A. Schiaffino, Compactness methods for a class of semilinear equations of evolution, *Nonlinear Anal. 2* (1978), 179-188.

258. A. Schiaffino and A. Tesei, On the asymptotic stability for abstract Volterra integro-differential equations, *Rend. Acc. Naz. Lincei, VIII, 67* (1979).

259. I. Segal, Nonlinear semigroups, *Ann. Math. 78* (1963), 339-364.

260. F. R. Sharpe and A. J. Lotka, A problem in age distribution, *Phil. Mag. 21* (1911), 435-438.

261. E. Sinestrari, Non-linear age-dependent population growth, *J. Math. Biol. 128* (1980), 1-15.

262. J. W. Sinko and W. Streifer, A new model for age-size structure of a population, *Ecology 48* (1967), 910-918.

263. J. W. Sinko and W. Streifer, A model for populations reproducing by fission, *Ecology 52* (1971), 330-335.

264. M. Slemrod and E. F. Infante, An invariance principle for dynamical systems on Banach space, *Instability of Continuous Systems*, H. Lerpholz, ed., Springer-Verlag, Berlin, 1971, pp. 215-221.

265. D. Smith and N. Keyfitz, *Mathematical Demography, Selected Papers*, Biomathematics, Vol. 6, Springer-Verlag, Berlin-Heidelberg-New York, 1977.

266. C. O. A. Sowunmi, Female dominant age-dependent deterministic population dynamics, *J. Math. Biol. 3* (1976), 9-17.

267. K. E. Swick, A nonlinear age-dependent model of single species population dynamics, *SIAM J. Appl. Math. 22* (1977), 484-498.

268. K. W. Swick, Periodic solutions of a nonlinear age-dependent model of single species population dynamics, *SIAM J. Math. Anal. 11* (1980), 901-910.

269. A. Taylor, *Introduction to Functional Analysis*, John Wiley, New York, 1967.

270. A. Tesei, Stability properties for partial Volterra integro-differential equations (to appear).

271. H. Thieme, Asymptotic estimates of the solutions of nonlinear integral equations and asymptotic speeds of the spread of populations, *J. Reine Angew. Math. 306* (1979), 94-121.

272. H. R. Thieme, Renewal theorems for linear periodic Volterra integral equations (to appear).

273. R. J. Thompson, On some functional differential equations: Existence of solutions and difference approximations, *J. Differential equations 14* (1973), 57-69.

274. C. C. Travis, W. M. Post, D. L. De Angelis, and J. Perkowski, Analysis of compensatory Leslie matrix models for competing species, *Theoret. Population Biol. 18* (1980), 16-30.

275. C. C. Travis and G. F. Webb, Existence and stability for partial functional differential equations, *Trans. Amer. Math. Soc. 200* (1974), 395-418.

276. C. C. Travis and G. F. Webb, Partial differential equations with deviating arguments in the time variable, *J. Math. Anal. Appl. 56* (1976), 397-409.

277. C. C. Travis and G. F. Webb, Existence, stability, and compactness in the α-norm for partial functional differential equations, *Trans. Amer. Math. Soc. 240* (1978), 129-143.

278. R. Triggiani, On the stabilizability problem in Banach space, *J. Math. Anal. Appl. 52* (1975), 383-403.

279. E. Trucco, Mathematical models for cellular systems. The von Foerster equation, *Bull. Math. Biophysics 27* (1965), 285-305, 449-471.

280. P. F. Verhulst, Notice sur la loi que la population suit dans son accroissement, *Correspondance Mathématique et Physique 10* (1838), 113-121.

281. P. F. Verhulst, Recherches mathématiques sur la loi d'accroissement de la population, *Mém. Acad. Roy. Bruxelles 18* (1845).

282. P. F. Verhulst, Recherches mathématiques sur la loi d'accroissement de la population, *Mém. Acad. Roy. Bruxelles 20* (1847).

283. R. Villella-Bressan, On suitable spaces for studying functional equations using semigroup theory, *Proceedings Conf. Nonl. Differential Equations*, Trento, Italy, Academic Press, 1981.

284. R. Villella-Bressan, Flow-invariant sets for functional differential equations, Research Notes in Math., *Pitman A.P.P. 48* (1981), 213-229.

285. R. Villella-Bressan, Functional equations of delay type in $L^1$ spaces (to appear).

286. R. Villella-Bressan and G. F. Webb, Nonautonomous functional equations, and nonlinear evolution operators (to appear).

287. R. B. Vinter, On the evolution of the state of linear differential delay equations in $M^2$: Properties of the generator, *J. Inst. Math. Appl. 21* (1978), 13-23.

288. R. B. Vinter, Semigroups on product spaces with application to initial value problems with non-local boundary conditions (to appear).

289. H. Von Foerster, Some remarks on changing populations, *The Kinetics of Cellular Proliferation*, Grune and Stratton, 1959. pp. 382-407.

290. J. A. Walker, *Dynamical Systems and Evolution Equations. Theory and Applications*, Plenum Press, 1980.

291. J. A. Walker and E. F. Infante, Some results on the precompactness of positive orbits of dynamical systems, *J. Math. Anal. Appl. 51* (1975), 56-67.

292. P. Waltman, Deterministic threshold models in the theory of epidemics, *Lecture Notes in Biomathematics*, Springer-Verlag, Berlin, 1974.

293. F. J. S. Wang, Stability of an age-dependent population, *SIAM J. Math. Anal. 11* (1980), 683-689.

294. G. F. Webb, Autonomous nonlinear functional differential equations and nonlinear semigroups, *J. Math. Anal. Appl 46* (1974), 1-12.

295. G. F. Webb, Functional differential equations and nonlinear semigroups in $L^p$ spaces, *J. Differential Equations 20* (1976), 71-89.

296. G. F. Webb, Asymptotic stability for abstract nonlinear functional differential equations, (1976), 225-230.

297. G. F. Webb, Linear functional differential equations with $L^2$ initial functions, *Funkcial. Ekvac. 19* (1976), 65-77.

298. G. F. Webb, Volterra integral equations as functional differential equations on infinite intervals, *Hiroshima Math. J. 7* (1977), 61-70.

299. G. F. Webb, Volterra integral equations and nonlinear semigroups, *Nonl. Anal. 1* (1977), 415-427.

300. G. F. Webb, An abstract semilinear Volterra integrodifferential equation, *Proc. Amer. Math. Soc. 69* (1978), 255-260.

301. G. F. Webb, Compactness of bounded trajectories of dynamical systems in infinite dimensional spaces, *Proc. Roy. Soc. Edin. 84A* (1979), 19-33.

302. G. F. Webb, An age-dependent epidemic model with spatial diffusion, *Arch. Rat. Mech. Anal. 75* (1980), 91-102.

303. G. F. Webb, Nonlinear semigroups and age-dependent population models, *Ann. Mat. Pura Appl. CXXIXX* (1981), 43-55.

304. G. F. Webb, A genetics model with age dependence and spatial diffusion, *Proc. Conf. Differential Equations and Applications*, Claremont College, Academic Press, 1981.

305. G. F. Webb, Nonlinear age-dependent population dynamics with continuous age distributions, *Evolution Equations and Their Applications*, Research Notes in Mathematics 68, Pitman, Boston-London-Melbourne, 1982, pp. 274-294.

306. G. F. Webb, Diffusive age-dependent population models and an application to genetics, *Math. Biosci. 61* (1982), 1-16.

307. G. F. Webb, A recovery-relapse epidemic model with spatial diffusion, *J. Math. Biol. 14* (1982), 177-194.

308. G. Webb, Nonlinear age-dependent population dynamics in $L^1$, *J. Integral Equations 5*, No 4 (1983), 309-328.

309. G. F. Webb, The semigroup associated with nonlinear age-dependent population dynamics, *Int. J. Computers Appl. Math. 9* (1983), 487-497.

310. M. Witten, Modelling cellular systems and aging process: II. Some thoughts on describing an asynchronously dividing cellular system, *Nonlinear Phenomena in Mathematical Sciences*, Academic Press, 1982, pp. 1023-1035.

311. M. Witten, Modelling cellular systems and aging processes: I. Mathematics of cell system models - A review, to appear in Mech. Aging and Development.

312. K. Yosida, *Functional Analysis*, Second Edition, Springer-Verlag, Berlin-Heidelberg-New York, 1968.

313. J. Zabczyk, A note on $C_0$-semigroups, *Bull. Acad. Polon. Sci. 23* (1975), 895-898.

314. J. Zabczyk, On semigroups corresponding to non-local boundary conditions with applications to system theory, Control Theory Centre, Report 49, University of Warwick, 1976.

315. J. Zabczyk, On decomposition of generators, *SIAM J. Control and Optimization 16* (1978), 523-534.

# Notation

# Index